further
advanced chemistry

B EARL
Head of Chemistry and Network Manager at
St Aidan's Church of England High School, Harrogate

L D R WILFORD
Head of Science Faculty and Assistant Head at
St Aidan's Church of England High School, Harrogate

Hodder Murray

A MEMBER OF THE HODDER HEADLINE GROUP

Titles in this series:
Introduction to Advanced Biology 978 0 7195 7671 3
Introduction to Advanced Chemistry 978 0 7195 8587 6
Introduction to Advanced Physics 978 0 7195 8588 3
Further Advanced Chemistry 978 0 7195 8608 8
Further Advanced Physics 978 0 7195 8609 5

The cover image shows a computer-generated representation of part of a molecule depicting its arrangement of atoms (balls). The rods holding the balls together represent the chemical bonds between the atoms.

© B Earl & L D R Wilford 2001

First published in 2001
by Hodder Murray, an imprint of Hodder Education,
a member of the Hodder Headline group
338 Euston Road
London NW1 3BH

Reprinted 2006

Layouts by Eric Drewery
Illustrations by Barking Dog Art

Typeset in 10/12pt Gill Sans by Wearset, Boldon, Tyne and Wear
Printed and bound in Dubai

A catalogue entry for this title is available from the British Library

ISBN 978 0 7195 8608 8

Contents

CONTENTS

Introduction

This is the second of two textbooks written especially for the new Advanced level General Certificate of Education: Chemistry specifications. The book covers the A2 components and is ideal for your second year course. The *Specification-matching matrix* below shows which chapters you need to study when preparing for the Module Tests from your particular Awarding Body.

This book follows on from *Introduction to Advanced Chemistry* which was written to help you make the step from GCSE Double Award Science. In both books the key points to be covered are listed in the *Starting points* at the beginning of each chapter, and the *Checklists* at the end of each chapter include definitions of all the key terms that have been introduced. These can be used to help you establish that you have covered and understood the main ideas.

Throughout the text there are questions to test whether you have understood what you have read. At the end of each chapter there are longer study questions, mainly taken from past examination papers. Try to answer as many as you can, because asking and answering questions is at the heart of your study of Chemistry. If you get really stuck, there are answers given to the shorter questions and numerical questions at the end of the book!

The new Key Skills qualification can be obtained through the work you do in Chemistry. To gain the qualification you have to show your abilities in a range of skills including Communication, Information Technology and Application of Number. There are a number of opportunities to develop these skills throughout the book, and the specific *Skills task* exercise at the end of each chapter can be used to provide evidence of your achievements.

L D R Wilford & B Earl

Specification-matching matrix

Board	AQA					Edexcel				OCR						WJEC			
Module*	1	2	3	4	5	1	2	4	5	2811	2812	2813	2814	2815	2816	CH1	CH2	CH4	CH5
Chapter 1				✓					✓						✓				✓
Chapter 2				✓			✓								✓				✓
Chapter 3					✓			✓					✓						✓
Chapter 4				✓			✓						✓						✓
Chapter 5			✓				✓								✓				✓
Chapter 6				✓			✓						✓						✓
Chapter 7							✓												✓
Chapter 8					✓			✓					✓						✓
Chapter 9			✓				✓					✓						✓	
Chapter 10			✓				✓					✓						✓	
Chapter 11			✓				✓					✓						✓	
Chapter 12			✓						✓			✓						✓	
Chapter 13			✓						✓			✓						✓	
Chapter 14			✓						✓					✓				✓	
Chapter 15			✓						✓			✓						✓	

* Modules in **bold** make up the AS course

Acknowledgements

The authors would like to thank Irene, Katharine, Michael and Barbara for their never-ending patience and encouragement during the production of this A2 textbook. In addition, thanks are given to Mr Dennis Richards, Headteacher at St Aidan's Church of England High School, Harrogate, for his support throughout. Also, we wish to thank Tim Jackson, Michelle Sadler and other editorial staff, and all others involved at John Murray for all their hard work on our behalf, and for their encouragement throughout the development of this new generation of Chemistry textbook.

Exam questions have been reproduced with kind permission from the following examination boards:

AQA: The Associated Examining Board (AEB)
         Northern Examinations and Assessment Board (NEAB)
London Examinations: A division of Edexcel Foundation
OCR

The accuracy and method of working in the answers provided in this book are the sole responsibility of the authors and have not been supplied or approved by the examination boards.

Thanks are due to the following for permission to reproduce copyright photographs:

Cover Laguna Design/Science Photo Library; **p.1** © Murray Robertson 1998–1999; **p.2** Fig. 1.1 *l* Rex Features, *r* John Townson/Creation, Fig. 1.2 Andrew Lambert; **p.3** Fig. 1.4 Empics; **p.5** Fig. 1.6 Andrew Lambert; **p.16** Fig. 2.1a John Townson/Creation, Fig. 2.1b Andrew Lambert, Fig. 2.1c Roger Scruton; **p.19** Fig. 2.2 Science Photo Library, Fig. 2.3 Andrew Lambert; **p.20** Fig. 2.4 *both* Andrew Lambert, Fig. 2.5 Martin Bond/Science Photo Library; **p.21** Fig. 2.6b Empics, Fig. 2.7 Science Photo Library; **p.22** Fig. 2.8a & b Roger Scruton; **p.23** Fig. 2.9 *both* Andrew Lambert; **p.24** Fig. 2.11 Oxford Scientific Films; **p.30** Fig. 3.1 *all* John Townson/Creation; **p.31** Fig. 3.2 Andrew Lambert, Fig. 3.3 *both* Andrew Lambert; **p.33** Fig. 3.4 *both* Andrew Lambert; **p.34** Fig. 3.5a–d Andrew Lambert, Fig. 3.6 *all* John Townson/Creation; **p.36** Fig. 3.7 Empics, Fig. 3.8 *both* Andrew Lambert; **p.43** Fig. 3.21 *all* John Townson/Creation; **p.46** Fig. 4.1a Andrew Lambert; **p.51** Fig. 4.5 Andrew Lambert; **p.56** Fig. 4.14 John Townson/Creation; **p.58** Fig. 4.16 *both* Andrew Lambert; **p.62** Fig. 5.1a & b *all* John Townson/Creation; **p.63** Fig. 5.2 Mary Evans Picture Library; **p.64** Fig. 5.3 Mary Evans Picture Library; **p.65** Fig. 5.4 Andrew Lambert; **p.66** Fig. 5.5 John Townson/Creation; **p.68** Fig. 5.6 Andrew Lambert; **p.69** Fig. 5.8 Andrew Lambert; **p.72** Fig. 5.11 Science Photo Library/BISP LECA, Fig. 5.12 *both* John Townson/Creation; **p.76** Fig. 5.14 Andrew Lambert; **p.83** © Murray Robertson 1998–1999; **p.86** Figs. 6.6 & 6.7 Andrew Lambert; **p.87** Fig. 6.8, 6.9 & 6.10 Andrew Lambert; **p.88** Fig. 6.11 Andrew Lambert, Fig. 6.12 Trip Photo Library, Fig. 6.13 Andrew Lambert; **p.89** Fig. 6.14 Robert Harding, Fig. 6.15 Andrew Lambert; **p.90** Fig. 6.16 Robert Harding; **p.91** Fig. 6.17 Andrew Lambert; **p.93** Fig. 6.21 Andrew Lambert; **p.94** Fig. 6.22, 6.23 & 6.24a Andrew Lambert, Fig. 6.24b *both* John Townson/Creation; **p.98** Fig. 7.1a *l* Spectrum Colour Library, *r* Manfred Kage/Science Photo Library, Fig. 7.1b Richard Megna/Fundamental/Science Photo Library, Fig. 7.1c Alfred Pasieka/Science Photo Library, Fig. 7.1d Colorsport; **p.99** Fig. 7.2 *t* Andrew Lambert, *bl* & *br* John Townson/Creation, Fig. 7.3 Courtesy of Pilkington Plc, Fig. 7.4 Adam Hart-Davis/Science Photo Library, Fig. 7.5a Scala/Museo dell Ermitage, Fig. 7.5b Andrew Lambert; **p.101** Fig. 7.7 Trip Photo Library; **p.102** Fig. 7.9 *all* John Townson/Creation, Fig. 7.10 John Townson/Creation; **p.103** Fig. 7.11a Roger Scruton, Fig. 7.11b Ace Photo Library, Fig. 7.12 *both* Andrew Lambert; **p.104** Fig. 7.13 Stock Market, Fig. 7.15a & b John Townson/Creation; **p.105** Fig. 7.16a Trip Photo Library, Fig. 7.16b Robert Harding, Fig. 7.16c Phillip Hayson/Science Photo Library, Fig. 7.17 *l* Kaj B Svensson/Science Photo Library, *c* Roberto de Gugiemo/Science Photo Library, *r* Natural History Museum; **p.106** Fig. 7.19a & b Andrew Lambert; **p.109** Fig. 8.1 John Townson/Creation, Fig. 8.2a Spectrum Colour Library, Fig. 8.2b BSIP, Beranger/Science Photo Library; **p.110** Fig. 8.3 Ace Photo Library, Fig. 8.4 Johnson Matthey; **p.112** Fig. 8.9 Andrew Lambert; **p.113** Fig. 8.10 Ace Photo Library; **p.114** Fig. 8.11 Andrew Lambert; **p.115** Fig. 8.12 Andrew Lambert; **p.116** Fig. 8.14b *both* John Townson/Creation, Fig. 8.15a Damien Lovegrove/Science Photo Library; **p.119** Fig. 8.21b John Townson/Creation, Fig. 8.22 *both* Andrew Lambert; **p.120** Fig. 8.26 Andrew Lambert; **p.121** Fig. 8.28 Andrew Lambert; **p.122** Fig. 8.29b Trip Photo Library, Fig. 8.30 John Townson/Creation; **p.123** Fig. 8.31 *all* Andrew Lambert, Fig. 8.32 Andrew Lambert; **p.124** Fig. 8.33 Roger Scruton; **p.125** Fig. 8.34 Corus Teesside Technology Centre, Fig. 8.35 Andrew Lambert; **p.127** Fig. 8.37 Andrew Lambert; **p.128** Fig. 8.38 Andrew Lambert; **p.129** Fig. 8.39 John Townson/Creation; **p.131** Figs. 8.41 & 8.42 Andrew Lambert; **p.135** © Murray Robertson 1998–1999; **p.136** Fig. 9.1a–c John Townson/Creation; **p.138** Fig. 9.4 *all* John Townson/Creation; **p.139** Fig. 9.7 Andrew Lambert; **p.140** Fig. 9.10 Malcolm Fielding, Johnson Matthey PLC/Science Photo Library; **p.141** Fig. 9.11 *both* Andrew Lambert, Fig. 9.12 & 9.13 Andrew Lambert; **p.142** Fig. 9.14 Andrew Lambert; **p.145** Fig. 10.2a Oxford Scientific Films, Fig. 10.2b John Townson/Creation, Fig. 10.2c Colorsport; **p.146** Fig. 10.4 Andrew Lambert, Fig. 10.6 John Townson/Creation; **p.147** Fig. 10.7 Shout Picture Library; **p.152** Fig. 10.12 Hattie Young/Science Photo Library; **p.154** Fig. 10.16 John Townson/Creation, Fig. 10.18 *both* John Townson/Creation; **p.157** Fig. 10.22 John Townson/Creation; **p.160** Fig. 11.1 Crown Copyright/Health & Safety Lab/Science Photo Library, Fig. 11.2 Empics, Fig. 11.3 John Townson/Creation; **p.163** Fig. 11.10 John Townson/Creation; **p.164** Fig. 11.12 *l* Ace Photo Library, *r* John Townson/Creation; **p.166** Fig. 11.13 Chemical Design/Science Photo Library; **p.170** Fig. 11.20 Andrew Lambert; **p.171** Fig. 11.24 Chris Priest & Mark Clarke/Science Photo Library, Fig. 11.25 J.C. Revy/Science Photo Library; **p.175** Fig. 12.1 *all* John Townson/Creation, Fig. 12.2b Science Photo Library; **p.176** Fig. 12.5 Mary Evans Picture Library; **p.180** Fig. 12.16 Andrew Lambert, Fig. 12.19b Dick Luria/Science Photo Library; **p.181** Fig. 12.22b John Townson/Creation; **p.182** Fig. 12.25b *both* John Townson/Creation, Fig. 12.29 Andrew Lambert; **p.183** Fig. 12.34 John Townson/Creation; **p.185** Fig. 12.39b Andrew Lambert; **p.188** Fig. 13.1a, b & d John Townson/Creation, Fig. 13.1c Enak Ltd; **p.189** Fig. 13.3 *all* John Townson/Creation, Fig. 13.4 *both* John Townson/Creation; **p.191** Fig. 13.10 Lawrence Livermore/National Laboratory/Science Photo Library; **p.192** Fig. 13.12 Linpac Ltd, Fig. 13.13 Mawson Triton; **p.193** Fig. 13.16 John Townson/Creation; **p.194** Fig. 13.17 Science Photo Library; **p.195** Fig. 13.21 Sidney Moulds/Science Photo Library, Fig. 13.24 Colorsport; **p.199** Fig. 14.1 Science Photo Library, Fig. 14.2 Joseph Nettis/Science Photo Library; **p.201** Fig. 14.7 John Townson/Creation; **p.211** Fig. 14.24 Andrew Lambert, Fig. 14.25 Colin Cuthbert/Science Photo Library; Fig. 14.26 Andrew Lambert; **p.212** Figs. 14.27 & 14.28 Andrew Lambert, Fig. 14.29 Stevie Grand/Science Photo Library, Fig. 14.30 Andrew Lambert; **p.213** Fig. 14.32 Andrew Lambert; **p.214** Fig. 14.34 Andrew Lambert; **p.215** Fig. 14.35 Ed Young/Science Photo Library; **p.216** Fig. 14.38 GlaxoSmithKline, Fig. 14.39 Simon Fraser/Searle Pharmaceuticals/Science Photo Library; **p.217** Fig. 14.40 C J Clegg, Fig. 14.41a Maximilian Stock Ltd/Science Photo Library, Fig. 14.41b Richardson/Custom Medical Stock Photo/Science Photo Library, Fig. 14.42 R. Maisonneuve, Publiphoto Diffusion/Science Photo Library; **p.221** Fig. 15.1a Roger Scruton, Fig. 15.1b John Townson/Creation; **p.222** Fig. 15.2 Geoff Tompkinson/Science Photo Library; **p.223** Fig. 15.4a Mary Evans Picture Library, Fig. 15.4b Science & Society Picture Library, Fig. 15.5a St. Bartholomew's Hospital, London/Science Photo Library; **p.225** Fig. 15.7 Empics; **p.227** Fig. 15.10b Perkins-Elmer Ltd; **p.230** Fig. 15.13 *both* Shout Picture Library, Fig. 15.14a Geoff Tompkison/Science Photo Library, Fig. 15.14b Alfred Pasieka/Science Photo Library; **p.231** Fig. 15.16a *l* James King-Holmes/Science Photo Library, *tr* & *br* John Townson/Creation; **p.235** Fig. 15.21 Dr Jurgen Scriba/Science Photo Library; **p.236** Fig. 15.24 Holt Studios, Fig. 15.25 Julian Baum/Science Photo Library, Fig. 15.26 Roger Scruton.

l = left, *r* = right, *t* = top, *b* = bottom, *c* = centre

The publishers have made every effort to contact copyright holders. If any have been inadvertently overlooked they will be pleased to make the necessary arrangements at the earliest opportunity.

1
PHYSICAL CHEMISTRY

Seaborgium is a radioactive, synthetic element discovered in 1974. It does not occur naturally.

1 Kinetics

STARTING POINTS ● The rate of reaction can be obtained by following some property that changes as the reaction proceeds, for example, a colour change or a change in pH.
● Methods involved in the determination of reaction rate include techniques such as colorimetric analysis, gas collection and titration techniques.
● Theories of reaction rates include collision theory and transition state theory. These theories help to explain the processes involved in chemical reactions.
● Catalysts are substances that provide a chemical reaction with an alternative route that has a lower activation energy.
● Catalysts have widespread uses throughout the chemical industry.

Figure 1.1
The chemical reaction used by this rocket to put the Space Shuttle into orbit is a very fast one, whilst the production of the green copper carbonate/hydroxide on this roof is caused by a slow reaction between the atmosphere and the copper. In both cases, it would be useful to understand more about the reaction on the molecular level.

So far in your study of reaction rates or kinetics you have looked at rates of different types of chemical reaction and the factors which affect those rates (see *Introduction to Advanced Chemistry*, Chapter 8). In particular you have seen that the rate of almost all chemical reactions is affected by a change in the concentration of the reactants (Figure 1.2).

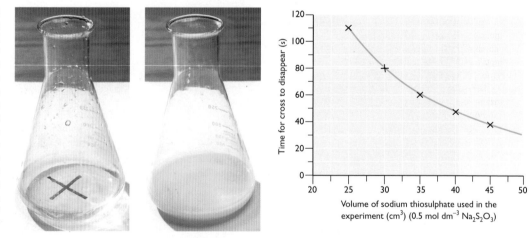

Figure 1.2
In the reaction between sodium thiosulphate and hydrochloric acid a yellow precipitate of sulphur is produced, which obscures the cross. The sample data in the graph shows the effect of different concentrations of sodium thiosulphate on the time taken for the cross to disappear. What is the effect of increasing the concentration of sodium thiosulphate on the rate of this reaction?

1 Give the main methods of altering the rate of a chemical reaction.

In this chapter we are particularly interested in seeing how the information derived from such studies can be used to obtain useful information about what is happening on a molecular level in chemical reactions. This will give us an insight into the mechanism of the reaction.

The rate equation

The rate of a chemical reaction can be found by following a property that changes as the reaction proceeds:

$$\text{rate of reaction} = \frac{\text{change in property}}{\text{time taken}}$$

For example, for concentration

$$\text{rate of reaction} = \frac{\text{change in concentration}}{\text{time taken}}$$

The reaction rate in this case is measured in units of $mol\,dm^{-3}\,s^{-1}$.

By analysing the reaction mixture at suitable time intervals, it is possible to determine the concentration of both reactants and products. We can then obtain a measure of the reaction rate.

The method used in the analysis depends on the reaction being studied. Earlier in your study of chemistry you will have used techniques such as

- colorimetry
- titration
- gas collection (Figure 1.3)

to follow chemical reactions.

Figure 1.3
a Gas collection can be used to follow chemical reactions.
b Graphs of volume of gas collected against time for various initial concentrations of hydrogen peroxide. As the concentration increases, so does the initial slope of the line on the graph.

a

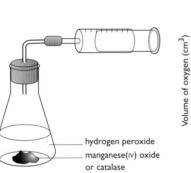

hydrogen peroxide
manganese(IV) oxide or catalase

b

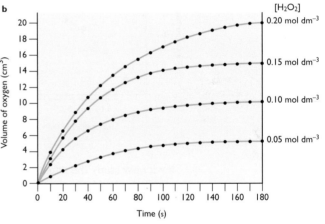

Figure 1.4
In humans hydrogen peroxide is toxic. It is a common by-product of metabolism but must be removed. Catalase is the fastest acting known enzyme and is present in the body to catalyse the decomposition of hydrogen peroxide.

Reactions are generally fastest in their early stages. This is indicated on Figure 1.3b by the slope of the various curves at the beginning of the reaction. Here the reaction is the breakdown of hydrogen peroxide into water and oxygen, catalysed by manganese(IV) oxide or the enzyme, catalase:

$$2H_2O_2(aq) \xrightarrow{\text{catalase}} 2H_2O(l) + O_2(g)$$

As the reaction proceeds the slope of the curve decreases, indicating that the reaction is slowing down. When the curve becomes horizontal the reaction has reached completion. By changing the *initial* concentration of hydrogen peroxide that is used, a collection of graphs can be produced showing how the rates change with change in *initial* concentration keeping the quantity of the catalyst constant (see Figure 1.3b). As the concentration of the hydrogen peroxide increases, so does the gradient of the graph at the start of each of the reactions.

The rate of reaction at the start of a reaction is called the **initial rate**. The initial rate of the reaction can be obtained by drawing a tangent to the curve at $t = 0$ and measuring its gradient. If we now plot a graph of initial rate (the slope of each of the tangents at $t = 0$ for the different concentrations of hydrogen peroxide shown in Figure 1.3b) against the concentration of hydrogen peroxide, then the graph shown in Figure 1.5 is produced. The straight line shown in Figure 1.5 indicates that the rate of this reaction is directly proportional to the concentration of hydrogen peroxide used:

$$\text{rate} \propto [\text{H}_2\text{O}_2]$$

or

$$\text{rate} = k[\text{H}_2\text{O}_2]$$

where k is a constant.

Figure 1.5
A graph showing how the initial rate of reaction for the catalysed decomposition of hydrogen peroxide depends upon $[\text{H}_2\text{O}_2]$.

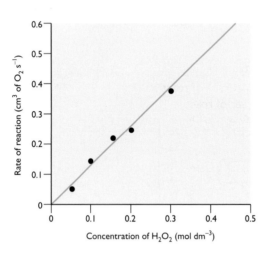

If we now study the effect of varying the concentration of the catalyst, catalase, on the decomposition of hydrogen peroxide solutions of the same concentration then a set of graphs similar to those shown in Figure 1.3b is obtained. This then leads to a graph similar to that shown in Figure 1.5 if *initial rate* is now plotted against [catalase]. Therefore, the rate of decomposition of hydrogen peroxide is directly proportional to the concentration of the catalase used:

$$\text{rate} \propto [\text{catalase}]$$

or

$$\text{rate} = k[\text{catalase}]$$

If we now combine the equation involving catalase with that involving hydrogen peroxide, we obtain an equation that indicates how the rate of the decomposition of hydrogen peroxide depends on the concentration of both of the reagents used:

$$\text{rate} = k[\text{H}_2\text{O}_2]\,[\text{catalase}]$$

where k is the **rate constant** and the overall equation is called the **rate equation**. (It should be noted that the higher the activation energy for a reaction then the lower will be the value of k. Also note that the values of k for each of the processes discussed above will be different.) The above rate equation was obtained from *actual* experiments. Provided we can carry out actual experiments to see how the rate of a chemical process is dependent on the concentrations of the reactants used, it is possible to write a rate equation for any chemical reaction. This rate equation can then be used to give us essential information about the mechanism of the reaction. (For a further discussion of rate equations and reaction mechanisms see page 9.)

Generally, for the reaction

$$A + B \rightarrow C + D$$

experiments may show that the rate of reaction is proportional to $[A]^x$ and $[B]^y$

$$\text{rate} \propto [A]^x$$

and

$$\text{rate} \propto [B]^y$$

The overall equation is therefore:

$$\text{rate} = k[A]^x[B]^y$$

where the powers x and y are the individual **orders of reaction** with respect to A and B. The total or **overall order of reaction** for this general case is given by $x + y$. The order of reaction can be defined as the power to which the concentration terms are raised in the rate equation.

In the catalase decomposition reaction of hydrogen peroxide, the rate equation determined *experimentally* is:

$$\text{rate} = k[H_2O_2]\,[\text{catalase}]$$

The order of reaction with respect to hydrogen peroxide is 1 and the order of reaction with respect to catalase is 1. We say that the reaction is first order with respect to both hydrogen peroxide and catalase. The overall order of reaction is $1 + 1 = 2$. We say that overall the reaction is second order.

A reaction may be zero order with respect to one of the reactants. For example, in the reaction between iodine (yellow/brown) and propanone in the presence of an acid catalyst:

$$I_2(aq) + CH_3COCH_3(l) \xrightarrow{H^+} CH_3COCH_2I(aq) + HI(aq)$$

the rate equation *determined by experiment* is:

$$\text{rate} = k[CH_3COCH_3]\,[H^+]$$

The rate equation indicates that the reaction is first order with respect to both propanone and hydrogen ions. However, it is zero order with respect to iodine, whose concentration term does not appear in the rate equation.

---

2 In the following rate equation state:
  a the order of reaction with respect to $HCrO_4^-$, HSO_3^- and H^+
  b the overall order of reaction.

$$\text{rate} = k[HCrO_4^-]\,[HSO_3^-]\,[H^+]$$

3 In the following rate equation suggest:
  a the order of reaction with respect to $HCOOCH_3$ and H^+
  b the overall order of reaction.

$$\text{rate} = k[HCOOCH_3]\,[H^+]$$

Figure 1.6
This data-logger system attached to a colorimeter allows direct collection and plotting of the absorption data for the reaction of iodine with propanone.

Calculation of order of reaction using the initial rates for pairs of experiments

It is possible to determine the order of reaction with respect to each reactant by comparing the initial concentration and the initial rates for pairs of experiments. For example, for the general reaction:

$$2A + B \rightarrow C + D$$

the initial rates measured for the reaction are shown in Table 1.1.

Table 1.1

Experiment	[A] $(mol\,dm^{-3})$	[B] $(mol\,dm^{-3})$	Initial rate $(mol\,dm^{-3}\,s^{-1})$
1	0.130	0.20	1.50×10^{-5}
2	0.130	0.40	6.00×10^{-5}
3	0.065	0.40	3.00×10^{-5}
4	0.065	0.20	0.75×10^{-5}

Using experiments 1 and 2, [A] stays the same but [B] is doubled. The initial rate quadruples from $1.50 \times 10^{-5}\,mol\,dm^{-3}\,s^{-1}$ to $6.00 \times 10^{-5}\,mol\,dm^{-3}\,s^{-1}$. This suggests that:

$$\text{rate} \propto [B]^2$$

This is confirmed by considering experiments 4 and 3, where the initial rate is again quadrupled when [B] is doubled. Therefore, the order with respect to B is 2.

Using experiments 4 and 1, [B] stays the same but [A] is doubled. The initial rate doubles from $0.75 \times 10^{-5}\,mol\,dm^{-3}\,s^{-1}$ to $1.50 \times 10^{-5}\,mol\,dm^{-3}\,s^{-1}$. This suggests that:

$$\text{rate} \propto [A]$$

This is confirmed by considering experiments 3 and 2, where the initial rate is again doubled when [A] is doubled. Therefore, the order with respect to A is 1.

Therefore since:

$$\text{rate} \propto [B]^2$$

and

$$\text{rate} \propto [A]$$

then the overall rate equation for the general reaction is

$$\text{rate} \propto [A]\,[B]^2$$

that is

$$\text{rate} = k[A]\,[B]^2$$

4 Describe how you could use a colorimeter to follow the reaction between iodine and propanone.

5 Determine the order of reaction with respect to NO and H_2, write the rate equation and calculate the value of the rate constant, including units, from the experimental data given for the following reaction:

$$2NO(g) + 2H_2(g) \rightarrow N_2(g) + 2H_2O(g)$$

Experiment	[NO] $(mol\,dm^{-3})$	[H$_2$] $(mol\,dm^{-3})$	Rate $(mol\,dm^{-3}\,s^{-1})$
1	1.25×10^{-2}	1.0×10^{-2}	1.2×10^{-6}
2	2.50×10^{-2}	2.0×10^{-2}	9.6×10^{-6}
3	1.25×10^{-2}	2.0×10^{-2}	2.4×10^{-6}
4	5.00×10^{-2}	2.0×10^{-2}	38.4×10^{-6}
5	2.50×10^{-2}	4.0×10^{-2}	19.2×10^{-6}

Using this rate equation it is possible to find a value for the rate constant, k. Simply substitute a set of values from a particular experiment into the overall rate equation. Using the values from experiment 1:

$$1.5 \times 10^{-5}\,mol\,dm^{-3}\,s^{-1} = k\,(0.130\,mol\,dm^{-3})\,(0.20\,mol\,dm^{-3})^2$$

$$\frac{1.50 \times 10^{-5}\,mol\,dm^{-3}\,s^{-1}}{(0.130\,mol\,dm^{-3})\,(0.20\,mol\,dm^{-3})^2} = k$$

$$k = 2.88 \times 10^{-3}\,dm^6\,mol^{-2}\,s^{-1}$$

Using concentration/time graphs to determine the order of reaction

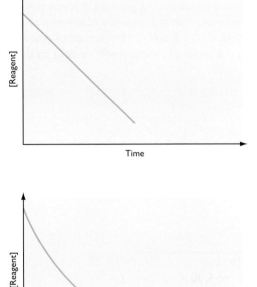

Figure 1.7
A zero-order reaction produces a graph of this type.

It is sometimes possible to tell the order of reaction with respect to a reactant by looking at the *concentration/time* graph. If the reaction is zero order with respect to a reagent then the graph produced is a straight line (see Figure 1.7). The data in Figure 1.7 comes from a set of experiments, followed over time.

In the case of first- and second-order reactions, the graph obtained is a curve. If the reaction is second order then a deeper curve is obtained for the graph of concentration/time (Figures 1.8 and 1.9).

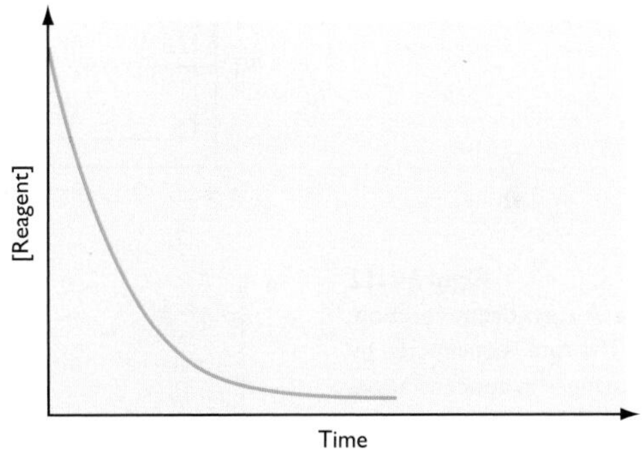

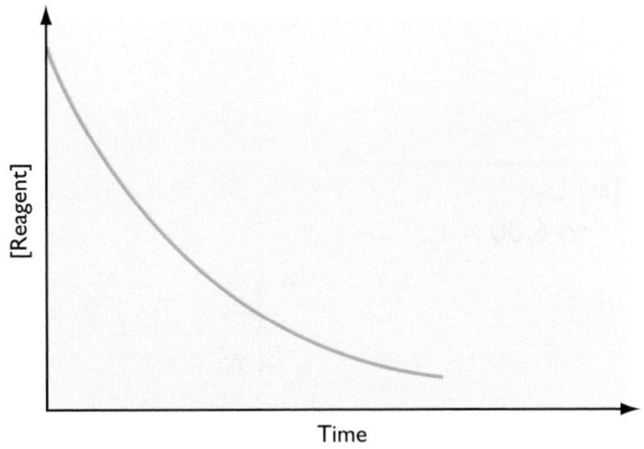

Figure 1.8 (left)
A first-order reaction produces a curve of this type.

Figure 1.9 (right)
A second-order reaction produces a deeper curve compared to the first-order reaction in Figure 1.8.

The reliability of this method for first- and second-order reactions is open to question since it relies on the interpretation of two 'slightly' different curves. This can be overcome, however, by looking at the **half-life** of a reaction.

Using half-lives

Figure 1.10
The graph shown has successive half-lives of approximately 25 seconds. This indicates that the reaction is first order with respect to $[H_2O_2]$.

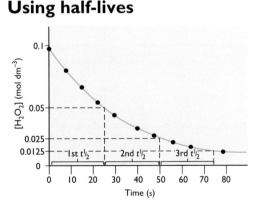

The half-life of a chemical reaction, $t_{1/2}$, is the time taken for the concentration of one of the reagents to be reduced by half. The first half-life for a reaction is taken from the start, the second is taken from the end of the first half-life, and so on. The graph in Figure 1.10 shows how the amount of hydrogen peroxide varies with time. If the reaction is first order with respect to hydrogen peroxide, then all the successive half-lives will, within the realms of experimental error, be the same.

The decomposition is first order with respect to $[H_2O_2]$, with a constant half-life of 25 seconds. If the reaction had been second order with respect to $[H_2O_2]$ then the curve obtained would have produced half-lives that would progressively increase. It should be noted that generally the half-life of a first-order reaction is independent of the initial concentration.

6 This question relates to an experiment between benzoyl chloride and phenylamine. Use the following data, relating to this reaction, to answer the questions below.

Time (min)	Concentration of phenylamine (mol dm^{-3})
0	0.0200
5	0.0133
10	0.0100
20	0.0069
30	0.0052

a Plot a graph of concentration of phenylamine against time.

b Obtain several half-lives for the reaction with respect to phenylamine.

c Use these half-life values to determine the order of reaction with respect to phenylamine.

Using rate/concentration graphs

It is possible to plot further useful graphs based on the concentration/time graphs shown in Figures 1.7, 1.8 and 1.9. To do this it is necessary to draw tangents to the concentration/time graphs at several time values and so obtain the gradient at these points (Figure 1.11). The gradient of the concentration/time graph is a measure of the rate of the reaction at that time.

Using data like that in Figure 1.11, a graph can be plotted of rate against concentration. The shape of this graph confirms the order of reaction with respect to the reagent being studied (Figure 1.12).

Figure 1.11
The rate of reaction at different time values is measured from the tangents taken from the concentration/time graph.

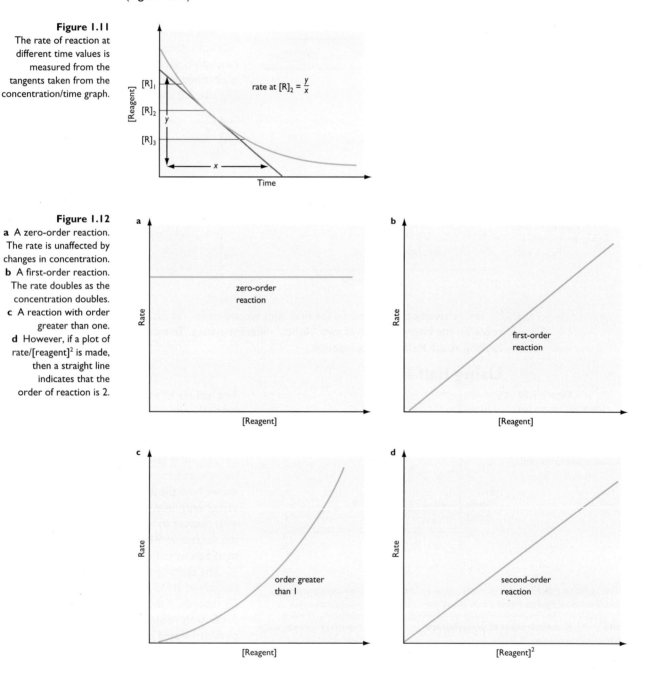

Figure 1.12
a A zero-order reaction. The rate is unaffected by changes in concentration.
b A first-order reaction. The rate doubles as the concentration doubles.
c A reaction with order greater than one.
d However, if a plot of rate/$[\text{reagent}]^2$ is made, then a straight line indicates that the order of reaction is 2.

Rate equations and reaction mechanisms

Rate studies allow us to investigate reactions on a molecular level. The great majority of reactions occur in more than one step and the separate steps that take us from reactants to products are called, collectively, the **reaction mechanism**. If we know the rate equation for a particular reaction we can make links to the reaction mechanism for that reaction, since the rate equation gives us information about the *slowest* step in the reaction mechanism, that is, the **rate-determining step**, sometimes known as the **rate-limiting step**. As an example, let us consider the reaction between 2-bromo-2-methylpropane, a tertiary halogenoalkane, and hydroxide ions from sodium hydroxide (see *Introduction to Advanced Chemistry*, Chapter 18, page 197).

$$CH_3-\underset{\underset{CH_3}{|}}{\overset{\overset{CH_3}{|}}{C}}-Br + OH^- \rightarrow CH_3-\underset{\underset{CH_3}{|}}{\overset{\overset{CH_3}{|}}{C}}-OH + Br^-$$

It is found from experimental data that the rate equation is:

$$rate = k[(CH_3)_3CBr]$$

This indicates that the reaction is first order with respect to $[(CH_3)_3CBr]$ but zero order with respect to $[OH^-]$. This suggests that OH^- is not involved in any of the steps up to and including the rate-determining or slow step for this reaction. The conclusion that follows is that the rate-determining step involves the heterolytic fission of the C—Br bond to produce the carbocation $(CH_3)_3C^+$, as shown below:

$$CH_3-\underset{\underset{CH_3}{|}}{\overset{\overset{CH_3}{|}}{C}}-Br \xrightarrow{slow} CH_3-\underset{\underset{CH_3}{|}}{\overset{\overset{CH_3}{|}}{C^+}} + Br^-$$

It has been shown that the reaction of $(CH_3)_3CBr$ with OH^- takes place in two steps. The first step is the slow step shown above and the second step involves the carbocation $(CH_3)_3C^+$ reacting quickly with OH^-:

$$CH_3-\underset{\underset{CH_3}{|}}{\overset{\overset{CH_3}{|}}{C^+}} + OH^- \xrightarrow{fast} CH_3-\underset{\underset{CH_3}{|}}{\overset{\overset{CH_3}{|}}{C}}-OH$$

The reaction of $(CH_3)_3CBr$ with OH^- is known as an **S$_N$1** or **substitution nucleophilic unimolecular** reaction. In other words, it is a reaction which takes place by **nucleophilic substitution** and whose rate is determined by the concentration of the tertiary halogenoalkane:

$$rate = k[\text{tertiary halogenoalkane}]$$

Unimolecular refers to the **molecularity**, that is the number of reacting species that take part in the rate-determining step. Molecularity is always a whole number. Rate-determining steps may be unimolecular, bimolecular or trimolecular.

There is a second type of nucleophilic substitution reaction whose rate depends on the concentration of both the halogenoalkane and the nucleophile. This is known as an **S$_N$2** reaction.

Primary halogenoalkanes undergo S$_N$2 reactions; for example, the reaction between 1-bromobutane and hydroxide ions. It is found from experimental data that the rate equation is:

$$rate = k[CH_3CH_2CH_2CH_2Br][OH^-]$$

This indicates that the reaction is first order with respect to [CH₃CH₂CH₂CH₂Br] and [OH⁻]. This suggests that both OH⁻ and CH₃CH₂CH₂CH₂Br are involved in the slow step for this reaction. In this reaction the OH⁻ attaches itself to the electropositive $C^{\delta+}$ atom at the same time as the halogen leaves. An intermediate is produced known as the **activated complex**. This complex arises if the particles involved have enough energy on collision to overcome repulsions between the outer electrons of the two species reacting.

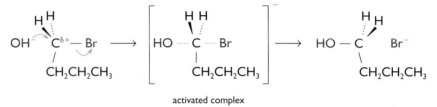

activated complex

Many reactions proceed through a **transition state**. For example, in the decomposition of hydrogen iodide a transition state complex is created (Figure 1.13) and if there is sufficient kinetic energy available on collision then this complex will decompose to produce products.

Figure 1.13
The transition state complex occurs at the peak of the reaction profile.

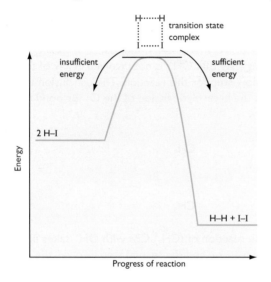

Effect of temperature

According to **collision theory**, before two chemicals can react together to form products the particles involved must collide. During this collision bonds are broken or made, which results in the rearrangement of the atoms present and the formation of the product. Collisions in which products arise are said to be **effective collisions** (Figure 1.14).

Figure 1.14
Effective collisions between methane and oxygen molecules produce carbon dioxide and water.

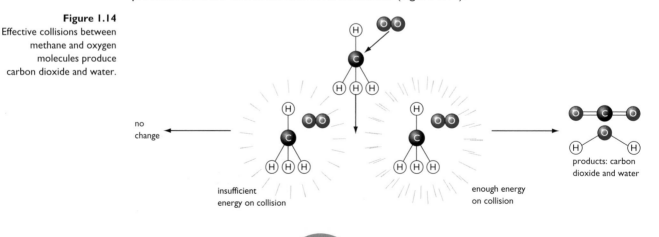

Not all collisions result in the formation of products, otherwise all chemical reactions would be over in a fraction of a second. In practice only a very small proportion of collisions are effective. There are two main reasons for this:

- the particles must have a certain minimum amount of energy on collision – this is known as the **activation energy** (Figure 1.15)
- the particles must approach one another in the correct orientation – this is called the **steric factor**.

Figure 1.15
The minimum energy required to form products is the activation energy.

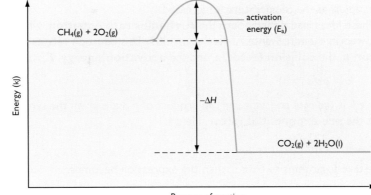

These ideas can be used to explain the effects of the factors, such as temperature, that change the rate of chemical reactions.

At higher temperatures reactant particles will, on average, have more kinetic energy. This results in the particles having a greater velocity, which results in a greater number of collisions. Also, when they collide more particles will have more than the activation energy for that reaction. Both these factors lead to more effective collisions and a faster rate of reaction.

At any moment some of the particles will have high energies, some will have low energies, but most of the particles will have energies close to the average energy of the system. The **Maxwell–Boltzmann distribution** curve shows the distribution of energies amongst reacting particles at a particular temperature, T_1 (measured in kelvin, K) (see Figure 1.16). It should be appreciated that originally the graph shown in the diagram was related to colliding particles in a pure gas. It has more recently been applied with some success, however, to systems involving chemical reactions.

Figure 1.16
The Maxwell–Boltzmann distribution curve at temperature T_1.

The area under the graph in Figure 1.16 is proportional to the number of particles in the reaction mixture. The curve shows why only a certain number of particles are able to undergo successful collisions. It is those that have more than the activation energy (E_a) for the reaction.

At a higher temperature, T_2, a greater proportion of the particles in the reaction mixture now have an energy greater than the activation energy and thus more successful collisions occur, which increases the rate of the reaction (Figure 1.17). As a guide it is found that increasing the temperature of a reaction mixture by 10 K will roughly double the rate of the reaction. It should be pointed out that this applies particularly to reactions with an activation energy of about 55 kJ mol^{-1}.

Figure 1.17
Maxwell–Boltzmann distribution curves for temperatures T_1 and T_2.

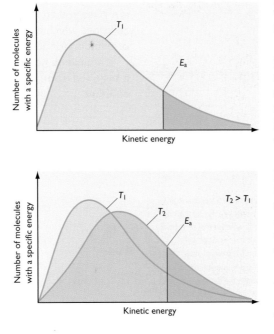

The Arrhenius rate equation

Using the Maxwell–Boltzmann distribution it can be suggested that the fraction of the molecules with energy greater than the activation energy, E_a, is given by the exponential expression, $e^{-E_a/RT}$, where R is the gas constant, T is the temperature in kelvin and 'e' is the base of natural logarithms ($= 2.718$). The fact that it is an exponential relationship means that a small rise in the temperature results in a large increase in the number of molecules with an energy greater than the activation energy, and this is shown by the area which has been shaded in Figure 1.17. Therefore, it can be seen that the fraction of molecules with an energy greater than E_a increases very rapidly as the temperature is increased.

These ideas lead directly onto the **Arrhenius rate equation**, which was first predicted by the Swedish chemist Svante Arrhenius in 1889. This equation relates three factors, the **steric factor**, p, the **collision factor**, z, and the **activation energy**, E_a. It is usually written:

$$k = pz(e^{-E_a/RT})$$

where k is the rate constant and the product of $p \times z$ is given the symbol A and is often referred to as the **pre-exponential factor**. Hence:

$$k = A(e^{-E_a/RT})$$

If we take logarithms to base e, then the expression becomes:

$$\ln k = \ln A + \ln e^{-E_a/RT}$$

or

$$\ln k = \ln A - E_a/RT$$

If this expression is compared with:

$$y = mx + c$$

which is the general equation for a straight line, then a graph of $\ln k$ against $1/T$ should be a straight line with a gradient of $-E_a/R$ (Figure 1.18).

Figure 1.18
A straight line graph produced by plotting $\ln k$ against $1/T$. The gradient is $-E_a/R$ and the intercept is $\ln A$.

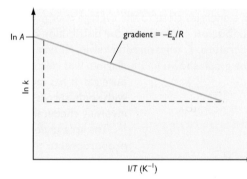

It is possible, therefore, from a plot of $\ln k$ against the reciprocal of the absolute temperature ($1/T$) to obtain a value of the activation energy from the gradient, $-E_a/R$. The intercept on the $\ln k$ axis of this graph is $\ln A$ (the pre-exponential factor). This value of $\ln A$ gives you an idea of the number of collisions which have taken place in the correct orientation for a reaction to take place.

It should be noted that a reaction which has a large activation energy will have a small rate constant.

7 Hydrogen iodide decomposes according to the following equation:

$$2HI(g) \rightarrow H_2(g) + I_2(g)$$

The following values for the rate constant, k, were obtained at different temperatures:

Temperature (K)	Rate constant, k (mol dm^{-3} s^{-1})
500	3.77×10^{-9}
600	6.68×10^{-6}
700	1.16×10^{-3}
800	7.73×10^{-2}

a Define the terms rate constant, k, and activation energy, E_a, and write the expression which relates k to E_a.
b Plot a graph of $\ln k$ against $1/T$.
c Use the graph you have plotted in part **b** to determine a value for the activation energy, E_a, for the decomposition of hydrogen iodide ($R = 8.31 \text{ J K}^{-1} \text{mol}^{-1}$).

● **Key skills** ICT
- Use of data-logging equipment to follow the rate of reaction by monitoring pH, transmittance of light through solutions, concentration or conduction, followed by subsequent processing of data.

Number
- Processing of data from reactions to obtain the order of reaction with respect to individual reactants, the overall rate equation and the activation energy for the reaction.
- Scaling of graphs.

● **Skills task** Discuss the following.

- Radioactive decay is a first-order process.
- Why does a reaction not occur after every collision between two reactant particles?
- Collision frequency decreases as the pressure of a gas is lowered.

CHECKLIST After studying Chapter 1 you should know and understand the following terms.

- ● **Initial rate:** The rate of a reaction at its start.
- ● **Rate equation:** An equation which shows how changes in concentration affect the rate of a reaction. The rate equation cannot be deduced from the balanced chemical equation; it has to be found by experiment.
- ● **Rate constant:** A constant for a given reaction at a particular temperature. Symbol: k.
- ● **Reaction mechanism:** The series of separate steps that describe a chemical reaction from start to finish, in terms of the bonds which break and the new bonds which form.
- ● **Rate-determining step:** The slowest step in the reaction mechanism which, therefore, controls the overall rate of the reaction.
- ● **Transition state:** The state of the reacting species when they are at the highest point of the activation energy barrier for a reaction step.
- ● **Collision theory:** A theory used to account for the effects of concentration, temperature and catalysts on reaction rates. It is assumed that chemical reactions occur by collision of the particles, with bond breaking and bond forming taking place.
- ● **Maxwell–Boltzmann distribution:** Shows the distribution of energy amongst particles at a particular temperature.
- ● **Arrhenius equation:** $k = Ae^{-E_a/RT}$ relates three factors, the steric factor, the collision factor and the activation energy, to the rate constant, k.

Examination questions

1 a A chemical reaction is first order with respect to compound **X** and second order with respect to compound **Y**.

i Write the rate equation for this reaction.
ii What is the overall order of this reaction?
iii By what factor will the rate increase if the concentrations of **X** and **Y** are **both** doubled? (4)

b The table below shows the initial concentrations of two compounds, **A** and **B**, and also the initial rate of the reaction that takes place between them at constant temperature.

Experiment	[A] (mol dm^{-3})	[B] (mol dm^{-3})	Initial rate (mol dm^{-3} s^{-1})
1	0.2	0.2	3.5×10^{-4}
2	0.4	0.4	1.4×10^{-3}
3	0.8	0.4	5.6×10^{-3}

i Determine the overall order of the reaction between **A** and **B**. Explain how you reached your conclusion.
ii Determine the order of reaction with respect to compound **B**. Explain how you reached your conclusion.
iii Write the rate equation for the overall reaction.
iv Calculate the value of the rate constant, stating its units. (7)

AQA, A level, Specimen Paper 6421, 2001/2

2 Hydrogen peroxide decomposes in the presence of metal oxide catalysts to form oxygen and water according to the following equation.

$$2H_2O_2 \rightarrow 2H_2O + O_2$$

You are required to plan an experiment to compare which of two metal oxides is the more effective catalyst by measuring the amount of oxygen formed during a 2 minute period immediately after a solution of hydrogen peroxide is added to a catalyst.

a Calculate the volume of oxygen formed (measured at 20 °C and 100 kPa) when 100 cm^3 of 1.00 mol dm^{-3} hydrogen peroxide decomposes. (5)

b Describe, giving details of the apparatus and procedure, how you could carry out experiments to compare the two metal oxides. Use your answer to part **a** to help you choose a suitable scale of apparatus and amount of hydrogen peroxide solution to use. (10)

AQA, A level, Specimen Paper 6421, 2001/2

3 a The product of bromination of ethane is bromoethane. This reacts with potassium cyanide in a solution of ethanol and water. The rate of this reaction was studied and the results are given below.

Experiment	[CN$^-$] (mol dm^{-3})	[C$_2$H$_5$Br] (mol dm^{-3})	Initial rate (mol dm^{-3} s^{-1})
1	0.060	0.020	1.0×10^{-5}
2	0.060	0.040	2.0×10^{-5}
3	0.120	0.020	2.0×10^{-5}

Deduce, showing your reasoning, the rate equation. (3)

b Two routes can be suggested for the reaction in part **a**.

Route I

$$CN^- + CH_3CH_2Br \rightarrow \left[\begin{array}{c} H \quad\quad CH_3 \\ NC \cdots C \cdots Br \\ H \end{array} \right]^-$$

$$\rightarrow CH_3CH_2CN + Br^-$$

Route II

$$CH_3CH_2Br \xrightarrow{slow} CH_3CH_2^+ + Br^-$$

then

$$CH_3CH_2^+ + CN^- \xrightarrow{fast} CH_3CH_2CN$$

i Explain which route is consistent with the rate equation in part **a**. (1)
ii This exothermic reaction is catalysed by iodide ions. Draw the enthalpy level diagram for both the uncatalysed and the catalysed reaction, labelling each clearly. (3)

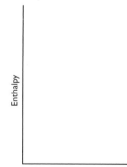

Edexcel, AS/A level, Module 2, June 2000

4 Iodine reacts with propanone in the presence of an acid catalyst according to the equation:

$$CH_3COCH_3 + I_2 \xrightarrow{H^+} CH_3COCH_2I + HI$$

Data concerning an experiment to determine the rate equation for this reaction are given in the following table:

Relative concentrations			Relative rate
[CH_3COCH_3]	[I_2]	[H^+]	
1	1	1	2
1	2	1	2
2	1	1	4
1	1	2	4

a i State why a rate equation cannot be written from a knowledge only of the chemical equation representing the reaction. (2)

ii Use the data to deduce the order of reaction with respect to

propanone   iodine   hydrogen ions (3)

iii What does the value for iodine tell you about the part that iodine plays in the rate determining step of the reaction? (1)

iv Write the rate equation for the reaction. (1)

b Draw an enthalpy level diagram for the reaction, showing on it both catalysed and uncatalysed pathways. (3)

London, AS/A level, Module 2, June 1999

5 Nitrogen oxides such as nitrogen monoxide, NO, and nitrogen dioxide, NO_2, are formed unintentionally by man and cause considerable harm to the environment.

a The oxidation of nitrogen monoxide in car exhausts may involve the following reaction:

$$NO(g) + CO(g) + O_2(g) \rightarrow NO_2(g) + CO_2(g)$$

This reaction was investigated in a series of experiments. The results are shown in the table below.

Experiment	[NO(g)] (mol dm^{-3})	[CO(g)] (mol dm^{-3})	[O$_2$(g)] (mol dm^{-3})	Initial rate (mol dm^{-3} s^{-1})
1	1.00×10^{-3}	1.00×10^{-3}	1.00×10^{-1}	0.44×10^{-3}
2	2.00×10^{-3}	1.00×10^{-3}	1.00×10^{-1}	1.76×10^{-3}
3	2.00×10^{-3}	2.00×10^{-3}	1.00×10^{-1}	1.76×10^{-3}
4	2.00×10^{-3}	2.00×10^{-3}	4.00×10^{-1}	7.04×10^{-3}

i For each reactant, NO, CO and O_2, deduce the order of reaction. Show your reasoning.

ii Deduce the rate equation and calculate the rate constant for this reaction.

iii Suggest, with a reason, what would happen to the value of the rate constant, k, as the car's exhaust gets hotter. (11)

b State **two** environmental consequences of nitrogen oxides. (2)

c Not all nitrogen compounds are harmful: some, such as nitrogen fertilisers, are beneficial to man.

A nitrogen fertiliser, **D**, was analysed in the laboratory and was shown to have the composition by mass Na, 27.1%; N, 16.5%; O, 56.4%. On heating, 3.40 g of **D** was broken down into sodium nitrite, $NaNO_2$, and oxygen gas.

Showing your working, suggest an identity for the fertiliser, **D**, and calculate the volume of oxygen that was formed. (Under the experimental conditions, 1 mole of gas molecules occupies 24 dm^3.) (4)

OCR, A level, Specimen Paper A7882, Sept 2000

6 The kinetics of a reaction are used to clarify reaction mechanisms. An experiment to determine the kinetics of the substitution reaction between 2-chloro-2-methylpropane and sodium hydroxide uses **equal initial** concentrations of these substances in aqueous ethanol solvent. A mixture was maintained at 25 °C, and samples taken at intervals. The samples are quenched in about twice their volume of cold propanone, and the concentration of sodium hydroxide is found.

Time (min)	0	7	15	27	44	60
Concentration (mol dm^{-3})	0.080	0.065	0.054	0.041	0.028	0.020

a Show by means of a suitable graph that the reaction is first order. (4)

b As performed, the results cannot distinguish between the rate laws

rate = k[OH$^-$] and **rate = k[halogenoalkane]**

Outline a further experiment which must be performed to enable the distinction to be made, showing how the new data would be used to establish the rate law. (3)

c The reaction is in fact first order with respect to the halogenoalkane. Write the mechanism for the substitution reaction, identifying the rate-determining step. (4)

d Nucleophilic substitution is usually accompanied by elimination as a competing reaction. Write the name and structural formula of the product of the elimination reaction with 2-chloro-2-methylpropane and state the conditions which favour elimination over substitution. (3)

London, A level, Synoptic Paper (CH6), June 1999

2 Further chemical equilibria

STARTING POINTS
● An equilibrium is established when the rate of the forward reaction equals the rate of the back reaction in a closed system.
● Any chemical equilibrium can be described by an equilibrium constant.
● The position of chemical equilibria can be affected by changing the conditions under which it is established.
● The effects of changing the conditions can be summarised by Le Chatelier's principle.
● Catalysts do not affect the position of a chemical equilibrium. They only affect the rate at which the equilibrium is established.

Figure 2.1 (below)
a The chemicals in the fireworks create the colourful display as they react to completion.
b When sodium chloride solution is added to silver nitrate solution a white precipitate of silver chloride is produced as the reaction goes to completion.
c The exothermic reaction of carbon burning completely to carbon dioxide creates the glow and flames as the carbon burns.

From your previous study of Chemistry you will be aware that some reactions go almost to completion (Figure 2.1), for example reactions of ions in solution:

$$Ag^+(aq) + Cl^-(aq) \rightarrow AgCl(s)$$

Other reactions never reach completion because the products react together to re-form the reactants; they are **reversible reactions**. For example:

$$H_2(g) + I_2(g) \rightleftharpoons 2HI(g)$$

$$CH_3COOH(l) + C_2H_5OH(l) \rightleftharpoons CH_3COOC_2H_5(l) + H_2O(l)$$

Reactions such as those above are dynamic and eventually establish an **equilibrium**. In the reaction mixture after any period of time there will be a quantity of all the chemicals present. Most organic reactions are in equilibrium.

a b c
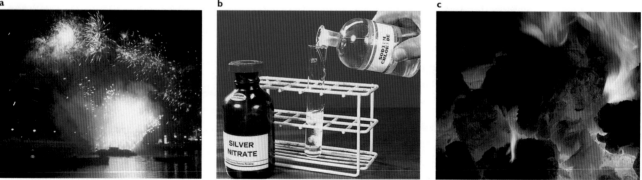

Characteristics of the equilibrium state

- Chemical equilibrium can only be attained in a **closed system**, one that cannot exchange matter with its surroundings.
- Chemical equilibria can be approached from either direction. A given equilibrium composition for a particular reaction can be made starting from substances from either the left-hand side or right-hand side of the equation. Changes of this kind are **reversible** reactions.
- Chemical equilibrium is **dynamic**, not static. This means that the forward and back reactions still take place once equilibrium has been established but they occur at equal rates. The overall result is that the amounts of reactants and products remain constant.
- **Homogeneous equilibria** are those which contain all the reactants and products in the same physical state.
- **Heterogeneous equilibria** contain reactants and products in different physical states.

The equilibrium constant

The amount of products and reactants in a particular equilibrium process can be measured by the **equilibrium constant**, given the symbol K_c.

Consider the reaction between hydrogen gas and iodine vapour at 698K.

$$H_2(g) + I_2(g) \rightleftharpoons 2HI(g)$$

The equilibrium concentration of each of the chemicals in the above equilibrium process was measured in three different runs of the experiment using different starting amounts of hydrogen gas and iodine vapour heated in a sealed vessel. In order to analyse the equilibrium mixture it was 'frozen' by cooling the reaction vessel rapidly, breaking open the tube and estimating the equilibrium concentration of iodine with a standard sodium thiosulphate solution. Knowing the initial amounts of hydrogen and iodine then allowed the equilibrium concentrations of all three chemicals to be calculated. The results are shown in Table 2.1.

Table 2.1
Results obtained by heating hydrogen and iodine in a sealed vessel, at 698K.

$[H_2(g)]_{eqm}$ (mol dm^{-3})	$[I_2(g)]_{eqm}$ (mol dm^{-3})	$[HI(g)]_{eqm}$ (mol dm^{-3})
4.56×10^{-3}	0.74×10^{-3}	13.54×10^{-3}
3.56×10^{-3}	1.25×10^{-3}	15.59×10^{-3}
2.25×10^{-3}	2.34×10^{-3}	16.85×10^{-3}

From your previous study of Chemistry you will be aware that the equilibrium law enables us to write an equilibrium constant expression for any chemical equilibrium process. We can write an expression for K_c for the general reaction:

$$aA + bB \rightleftharpoons cC + dD$$

$$K_c = \frac{[C]^c [D]^d}{[A]^a [B]^b}$$

In the expression for K_c above, [ ] stands for '**concentration of**' in units of mol dm^{-3}. This is a general expression of the **equilibrium law**. It should be noted that K_c, the equilibrium constant, is constant at a given temperature. If the temperature changes, the value of K_c changes.

For the equilibrium for which we have data in Table 2.1, an expression for K_c can be written as:

$$K_c = \frac{[HI(g)]^2_{eqm}}{[H_2(g)]_{eqm} [I_2(g)]_{eqm}}$$

where $[H_2(g)]_{eqm}$, for example, is the equilibrium concentration of hydrogen gas.

If the equilibrium concentrations for each of the three experiments mentioned in Table 2.1 are substituted in this expression it is found that K_c for this particular equilibrium at 698K is about 54.

Table 2.2
Data obtained by heating hydrogen iodide in a sealed vessel.

$[HI(g)]_{eqm}$ (mol dm^{-3})	$[H_2(g)]_{eqm}$ (mol dm^{-3})	$[I_2(g)]_{eqm}$ (mol dm^{-3})	$K_c = \dfrac{[HI(g)]^2_{eqm}}{[H_2(g)]_{eqm} [I_2(g)]_{eqm}}$
3.53×10^{-3}	0.48×10^{-3}	0.48×10^{-3}	54.08
3.66×10^{-3}	0.50×10^{-3}	0.50×10^{-3}	53.58
8.41×10^{-3}	1.14×10^{-3}	1.14×10^{-3}	54.42

The above equilibrium can also be approached from the other direction by decomposition of HI(g). When some hydrogen iodide is placed in a sealed tube and heated to 698K and allowed to reach equilibrium with the hydrogen and iodine that is produced, the data in Table 2.2 were obtained.

In Table 2.2 the values for K_c for the three experiments have been calculated, again giving a value of approximately 54 (at 698K). Notice that for this particular reaction K_c has no units as they cancel out in the K_c expression. It is important to work out the units of K_c for each reaction.

If K_c has a value larger than 1 it indicates that there is a larger proportion of products than reactants in the equilibrium mixture. If K_c is smaller than 1 then there is a larger proportion of reactants than products in the equilibrium mixture.

Whenever an equilibrium constant is quoted it must be accompanied by an equation for the equilibrium process being considered and the temperature at which equilibrium is reached must also be given.

Calculations involving K_c

K_c can be used to determine the concentration of a particular reactant or product in an equilibrium mixture.

Example N_2O_4 gas dissociates in a reversible reaction to form red/brown NO_2 gas. At a particular temperature K_c for this dissociation is $1.0 \times 10^{-5}\,mol\,dm^{-3}$ and the concentration of NO_2 gas, determined by colorimetry, was $1.6 \times 10^{-3}\,mol\,dm^{-3}$. What was the equilibrium concentration of N_2O_4 gas?

In answering any question involving equilibrium constants it is always very important to write down a balanced chemical equation for the process.

$$N_2O_4(g) \rightleftharpoons 2NO_2(g)$$

equilibrium concentration $\quad\quad x \quad\quad 1.6 \times 10^{-3}$
(mol dm^{-3})

$$K_c = \frac{[NO_2(g)]^2_{eqm}}{[N_2O_4(g)]_{eqm}}$$

$$1.0 \times 10^{-5}\,mol\,dm^{-3} = \frac{(1.6 \times 10^{-3}\,mol\,dm^{-3})^2}{x}$$

$$x = \frac{(1.6 \times 10^{-3}\,mol\,dm^{-3})^2}{1.0 \times 10^{-5}\,mol\,dm^{-3}} = 0.256\,mol\,dm^{-3}$$

Therefore,

$$[N_2O_4(g)]_{eqm} = 0.256\,mol\,dm^{-3}$$

In another form of question, you must be able to use the balanced chemical equation to calculate K_c given the initial concentrations of the reactants and the equilibrium concentrations of a reactant.

Example   When hydrogen and iodine react together they form hydrogen iodide. 10 mol of hydrogen and 5 mol of iodine were allowed to react and to reach equilibrium, producing 9 mol of hydrogen iodide. Calculate the value of K_c at this temperature.

Again, write down a balanced chemical equation for the process.

$$H_2(g) \quad + \quad I_2(g) \rightleftharpoons 2HI(g)$$

initial amount (mol) $\quad\quad 10 \quad\quad\quad 5 \quad\quad\quad -$

Since there is a $1:2$ mole relationship between the H_2/I_2 and HI we can calculate the equilibrium amounts of H_2 and I_2.

equilibrium amount (mol) $(10 - 9/2)\,(5 - 9/2)\,9$

$$K_c = \frac{[HI(g)]^2_{eqm}}{[H_2(g)]_{eqm}\,[I_2(g)]_{eqm}}$$

In order to calculate a value for K_c we need to use the concentrations of the chemicals involved. At present we have only calculated the amount of each, in moles, not their concentrations in mol dm^{-3}. To do this we need to incorporate the volume of the reaction vessel into the expression. In this example we are not given the volume so we assume that it is V dm^3. Concentrations can now be used in the expression:

$$K_c = \frac{(9/V\,mol\,dm^{-3})^2}{5.5/V\,mol\,dm^{-3} \times 0.5/V\,mol\,dm^{-3}} = 29.5\,\text{(no units)}$$

On this occasion the volume cancels, but this is not the case generally.

1 Write down K_c expressions for each of the following equilibria, giving the units in each case.
  a $PCl_5(g) \rightleftharpoons PCl_3(g) + Cl_2(g)$
  b $CH_3COOH(l) + C_2H_5OH(l) \rightleftharpoons CH_3COOC_2H_5(l) + H_2O(l)$
  c $C_2H_4(g) + H_2O(g) \rightleftharpoons C_2H_5OH(g)$
2 A mixture of 2.5 mol of hydrogen and 2.5 mol of iodine was allowed to reach equilibrium in a vessel of volume 4 dm^3, at a particular temperature. The equilibrium mixture contained 4 mol of hydrogen iodide. Calculate the value of K_c at this temperature.
3 At a particular temperature $K_c = 3.0 \times 10^{-4}\,mol\,dm^{-3}$ for the equilibrium:
$2H_2S(g) \rightleftharpoons 2H_2(g) + S_2(g)$
The equilibrium mixture contained
$[H_2S(g)]_{eqm} = 5.0 \times 10^{-3}\,mol\,dm^{-3}$ and
$[S_2(g)]_{eqm} = 2.4 \times 10^{-3}\,mol\,dm^{-3}$.
Calculate the equilibrium concentration of hydrogen gas.

Changes in the equilibrium conditions

Figure 2.2
The French scientist Henri
Le Chatelier (1850–1936)

The French chemist Henri Le Chatelier (1850–1936) (Figure 2.2) studied chemical equilibria at great length. It is his principle which allows us to predict the direction in which a chemical equilibrium will move when conditions such as the concentrations of reactants and products, the temperature at which the equilibrium is obtained and pressure (for gaseous equilibria), are altered.

Le Chatelier's principle states:

If a chemical system at equilibrium undergoes changes in the equilibrium conditions under which it has been established, then the chemical system responds by minimising the effect of the changes.

By using this principle it is possible to change conditions in order to provide higher yields of a particular substance in the equilibrium mixture. In industry this makes the production of certain chemicals such as ammonia and sulphuric acid economically viable. It is important to note that the principle is purely qualitative and so gives us no idea about the quantities of certain chemicals in the equilibrium process.

Changing concentration

In the AS level course you will have used this principle to explain how increasing the concentration of a reactant in an equilibrium causes the position of equilibrium to move to the right, increasing the concentration of products in the reaction mixture. As a quick reminder consider the equilibrium:

$$BiCl_3(aq) + H_2O(l) \rightleftharpoons BiOCl(s) + 2HCl(aq)$$

Figure 2.3
The equilibrium moves to the right as water is added to the equilibrium mixture.

- If water is added to the above chemical system, once it has reached equilibrium, the system minimises the effect by moving to the right, producing more $BiOCl(s)$ and $HCl(aq)$ (Figure 2.3).
- If water is removed from the system the equilibrium will move to the left to produce more $H_2O(l)$.

In both of these examples, the position of equilibrium is changed. Importantly, however, when the system returns to equilibrium the value of K_c will be unchanged.

Changing pressure

Changes in pressure only affect chemical equilibria that contain gaseous reactants or products and where there are differences in the numbers of moles of gaseous reactants or products on each side of the equation.

For example, consider the equilibrium between colourless dinitrogen tetraoxide gas, N_2O_4, and red/brown nitrogen dioxide gas, NO_2.

$$2NO_2(g) \rightleftharpoons N_2O_4(g)$$
red/brown   colourless

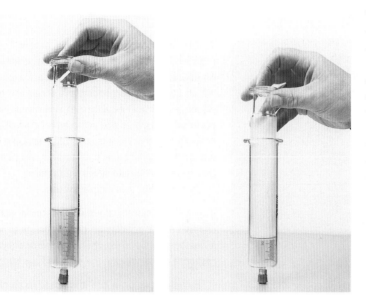

When the equilibrium pressure is increased the equilibrium moves in such a direction as to counter the increase in pressure. It does this by moving in the direction which will result in a reduction of the number of molecules, effectively reducing the pressure. In this case the equilibrium moves to the right, producing more of the colourless N_2O_4 gas (Figure 2.4). Once again, although the position of equilibrium has been changed, when the equilibrium is re-established at the higher pressure the value of K_c is unchanged.

Figure 2.4
An increase in pressure results in the formation of more colourless N_2O_4 gas from NO_2 gas.

Changing temperature

If the temperature at which equilibrium is established is raised, the system moves in such a way as to oppose the change. It does this by moving in the direction of the endothermic reaction, the reaction which takes in heat energy. A decrease in temperature results in the exothermic reaction being favoured, the reaction that would heat up the system by giving out heat energy.

Consider the equilibrium process between sulphur dioxide, oxygen and sulphur trioxide, which is one of the steps in the industrial manufacture of sulphuric acid by the Contact process (Figure 2.5).

$$2SO_2(g) + O_2(g) \rightleftharpoons 2SO_3(g) \qquad \Delta H = -197\,kJ\,mol^{-1}$$

A decrease in temperature would result in the forward reaction being favoured and the direction of equilibrium being moved to the right.

In industry a temperature of 450 °C is used. You may think that this is not necessarily a low temperature, which would favour the production of sulphur trioxide. You would be right. It is, however, a compromise temperature that results in sufficient sulphur trioxide being produced at a fast enough rate.

Figure 2.5
In the manufacture of sulphuric acid, once the SO_3 has been produced it is then dissolved in 98% sulphuric acid to produce oleum. This is then diluted with more 98% sulphuric acid to produce the concentration of acid required.

Using a catalyst

A catalyst has no effect on the position of a chemical equilibrium. It does, however, decrease the time required for the equilibrium to be established. It does this by providing an alternative pathway for the reaction with a lower activation energy for both the forward and back reactions (Figure 2.6a). This has the effect of increasing the rate of both reactions to the same extent.

Figure 2.6
a A catalyst does not affect the position of equilibrium but increases the rate at which it is established by reducing the activation energy of the forward and reverse reactions.
b The lower the bar, the easier it is for the pole vaulter to go over.

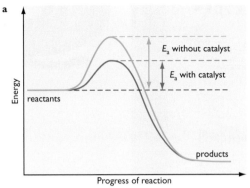

4 In which direction would the position of equilibrium move in each of the following reactions if the pressure was increased (by compressing the reaction mixture)?
a $PCl_5(g) \rightleftharpoons PCl_3(g) + Cl_2(g)$
b $H_2(g) + I_2(g) \rightleftharpoons 2HI(g)$
c $COCl_2(g) \rightleftharpoons CO(g) + Cl_2(g)$
d $N_2(g) + 3H_2(g) \rightleftharpoons 2NH_3(g)$

Gaseous equilibria

When dealing with chemical equilibria which involve only gases it is often more convenient to write an expression for the equilibrium constant in terms of **partial pressures** instead of in terms of concentration.

Partial pressure is a different way of expressing the concentration of a particular gas in a mixture of gases. The partial pressure of a gas is the pressure contributed by that individual gas; it is the pressure that gas would exert if it alone occupied the reaction vessel. The sum of all the partial pressures of all the gases in a mixture of gases which do not react with one another is equal to the total pressure of the system – this is known as **Dalton's law** (Figure 2.7).

The partial pressure of any individual gas in a mixture of gases can be found using the general relationship:

partial pressure of a gas = mole fraction of that gas $\times$ total pressure

The mole fraction of an individual gas in a mixture is given by the relationship:

$$\text{mole fraction of a gas} = \frac{\text{number of moles of that gas}}{\text{total number of moles of all gases}}$$

Figure 2.7
John Dalton (1766–1844): Dalton's law states that the sum of the partial pressures of the individual gases in a mixture is equal to the total pressure.

When the equilibrium constant is expressed in terms of partial pressures instead of concentrations it is given the symbol, K_p.

For the general reaction:

$$aA(g) + bB(g) \rightleftharpoons cC(g) + dD(g)$$

$$K_p = \frac{p(C(g)_{eqm})^c \, p(D(g)_{eqm})^d}{p(A(g)_{eqm})^a \, p(B(g)_{eqm})^b}$$

where p = partial pressure.

Figure 2.8
Ammonia gas is a very important raw material for the chemical industry. It is used to make **a** fertilisers and nitric acid and **b** some polyamides and explosives.

a

b

Consider, as an example, the gaseous equilibrium which is established when nitrogen and hydrogen react to produce ammonia gas in the Haber process (Figure 2.8).

$$N_2(g) + 3H_2(g) \rightleftharpoons 2NH_3(g)$$

In an equilibrium mixture of ammonia, hydrogen and nitrogen the total pressure was 2×10^5 Pa and the mixture contained 1 mol of ammonia, 3.6 mol of hydrogen and 13.5 mol of nitrogen. Calculate the equilibrium constant for this equilibrium in terms of partial pressure, K_p.

$$\text{total number of moles of gas} = 1 + 3.6 + 13.5$$
$$= 18.1 \text{ mol}$$

partial pressure of $NH_3(g)$, $p(NH_3(g)_{eqm}) = \dfrac{1}{18.1} \times 2 \times 10^5$

$$= 1.1 \times 10^4 \text{ Pa}$$

partial pressure of $H_2(g)$, $p(H_2(g)_{eqm}) = \dfrac{3.6}{18.1} \times 2 \times 10^5$

$$= 4.0 \times 10^4 \text{ Pa}$$

partial pressure of $N_2(g)$, $p(N_2(g)_{eqm}) = \dfrac{13.5}{18.1} \times 2 \times 10^5$

$$= 14.9 \times 10^4 \text{ Pa}$$

$$K_p = \frac{p(NH_3(g)_{eqm})^2}{p(N_2(g)_{eqm}) \, p(H_2(g)_{eqm})^3}$$

$$= \frac{(1.1 \times 10^4)^2 \text{ Pa}^2}{(14.9 \times 10^4) \text{ Pa} \times (4 \times 10^4)^3 \text{ Pa}^3}$$

$$= 1.27 \times 10^{-11} \text{ Pa}^{-2}$$

Neither K_p expressions, nor K_c expressions, include any solids that might be present in the equilibrium mixture.

5 Write down K_p expressions for the following gaseous equilibria. In each case state the units of K_p.
 a $2NO(g) \rightleftharpoons N_2(g) + O_2(g)$
 b $2SO_2(g) + O_2(g) \rightleftharpoons 2SO_3(g)$
 c $2CO(g) + O_2(g) \rightleftharpoons 2CO_2(g)$

6 Phosphorus(V) chloride decomposes on heating to give phosphorus(III) chloride and chlorine gas. The equation for this process is:

$PCl_5(g) \rightleftharpoons PCl_3(g) + Cl_2(g)$

At a pressure of 1.0×10^6 Pa and a temperature of 550 °C the amount of each gas present at equilibrium was 0.3 mol of phosphorus(V) chloride, 0.7 mol of phosphorus(III) chloride and 0.7 mol of chlorine. Calculate the value of K_p.

7 In the following equilibrium:

$2NO_2(g) \rightleftharpoons 2NO(g) + O_2(g)$

the amount of each gas present at equilibrium was 0.95 mol of $NO_2(g)$, 0.06 mol of $NO(g)$ and 0.06 mol of $O_2(g)$.

Given that $K_p = 7.0 \times 10^{-6}$ Pa, what was the total pressure needed to achieve this equilibrium composition?

Partition coefficients

Partition coefficients are special cases of equilibrium constants. They describe the equilibrium set up when a solute is added to two immiscible liquids.

For example, water and hexane do not mix. When they are both added to a flask or beaker, two separate layers are formed, with the hexane layer being the top layer. If now some iodine crystals are added to the beaker and the contents are shaken some of the iodine is seen to dissolve in the hexane layer and some in the water layer (Figure 2.9).

Figure 2.9
Iodine dissolves in both the hexane and water layers.

Analysis of the iodine concentration in each layer can be carried out by separating the two layers using a separating funnel and then by titration with sodium thiosulphate. At a given temperature the ratio of the iodine concentrations in the two layers is a constant, showing that an equilibrium has been established.

$$I_2(aq) \rightleftharpoons I_2(hexane)$$

At the point where the two layers meet iodine is continually moving between the two solvents. The rate in both directions is the same (Figure 2.10). It is said that the iodine has been **partitioned** between the two layers.

Figure 2.10
Iodine molecules interchange between the two layers – a dynamic equilibrium is set up.

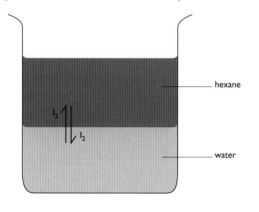

An expression for the equilibrium constant can be written in much the same way as you have seen in previous examples. In this particular case the equilibrium constant is the partition coefficient and is given the symbol K, followed by letters to identify the two solvents, K_{hw} in this case.

$$K_{hw} = \frac{[I_2(hexane)]_{eqm}}{[I_2(aq)]_{eqm}}$$

In this particular case iodine is much more soluble in hexane than in water so the value of the partition coefficient is greater than 1. This indicates that the position of equilibrium lies to the right.

Partition coefficient values are significant in various extraction processes, where it is important to know how well a particular solute dissolves in a particular solvent. Such extraction techniques include solvent extraction and chromatography. It is also important to know particular partition coefficients when developing pesticides and insecticides. Pesticides and insecticides need to be soluble in the fatty tissues of the animals that they are designed to kill but need to be much less soluble in water so that the chemicals are simply not washed away by rain.

Solubility products

Figure 2.11
The sea creatures from
which this coral reef is
made create high localised
conditions of calcium and
carbonate ions so that the
solubility product of
calcium carbonate is
exceeded. When this
happens, the calcium
carbonate precipitates on
the reef around the
creatures.

It is usual to be able to describe a particular substance as being either soluble or insoluble in water. However, it is very unusual for any substance to be totally insoluble in water. Most substances dissolve to some extent, even if it is to a very small degree; calcium carbonate is such a substance. Substances like calcium carbonate are said to be **sparingly soluble**.

When calcium carbonate is placed in water some aqueous calcium and carbonate ions are formed and an equilibrium is set up between these ions and the solid calcium carbonate:

$$CaCO_3(s) \rightleftharpoons Ca^{2+}(aq) + CO_3^{2-}(aq)$$

An equilibrium expression can therefore be written:

$$K_c = \frac{[Ca^{2+}(aq)]_{eqm}[CO_3^{2-}(aq)]_{eqm}}{[CaCO_3(s)]_{eqm}}$$

Unlike other equilibria, the addition of more solid calcium carbonate will not cause the position of equilibrium to move further to the right. This is because the solution is saturated at that particular temperature. So, because the position of equilibrium is not changed by the addition of more solid calcium carbonate it can be removed from the equilibrium constant expression. The expression is therefore re-written using K_{sp} instead of K_c:

$$K_{sp} = [Ca^{2+}(aq)][CO_3^{2-}(aq)]$$

where K_{sp} stands for the **solubility product** of calcium carbonate. At 298 K the value of the solubility product of calcium carbonate is $5 \times 10^{-9}\,mol^2\,dm^{-6}$. This is a very small value and it means that very few calcium and carbonate ions are present in solution and hence that calcium carbonate is not very soluble. Solubility products only apply to sparingly soluble substances. If the value of the solubility product is exceeded then the substance will precipitate. Table 2.3 shows some solubility product values for other sparingly soluble substances.

Table 2.3

Compound	K_{sp}	Units
PbI_2	7.0×10^{-9}	$mol^3\,dm^{-9}$
$AgCl$	2.0×10^{-10}	$mol^2\,dm^{-6}$
$CaSO_4$	2.4×10^{-5}	$mol^2\,dm^{-6}$
Ag_2CrO_4	2.4×10^{-12}	$mol^3\,dm^{-9}$
AgI	8.0×10^{-17}	$mol^2\,dm^{-6}$
$BaSO_4$	1.0×10^{-10}	$mol^2\,dm^{-6}$
PbS	1.2×10^{-28}	$mol^2\,dm^{-6}$
CuS	6.0×10^{-36}	$mol^2\,dm^{-6}$

It should be noted by looking at Table 2.3 that the units of the solubility product depend on the substance being considered.

Consider, for example, lead(II) iodide, PbI_2. If lead(II) iodide is placed in a beaker of water the following equilibrium is set up:

$$PbI_2(s) \rightleftharpoons Pb^{2+}(aq) + 2I^-(aq)$$

The K_{sp} expression is

$$K_{sp} = [Pb^{2+}(aq)][I^-(aq)]^2$$

The concentration of the iodide ions is squared as a result of the 2 mol of iodide ions which are formed if 1 mol of lead(II) iodide is dissolved. In this case K_{sp} has units of $mol^3\,dm^{-9}$ as it is the product of three concentration terms, each of which has units of $mol\,dm^{-3}$.

Solubility products can be used to predict whether a precipitate will form from a solution. The solubility product K_{sp} for lead(II) iodide $= 7.0 \times 10^{-9}\,mol^3\,dm^{-9}$. If two solutions of soluble salts that contain both lead(II) ions and iodide ions are mixed together such that their concentrations cause the solubility product to be exceeded then a precipitate will result.

Example Would a precipitate of lead(II) iodide be formed if $500\,cm^3$ of $0.1\,mol\,dm^{-3}$ lead(II) nitrate were added to $500\,cm^3$ of $0.05\,mol\,dm^{-3}$ sodium iodide?

- From Table 2.3 we know that $K_{sp}(PbI_2) = 7.0 \times 10^{-9}\,mol^3\,dm^{-9}$.
- The concentration of lead(II) ions on mixing will be halved as the volume of the final solution is doubled.

$$[Pb^{2+}(aq)] = 0.05\,mol\,dm^{-3}$$

- The concentration of iodide ions from the sodium iodide solution will also be halved.

$$[I^-(aq)] = 0.025\,mol\,dm^{-3}$$

When the solutions are mixed, therefore:

$$[Pb^{2+}(aq)][I^-(aq)]^2 = 0.05\,mol\,dm^{-3} \times (0.025)^2\,mol^2\,dm^{-6}$$
$$= 3.125 \times 10^{-5}\,mol^3\,dm^{-9}$$

This exceeds the solubility product of lead(II) iodide and therefore a precipitate will be formed.

The solubility of a sparingly soluble substance can be determined from the solubility product.

Example What is the solubility of silver chloride at 298 K given that K_{sp} of silver chloride is $2.0 \times 10^{-10}\,mol^2\,dm^{-6}$?

When silver chloride is added to water the following equilibrium is set up:

$$AgCl(s) \rightleftharpoons Ag^+(aq) + Cl^-(aq)$$

$$K_{sp} = [Ag^+(aq)][Cl^-(aq)]$$
$$= 2.0 \times 10^{-10}\,mol^2\,dm^{-6}$$

When 1 mol of silver chloride dissolves, 1 mol of silver ions and 1 mol of chloride ions are formed. The concentrations of the silver and chloride ions are, therefore, the same.

We can say, therefore:

$$\text{Let } x = [Ag^+(aq)] = [Cl^-(aq)]$$
$$\text{So, } x^2 = 2.0 \times 10^{-10}\,mol^2\,dm^{-6}$$
$$x = \sqrt{2.0 \times 10^{-10}\,mol^2\,dm^{-6}}$$
$$x = 1.4 \times 10^{-5}\,mol\,dm^{-3}$$

Since each mole of silver chloride that dissolves produces 1 mol of silver ions and 1 mol of chloride ions in solution we can say that the solubility of silver chloride is $1.4 \times 10^{-5}\,mol\,dm^{-3}$.

If the solubility of a sparingly soluble salt is known then its solubility product can be determined.

Example   What is the solubility product of silver chromate, Ag_2CrO_4, given that its solubility is $8.43 \times 10^{-5}\,mol\,dm^{-3}$?

The equation for the equilibrium is:

$$Ag_2CrO_4(s) \rightleftharpoons 2Ag^+(aq) + CrO_4^{2-}(aq)$$

$$K_{sp} = [Ag^+(aq)]^2 [CrO_4^{2-}(aq)]$$

When $8.43 \times 10^{-5}\,mol\,dm^{-3}$ of silver chromate dissolve:

- $[Ag^+(aq)]\quad = 2 \times 8.43 \times 10^{-5}\,mol\,dm^{-3}$
  $\qquad\qquad\quad = 1.686 \times 10^{-4}\,mol\,dm^{-3}$
- $[CrO_4^{2-}(aq)] = 8.43 \times 10^{-5}\,mol\,dm^{-3}$

$$K_{sp} = (1.686 \times 10^{-4})^2\,mol^2\,dm^{-6} \times 8.43 \times 10^{-5}\,mol\,dm^{-3}$$
$$\quad = 2.4 \times 10^{-12}\,mol^3\,dm^{-9}$$

The common ion effect

So far we have only considered the dissolving of sparingly soluble substances in water. In this section we look at what happens if a sparingly soluble substance is added to a solution that already contains one of its ions.

Earlier we calculated the solubility of silver chloride in water as being $1.4 \times 10^{-5}\,mol\,dm^{-3}$. What would be the effect on the solubility of silver chloride if it was added to a solution which already contained chloride ions at a concentration of $0.05\,mol\,dm^{-3}$?

$$AgCl(s) \rightleftharpoons Ag^+(aq) + Cl^-(aq)$$

$$K_{sp} = [Ag^+(aq)][Cl^-(aq)]$$
$$\quad = 2.0 \times 10^{-10}\,mol^2\,dm^{-6}$$

When silver chloride dissolves, only small amounts of silver ions and chloride ions are formed. In our previous calculation we said that the concentrations of the two ions were the same. This time we cannot make this assumption as there will be a much higher concentration of chloride ions than silver ions. In fact the concentration of chloride ions will be $0.05\,mol\,dm^{-3}$ plus those formed from the silver chloride dissolving. It is at this point we make an assumption to make the calculation slightly easier. The quantity of chloride ions coming from the dissolved silver chloride can be neglected, as it is going to be much smaller than the $0.05\,mol\,dm^{-3}$ already present in the solution, $[Cl^-(aq)] = 0.05\,mol\,dm^{-3}$.

$$K_{sp} = 2.0 \times 10^{-10}\,mol^2\,dm^{-6}$$
$$\quad = [Ag^+(aq)] \times 0.05\,mol\,dm^{-3}$$

$$[Ag^+(aq)] = \frac{2.0 \times 10^{-10}\,mol^2\,dm^{-6}}{0.05\,mol\,dm^{-3}}$$
$$\qquad\qquad = 4.0 \times 10^{-9}\,mol\,dm^{-3}$$

Since 1 mol of silver ions is formed for each mole of silver chloride dissolved, the solubility of silver chloride is $4.0 \times 10^{-9}\,mol\,dm^{-3}$. In pure water the solubility was $1.4 \times 10^{-5}\,mol\,dm^{-3}$. The silver chloride is, therefore, much less soluble in the chloride solution than in pure water. This effect is known as the **common ion effect**. A similar reduction in solubility would have occurred if the silver chloride had been dissolved in a solution already containing $0.05\,mol\,dm^{-3}$ silver nitrate.

8 Calculate the solubility of each of the following sparingly soluble salts given their solubility products.

	solubility product
a silver bromide	$5.0 \times 10^{-13}\,mol^2\,dm^{-6}$
b calcium hydroxide	$5.5 \times 10^{-6}\,mol^3\,dm^{-9}$
c lead(II) chloride	$2.0 \times 10^{-5}\,mol^3\,dm^{-9}$

9 Calculate the solubility products of the following sparingly soluble salts from their solubilities.

	solubility ($mol\,dm^{-3}$)
a cadmium carbonate, $CdCO_3$	1.6×10^{-7}
b calcium fluoride, CaF_2	2.2×10^{-4}
c chromium(III) hydroxide, $Cr(OH)_3$	1.4×10^{-8}

10 Calculate the solubility of silver bromide, AgBr, in $0.005\,mol\,dm^{-3}$ sodium bromide solution.

K_{sp} for silver bromide is $5.0 \times 10^{-13}\,mol^2\,dm^{-6}$.

Compare with your answer to Question 8a – what do you notice?

● **Key skills**   ICT

- Use of spreadsheets and graphical display to investigate the position of equilibrium with variation of conditions.

Number

- Using equilibrium concentrations to determine values of K_c and K_{sp}.

● **Skills task**   Find information to allow you to discuss the fact that seashells dissolve in deep sea water but not in shallow water.

CHECKLIST   After studying Chapter 2 you should know and understand the following terms.

● **Dynamic equilibrium:** An equilibrium during a chemical reaction in which the forward and back reactions occur at the same rate.

● **Equilibrium constant:** For the general reaction:

$$aA + bB \rightleftharpoons cC + dD$$

the equilibrium constant, $K_c = \dfrac{[C]^c [D]^d}{[A]^a [B]^b}$

where [ ] = concentration in $mol\,dm^{-3}$.

● **Homogeneous equilibria:** Equilibria processes which contain all the reactants and products in the same physical state.

● **Heterogeneous equilibria:** Equilibria processes which contain reactants and products in different physical states.

● **Le Chatelier's principle:** If a chemical system at equilibrium undergoes changes in the conditions under which it has been established then the chemical system responds by minimising the effects of the changes.

● **Partial pressure:** For a gas in a mixture, this is given by the product of the mole fraction of that gas and the total pressure of the system.

● **Mole fraction:** For an individual gas in a mixture this is given by the number of moles of that gas divided by the total number of moles of the gases present.

● **Dalton's law:** The total pressure of a system is given by the sum of the partial pressures of all the gases in the mixture.

● **Partition coefficients:** These describe the equilibrium established when a solute is added to two immiscible solvents.

● **Solubility products:** Equilibrium constants for sparingly soluble salts in equilibrium with solutions of their ions.

● **Common ion effect:** The effect observed on the position of equilibrium for a sparingly soluble salt when an ionic compound which contains one of the ions involved in the equilibrium is added.

Examination questions

I Ethane can be cracked at high temperatures to yield ethene and hydrogen, according to the equation:

$$C_2H_6(g) \rightleftharpoons C_2H_4(g) + H_2(g)$$

The standard enthalpy of formation of ethene is positive whereas that of ethane is negative.

a Discuss the effect on the equilibrium constant, K_p, of changes to
    i the temperature
    ii the pressure. (3)
b Calculate the value of the equilibrium constant, K_p, for this cracking reaction, given that 1.00 mol of ethane under an equilibrium pressure of 180 kPa at 1000 K can be cracked to produce an equilibrium yield of 0.36 mol of ethene. (7)

AQA, A level, Specimen Paper 6421, 2001/2

2 a When solid calcium nitrate is heated, brown fumes of nitrogen dioxide, NO_2, are seen and the solid remaining after decomposition is calcium oxide.
    i Write a balanced equation for the thermal decomposition of calcium nitrate. (2)
    ii Describe the changes you would see when cold water is added drop by drop to cold calcium oxide and give the chemical equation for the reaction. (3)
    iii State whether barium nitrate will decompose more easily or less easily than calcium nitrate on heating with a Bunsen burner. (1)
    iv Account for the trend in the thermal stability of the nitrates of the elements in Group 2. (3)
b The brown fumes in part **a** are not pure NO_2 but a mixture of N_2O_4 and NO_2.

$$N_2O_4(g) \rightleftharpoons 2NO_2(g)$$
pale yellow   dark brown

A transparent glass syringe was filled with the gaseous mixture of N_2O_4 and NO_2 and its tip sealed. When the piston of the syringe was rapidly pushed well into the body of the syringe, thereby compressing the gas mixture considerably, the colour of the gas became momentarily darker but then became lighter again.
    i Suggest why compressing the gases causes the mixture to darken. (1)
    ii Explain why the mixture turns lighter on standing. (2)
    iii Write an expression for the equilibrium constant, K_p, for this equilibrium. (1)
    iv 1.0 mole of N_2O_4 was allowed to reach equilibrium at 400 K. At equilibrium the partial pressure of N_2O_4 was found to be 0.15 atm.
    Given that the equilibrium constant K_p for this reaction is 48 atm, calculate the partial pressure of NO_2 in the equilibrium mixture. (3)

Edexcel, A level, Module 3, June 2000

3 The manufacture of sulphuric acid is based on the equilibrium reaction

$$2SO_2(g) + O_2(g) \rightleftharpoons 2SO_3(g) \qquad \Delta H = -197 \, kJ \, mol^{-1}$$

a The equilibrium is a **dynamic equilibrium**; explain the nature of such an equilibrium. (2)
b Give the expression for K_c for this equilibrium, giving consideration to the units. (2)
c State and explain the effects of the following on the equilibrium position:
    i an increase in temperature (2)
    ii the presence of a catalyst. (2)
d State and explain the effects of the following on the rate of attainment of equilibrium:
    i an increase in temperature (2)
    ii the presence of a catalyst. (2)

London, AS/A level, Module 2, June 1999

4 Ammonia is manufactured by passing hot nitrogen and hydrogen at high pressure over a catalyst containing iron.

In an experiment, 9.0 moles of nitrogen and 27 moles of hydrogen were put into an iron vessel of volume 10 dm³. This was then heated to 250 °C and allowed to reach equilibrium. It was found that two thirds of the nitrogen and hydrogen were converted into ammonia.

$$N_2(g) + 3H_2(g) \rightleftharpoons 2NH_3(g) \qquad \Delta H = -92.4 \, kJ \, mol^{-1}$$

a Give the expression for K_c. (1)
b Complete the table below.

	N_2	H_2	NH_3
Moles at start	9.0	27	0
Moles at equilibrium			
Concentration at equilibrium (mol dm⁻³)			

(2)
c Calculate the value of K_c at 250 °C, stating its units. (2)
d State the effect, if any, of an increase in temperature on the value of the **equilibrium constant**, giving a reason. (2)
e State the effect, if any, of an increase in pressure on the value of the **equilibrium constant**, giving a reason. (2)

London, AS/A level, Module 2, Jan 2000

5 New Zealand has no oilfields of its own and until recently relied completely on imported oil to meet its need for liquid fuels. The country does, however, have large reserves of natural gas (which is largely methane) and, since 1985, much of the petrol needed in New Zealand has been produced by chemical conversion of methane into liquid, hydrocarbon fuel.

The first stage in this process involves production of methanol from methane using the reactions in equations 1 and 2. Data about these reactions are shown in the table below.

$$CH_4(g) + H_2O(l) \rightleftharpoons CO(g) + 3H_2(g) \qquad \text{(equation 1)}$$

$$CO(g) + 2H_2(g) \rightleftharpoons CH_3OH(g) \qquad \text{(equation 2)}$$

	Reaction 1	Reaction 2
Conditions:		
Temperature (K)	1100	550
Pressure (atm)	5	100
Catalyst	Nickel	Copper/zinc
$\Delta H_{reaction}$ (kJ mol^{-1})	+206	−128

a Catalysts play a key role in increasing the rates of **reactions 1** and **2**.
i State **two** other ways in which the conditions above are chosen to increase the rates of **reactions 1** and **2**. (2)
ii For **each** answer to part **a i** explain why this choice of conditions leads to increased reaction rate. (4)

b The conditions used for **reactions 1** and **2** are chosen to give optimum yields at equilibrium from these reactions. For each reaction, explain why the conditions of temperature and pressure chosen give an optimum yield of products.
(In this question 1 mark is available for the quality of written communication.) (8)

c In the second stage of the process, methanol is converted into a mixture of hydrocarbons by **reactions 3** and **4**. (In equation 4, the mixture of hydrocarbons (petrol) is represented by octene, C_8H_{16}.) The reactions take place at 600 K in the presence of a zeolite catalyst.

$$2CH_3OH(g) \rightleftharpoons CH_3OCH_3(g) + H_2O(g) \quad \text{(equation 3)}$$

$$4CH_3OCH_3(g) \rightleftharpoons C_8H_{16}(g) + 4H_2O(g) \quad \text{(equation 4)}$$

i Write an expression for K_p for the equilibrium in equation 3 in terms of the partial pressures of the gases involved. (2)
ii Under the conditions used in the industrial process, **reaction 3** reaches equilibrium. Calculate the equilibrium partial pressure of methoxymethane (CH_3OCH_3) when the partial pressure of methanol at equilibrium is 0.142 atm. ($K_p = 9.00$) (3)

d Zeolites are crystalline aluminosilicate materials with structures containing a network of linked channels through which molecules can pass. The channels restrict the size of the hydrocarbon molecules produced and their passage out of the zeolite.

In the zeolite used as a catalyst for **reactions 3** and **4**, only molecules with up to 12 carbon atoms can be formed and pass through the channels. Despite this, up to 200 hydrocarbon compounds are present in the reaction product. This large number is due to the fact that, for most hydrocarbons, there are several ways in which the carbon and hydrogen atoms can be arranged for any given formula.
i Give **three** ways in which the arrangement of carbon and hydrogen atoms can give rise to different molecules of the same formula. (3)
ii Describe another industrial use for zeolites. (2)

OCR, A level, Specimen Paper A7887, Sept 2000

6 Ethanol, C_2H_5OH, is an important industrial chemical with about 200 000 tonnes manufactured in the UK each year. The usual method of manufacture is by the hydration of ethene with steam in the presence of a phosphoric acid catalyst at 550 K and a pressure of about 7000 kPa.

$$C_2H_4(g) + H_2O(g) \rightleftharpoons C_2H_5OH(g) \qquad \Delta H = -46 \text{ kJ mol}^{-1}$$

a i Predict, with justification, the optimum conditions for this reaction.
ii Explain why the actual conditions used may be different from the optimum conditions.
iii The boiling points of the three chemicals involved in this equilibrium are shown in the table below.

Compound	C_2H_4	H_2O	CH_3CH_2OH
Boiling point (°C)	−104	100	78

Suggest how the ethanol could be separated from the equilibrium mixture. (8)
b i Write an expression for K_p of this reaction and
ii explain, with a reason in each case, whether you would expect the value of K_p to alter if any of the external variables below were changed as indicated:
increase in temperature
increase in pressure
presence of catalyst (5)
c Alcohols such as ethanol can be used as alternative fuels to petrol. The combustion of ethanol tends to be more complete than the combustion of the alkanes present in petrol, partly because less oxygen is required for combustion.
i Use equations to compare the amount of oxygen required per gramme of fuel combusted.
ii Suggest why there is this difference between the amount of oxygen required per gramme for these two fuels. (5)

OCR, A level, Specimen Paper A7882, Sept 2000

3 Further redox chemistry

STARTING POINTS ● An oxidation process involves either the addition of oxygen, the removal of hydrogen or the loss of electrons.

● A reduction process involves either the removal of oxygen, the addition of hydrogen or the gain of electrons.

● Half-equations can be used to represent the oxidation and reduction processes in a redox reaction.

● Oxidation numbers can be assigned to an atom or ion to describe its relative state of oxidation or reduction.

● An increase in oxidation number is oxidation, a decrease in oxidation number is reduction.

● An oxidising agent is one which brings about the oxidation of another species.

● A reducing agent is one which brings about the reduction of another species.

● A redox titration is a method used to determine the concentration of a solution of a reducing agent or an oxidising agent.

Earlier in your Chemistry course you will have covered work associated with volumetric analysis, sometimes called titrimetric analysis because it involves titrations, and also redox processes. In this chapter we are going to extend these ideas and look at the use of redox reactions as applied to titrimetric analysis.

Figure 3.1
Redox titrations can be used to ensure consumer protection by checking these materials contain the right amounts or concentrations of the substances in them.

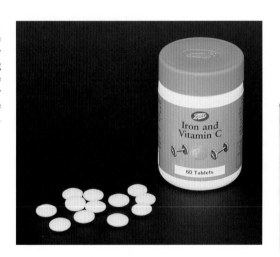

Redox titrations

Redox titrations can be used throughout the chemical industry in quality control laboratories to ensure consumer protection (Figure 3.1). A redox titration is one that involves both oxidation and reduction reactions. Three common redox techniques that are carried out involve the manganate(VII) ion, MnO_4^-(aq), the dichromate(VI) ion, $Cr_2O_7^{2-}$(aq), and the thiosulphate ion $S_2O_3^{2-}$(aq).

Potassium manganate(VII) titrations

Manganate(VII) titrations can be used to determine the concentration of a reducing agent in a solution. The manganate(VII) ion has an intense purple colour in concentrated solution. Even in dilute solutions a solution of manganate(VII) appears to have a strong pink colour (Figure 3.2).

Figure 3.2
The manganate(VII) ion has an intense purple colour in concentrated solution and is still strongly pink in dilute solution.

During the titration of a reducing agent, the manganate(VII) ion, MnO_4^-(aq), is reduced to the manganese(II) ion, Mn^{2+}(aq), which is very pale pink and appears colourless in dilute solution. The titration is carried out by adding a solution of the acidified manganate(VII) ions from a burette to a solution of the reducing agent so the pink colour of the manganate(VII) ion is removed as it is reduced. When all the reducing agent has reacted the next drop of manganate(VII) produces a faint pink tinge to the solution (Figure 3.3). This is the end-point of the titration. As a result of this colour change no further indicator is required in this type of titration.

A typical reducing agent which can be estimated in this way is the iron(II) ion, Fe^{2+}(aq). Thus a manganate(VII) titration can be used to determine the iron content of iron tablets that can be obtained from any pharmacy. The iron in iron tablets is usually in the form of hydrated iron(II) sulphate, although other iron compounds are sometimes used.

Figure 3.3
The pink colour of the manganate(VII) is removed by the reducing agent until the reducing agent has all reacted. The end-point is the formation of a faint pink colour.

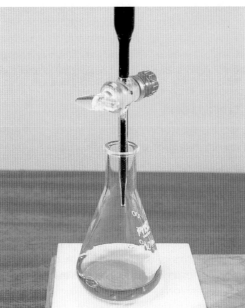

Example

4.00 g of powdered iron tablets, containing hydrated iron(II) sulphate, were dissolved in some dilute sulphuric acid and the solution made up to 250 cm^3. 25.00 cm^3 of this solution was titrated with 0.020 mol dm^{-3} potassium manganate(VII) solution, $KMnO_4$(aq). 14.30 cm^3 of the manganate(VII) solution was needed to give a faint pink colour. What was the percentage of iron in the iron tablets?

The reduction of the manganate(VII) ion can be shown by the equation:

$$MnO_4^-(aq) + 8H^+(aq) + 5e^- \rightarrow Mn^{2+}(aq) + 4H_2O(l)$$

As the iron(II) ions reduce the manganate(VII) ion, they are oxidised to iron(III) ions:

$$Fe^{2+}(aq) \rightarrow Fe^{3+}(aq) + e^-$$

By combining these two equations we obtain the overall equation for the reaction:

$$MnO_4^-(aq) + 8H^+(aq) + 5Fe^{2+}(aq) \rightarrow 5Fe^{3+}(aq) + Mn^{2+}(aq) + 4H_2O(l)$$

The number of moles of manganate(VII) ions reacting with the iron(II) ions is given by:

$$\text{moles of } MnO_4^-(aq) = 0.020 \, \text{mol dm}^{-3} \times \frac{14.30 \, \text{cm}^3}{1000}$$

$$= 2.86 \times 10^{-4} \, \text{mol}$$

From the overall equation 1 mol of MnO_4^-(aq) reacts with 5 mol of Fe^{2+}(aq), so the number of moles of Fe^{2+}(aq) in 25 cm^3 of solution is given by:

$$\text{moles of } Fe^{2+}(aq) = 5 \times 2.86 \times 10^{-4} \, \text{mol}$$
$$= 1.43 \times 10^{-3} \, \text{mol}$$

In the original solution there would be ten times this amount.

$$\text{moles of } Fe^{2+}(aq) = 10 \times 1.43 \times 10^{-3} \, \text{mol}$$
$$= 1.43 \times 10^{-2} \, \text{mol}$$

1 mol of Fe^{2+}(aq) has a mass of 56 g, therefore

$$\text{mass of Fe in 2.00 g of tablets} = 1.43 \times 10^{-2} \, \text{mol} \times 56 \, \text{g mol}^{-1}$$
$$= 0.80 \, \text{g}$$

$$\text{\% of Fe in the tablets} = \frac{0.80 \, \text{g}}{4.00 \, \text{g}} \times 100 = 20\%$$

I Some solid ammonium iron(II) sulphate, $FeSO_4.(NH_4)_2SO_4.6H_2O$, was made up to 500 cm^3 of solution using dilute sulphuric acid. 25.00 cm^3 of the solution was pipetted into a conical flask. 22.80 cm^3 of a 0.020 mol dm^{-3} solution of potassium manganate(VII) was needed for a complete reaction. Calculate (in mol dm^{-3}) the concentration of the ammonium iron(II) sulphate solution.

Manganate(VII) titrations can be used to estimate other reducing agents, such as ethanedioate ions, $C_2O_4^{2-}$(aq). However, they should not be used to estimate reducing agents in solutions which also contain chloride ions, Cl^-(aq), as the manganate(VII) is such a powerful oxidising agent itself that the chloride ions are oxidised to chlorine gas. This not only gives an inaccurate titration value but could be dangerous due to the production of toxic chlorine gas.

Dichromate(VI) titrations

Dichromate(VI) titrations, like manganate(VII) titrations, are used to estimate reducing agents. During the titration the dichromate(VI) ion is reduced according to the equation:

$$Cr_2O_7^{2-}(aq) + 14H^+(aq) + 6e^- \rightarrow 2Cr^{3+}(aq) + 7H_2O(l)$$

Solutions containing the dichromate(VI) ion are orange. During the titration of a reducing agent this colour is replaced by a green colour, indicating the presence of the chromium(III) ion, Cr^{3+}(aq) (Figure 3.4). This change in colour is used in breathalyser tubes to detect the presence of ethanol in breath. Solid orange potassium dichromate(VI) crystals in the breathalyser tube change to green as ethanol passes over them. The ethanol is oxidised by the dichromate(VI) ions.

Dichromate(VI) is not as powerful an oxidising agent as manganate(VII) and can be used with solutions which might contain the chloride ion, Cl^-(aq). As with manganate(VII), a suitable reducing agent to use as an example is the iron(II) ion, Fe^{2+}(aq).

Figure 3.4
Potassium dichromate(VI) is orange in colour and the end-point of the redox reaction is seen when a green colour is observed.

Example

A 0.90 g sample of steel was dissolved in concentrated sulphuric acid and the solution made up to 250 cm³ in a volumetric flask. 25 cm³ of the solution was pipetted into a conical flask and titrated against 0.010 mol dm⁻³ potassium dichromate(VI) solution. 23.65 cm³ of the dichromate(VI) solution was required to reach the end-point. Calculate the percentage of iron in the sample of steel.

The dichromate(VI) ion oxidises the iron(II) ions to iron(III) ions:

$$Fe^{2+}(aq) \rightarrow Fe^{3+}(aq) + e^-$$

The overall redox equation for the reaction is:

$$Cr_2O_7{}^{2-}(aq) + 14H^+(aq) + 6Fe^{2+}(aq) \rightarrow 6Fe^{3+}(aq) + 2Cr^{3+}(aq) + 7H_2O(l)$$

$$\text{moles of dichromate(VI) used} = 0.010\,mol\,dm^{-3} \times \frac{23.65\,cm^3}{1000}$$

$$= 2.365 \times 10^{-4}\,mol$$

$$\text{moles of iron(II) ions} = 6 \times 2.365 \times 10^{-4}\,mol$$
$$= 1.419 \times 10^{-3}\,mol$$

In the original 250 cm³ solution, there were ten times the number of moles of iron(II) ions.

$$\text{moles of iron(II) ions} = 10 \times 1.419 \times 10^{-3}\,mol$$
$$= 1.419 \times 10^{-2}\,mol$$

$$\text{mass of iron in the sample of steel} = 56\,g\,mol^{-1} \times 1.419 \times 10^{-2}\,mol$$
$$= 0.795\,g$$

$$\text{\% of iron in the sample of steel} = \frac{0.795\,g}{0.90\,g} \times 100$$

$$= 88.3\%$$

2 1.00 g of a video tape, containing iron, was dissolved in concentrated sulphuric acid and the volume was carefully made up to 250 cm³ with distilled water. 25.00 cm³ of this solution was titrated against 0.00100 mol dm⁻³ potassium dichromate(VI) solution. 22.50 cm³ of the potassium dichromate(VI) were required for a complete reaction. Calculate the percentage of iron in the video tape.

Thiosulphate titrations

Thiosulphate titrations can be used to estimate oxidising agents such as iodine, $I_2(aq)$. During such titrations, the thiosulphate ion, $S_2O_3^{2-}(aq)$, is oxidised to the tetrathionate ion, $S_4O_6^{2-}(aq)$:

$$2S_2O_3^{2-}(aq) \rightarrow S_4O_6^{2-}(aq) + 2e^-$$

and the aqueous iodine is reduced to aqueous iodide ions, $I^-(aq)$:

$$I_2(aq) + 2e^- \rightarrow 2I^-(aq)$$

The overall redox equation for the reaction is:

$$I_2(aq) + 2S_2O_3^{2-}(aq) \rightarrow 2I^-(aq) + S_4O_6^{2-}(aq)$$

Aqueous iodine is produced when iodide ions are oxidised by the addition of an oxidising agent. The iodine can then be titrated against a sodium thiosulphate solution using starch (or iodine indicator – thyodene) as indicator. With aqueous iodine the indicator forms a blue/black colour which changes to colourless when the aqueous iodine has reacted.

Unlike other titrations which use indicators, in this reaction it is usual to add the indicator after sufficient thiosulphate has been added to reduce the brown colour of the aqueous iodine to a pale straw colour. The indicator is added in this way because if it is added too soon specks of blue form which do not disappear at the end-point (Figure 3.5).

Figure 3.5
The different colours observed during a thiosulphate titration:
a aqueous iodine
b addition of the indicator at straw colour
c blue/black colour of indicator with aqueous iodine
d the end-point.

a   b   c   d

There are many occasions when this type of redox reaction can be used, for example in the determination of the 'available' chlorine in a commercial bleach. Bleaches such as Domestos and Parazone contain sodium chlorate(I), also known as sodium hypochlorite (Figure 3.6).

Figure 3.6
These commercial bleaches contain different amounts of sodium hypochlorite.

In acid solution, chlorate(I) ions react with chloride ions releasing chlorine:

$$OCl^-(aq) + Cl^-(aq) + 2H^+(aq) \rightarrow H_2O(l) + Cl_2(aq)$$

The chlorine released is referred to as 'available' chlorine; the larger the amount the stronger the bleaching action of the bleach. The 'available' chlorine so produced can be estimated by:

- initial displacement of iodine from potassium iodide by the chlorine produced:

$$Cl_2(aq) + 2I^-(aq) \rightarrow 2Cl^-(aq) + I_2(aq)$$

- titration of the displaced iodine using sodium thiosulphate.

Another example of how thiosulphate titrations can be used is in the determination of copper, for example, in a brass screw. A brass screw can be dissolved in concentrated nitric acid to form a solution of copper(II) ions. If an excess of potassium iodide is added to this solution, the iodide is oxidised to aqueous iodine as the copper(II) ion is reduced to form copper(I) iodide.

$$2Cu^{2+}(aq) + 4I^-(aq) \rightarrow 2CuI(s) + I_2(aq)$$

The aqueous iodine produced can then be titrated against thiosulphate solution.

Example In an experiment to determine the copper content of a brass screw, a screw of mass 3.75 g was dissolved in a beaker of concentrated nitric acid in a fume cupboard. The resulting solution, which contained copper(II) ions, $Cu^{2+}(aq)$, was made up to 250 cm^3 with distilled water. After mixing, 25 cm^3 of the solution was pipetted into a conical flask. Some sodium carbonate solution was added slowly to the copper(II) solution until a faint blue precipitate was obtained, indicating that the excess nitric acid had been neutralised. The precipitate was then dissolved by the addition of a small amount of ethanoic acid. To this now neutral solution of copper(II) ions, an excess of potassium iodide was added. (It is important to neutralise the excess acid as it would react with the sodium thiosulphate to produce sulphur dioxide and sulphur.)

The solution was then titrated against a $0.20\,mol\,dm^{-3}$ solution of sodium thiosulphate using an iodine indicator. 21.10 cm^3 was required to reach the end-point. What was the percentage of copper in the brass screw?

$$\text{moles of sodium thiosulphate used} = \frac{0.20\,mol\,dm^{-3} \times 21.10\,cm^3}{1000} = 4.22 \times 10^{-3}\,mol$$

From the equation:

$$I_2(aq) + 2S_2O_3^{2-}(aq) \rightarrow 2I^-(aq) + S_4O_6^{2-}(aq)$$

2 mol of thiosulphate ions react with 1 mol of aqueous iodine, therefore:

$$\text{number of moles of iodine} = \frac{1}{2} \times 4.22 \times 10^{-3}\,mol$$

$$= 2.11 \times 10^{-3}\,mol$$

Knowing that the iodine is produced by the reaction:

$$2Cu^{2+}(aq) + 4I^-(aq) \rightarrow 2CuI(s) + I_2(aq)$$

allows us to calculate the number of moles of $Cu^{2+}(aq)$:

$$\text{moles of } Cu^{2+}(aq) = 2 \times 2.11 \times 10^{-3}\,mol = 4.22 \times 10^{-3}\,mol$$

This is the number of moles of copper in 25 cm^3, so multiply by 10 for the original 250 cm^3:

$$\text{number of moles of copper} = 4.22 \times 10^{-3}\,mol \times 10 = 4.22 \times 10^{-2}\,mol$$

$$\text{the mass of copper is therefore} = 4.22 \times 10^{-2}\,mol \times 63.5\,g\,mol^{-1} = 2.68\,g$$

$$\text{\% of copper in the brass screw} = \frac{2.68\,g}{3.75\,g} \times 100 = 71.5\%$$

3 Swimming pools are kept sterile by the presence of chloric(I) acid, HOCl(aq). The amount of chloric(I) acid can be determined by adding potassium iodide to a sample of the water and then titrating the aqueous iodine, produced by the oxidation of the iodide ion by chloric(I) acid, with sodium thiosulphate solution.

$$HOCl(aq) + H^+(aq) + 2I^-(aq) \rightarrow I_2(aq) + Cl^-(aq) + H_2O(l)$$

A 50 cm³ sample of water was titrated with thiosulphate in the presence of an iodine indicator. 15.80 cm³ of a 0.001 mol dm⁻³ solution of sodium thiosulphate were needed for complete reaction. Calculate the concentration of the chloric(I) acid in the water sample.

Figure 3.7
It is necessary to keep swimming pools free from harmful bacteria. Adding sodium hypochlorite is a cheap way of doing this.

Oxidising and reducing power

When a redox reaction takes place, as in the examples given in the first part of this chapter, electrons move from one species (the reducing agent) to another (the oxidising agent). Take a look at the simple example below.

Example

When zinc metal is placed into a beaker containing copper(II) sulphate a reaction occurs (Figure 3.8). The ionic equation for the process is:

$$Zn(s) + Cu^{2+}(aq) \rightarrow Cu(s) + Zn^{2+}(aq)$$

Figure 3.8
Zinc reacts with copper(II) sulphate to form copper metal and zinc sulphate.

By separating out the two half-equations we can easily see the movement of the electrons:

$$Zn(s) \rightarrow Zn^{2+}(aq) + 2e^-$$
$$Cu^{2+}(aq) + 2e^- \rightarrow Cu(s)$$

Zinc, the reducing agent, loses electrons as it is oxidised and these combine with the copper(II) ions, the oxidising agent, as it is reduced to form copper metal.

A similar reaction to the one above takes place when magnesium metal is added to a beaker containing copper(II) ions.

$$Mg(s) + Cu^{2+}(aq) \rightarrow Cu(s) + Mg^{2+}(aq)$$

This time the magnesium is the reducing agent, but how good a reducing agent is it compared to zinc metal? To answer this question we have to do some measurements on the flow of electrons between the two half-cells. To do this we need to separate the two half-cells from one another but still allow for a complete electrical circuit. This is done by setting up an electrochemical cell.

Electrochemical cells

An electrochemical cell consists of two half-cells, one for each of the half-equations. Each half-cell is made up of the different species which are present in one of the half-equations.

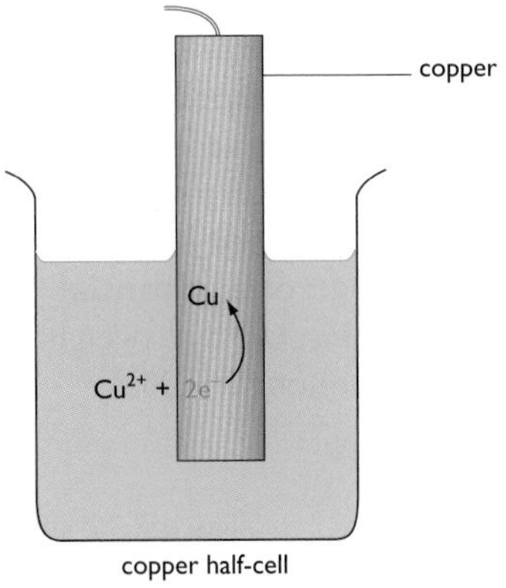

Figure 3.9
The copper half-cell consists of a piece of copper metal in contact with a solution of copper(II) ions.

copper

Cu

$Cu^{2+} + 2e^-$

copper half-cell

For example, the half-equation:

$$Cu^{2+}(aq) + 2e^- \rightleftharpoons Cu(s)$$

is made up of a piece of clean copper metal in contact with a solution containing aqueous copper(II) ions, for example, copper(II) sulphate solution. Notice that the equation above has now been written as a reversible reaction. This is because although we know that in the reaction with zinc or magnesium, copper(II) ions are reduced to copper metal, it could well be that in other reactions copper metal is oxidised to copper(II) ions. The half-cell for this half-equation is shown in Figure 3.9.

Figure 3.10 shows the electrochemical cell that represents the equation:

$$Zn(s) + Cu^{2+}(aq) \rightarrow Cu(s) + Zn^{2+}(aq)$$

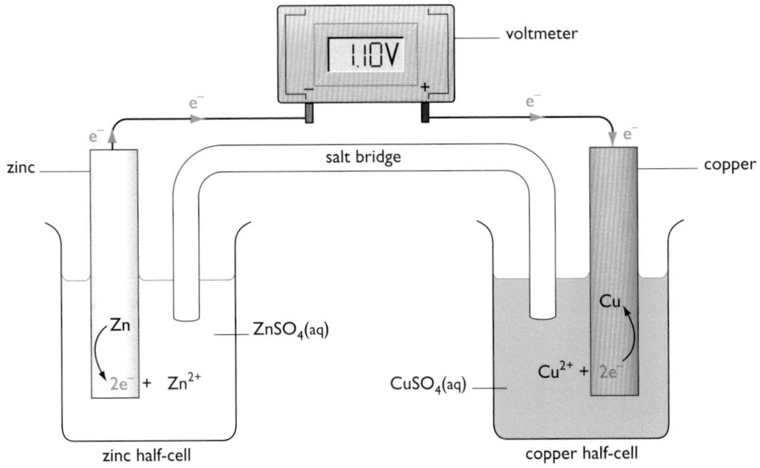

Figure 3.10
An electrochemical cell made by connecting a zinc half-cell to a copper half-cell.

voltmeter

1.10V

e^- e^- e^- e^-

salt bridge

zinc

copper

Zn

$ZnSO_4(aq)$

$2e^- + Zn^{2+}$

$CuSO_4(aq)$

Cu

$Cu^{2+} + 2e^-$

zinc half-cell

copper half-cell

In this example each of the pieces of metal act as electrodes, through which electrons may leave or enter each half-cell. The electrodes are connected to one another through wires and a **high-resistance voltmeter** that measures the potential difference between the two electrodes. A high-resistance voltmeter is used so that negligible current flows. The maximum potential difference recorded is known as the cell e.m.f., E_{cell}. The voltage produced under standard conditions, using concentrations of $1 \, mol \, dm^{-3}$, a temperature of 298 K and a pressure of $1.01 \times 10^5 \, Pa$ (1 atmosphere), is given the symbol, $E_{cell}^\ominus$. The use of standard conditions allows for consistent comparison of values. The connection between the two solutions is made using a **salt bridge**. A salt bridge, in its simplest form, is simply a piece of filter paper soaked in a saturated solution of a simple salt such as potassium nitrate. The salt bridge allows ions to move between the two solutions, thus completing the circuit. If the salt bridge dries out then the movement of ions is hampered and the circuit eventually is broken.

As can be seen from Figure 3.10 the maximum voltage recorded, under standard conditions, is 1.10 V. You will note that in the diagram the zinc half-cell is connected to the negative terminal of the voltmeter. Zinc is the more **negative half-cell** because it is the zinc metal which loses electrons and so there is a large negative charge on this electrode compared to the copper half-cell, the **positive half-cell**.

If the zinc half-cell in Figure 3.10 is replaced by a magnesium half-cell an $E_{cell}^\ominus$ of 2.71 V is recorded. What do these values tell us about the relative reducing powers of zinc and magnesium? Since the copper half-cell is present on both occasions we can deduce that the magnesium half-cell must be losing electrons more readily than the zinc half-cell and as such magnesium is a more powerful reducing agent than zinc.

Figure 3.11
The standard hydrogen half-cell is used as one half of an electrochemical cell used to determine standard electrode potentials for other half-cells.

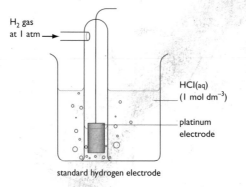

In the example just considered, the copper half-cell has been the 'reference' cell. It is usual however to use the **standard hydrogen half-cell** (Figure 3.11) as the reference cell to produce $E_{cell}^{\ominus}$ values, as it is not possible to measure the electrode potential of an isolated individual half-cell. The standard hydrogen half-cell consists of a special electrode composed of platinum onto which hydrogen gas adsorbs, thus ensuring it is in contact with the 1 mol dm^{-3} hydrogen ion solution.

The $E_{cell}^{\ominus}$ produced by an electrochemical cell formed from a standard hydrogen half-cell and another half-cell is known as the **standard electrode potential**. The standard electrode potentials of all other half-cells is measured against the hydrogen half-cell. The potential of the standard hydrogen half-cell is defined, by convention, as 0.00 V.

$$2H^+(aq) + 2e^- \rightleftharpoons H_2(g) \qquad E_{cell}^{\ominus} = 0.00\,V$$

Figure 3.12
The voltage produced by this electrochemical cell is the standard electrode potential of the magnesium half-cell.

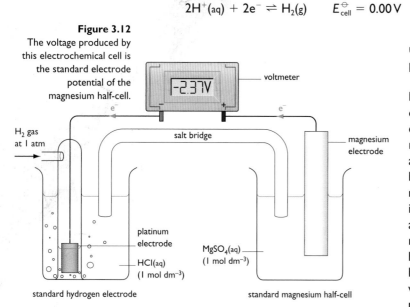

Figure 3.12 shows the electrochemical cell used to determine the standard electrode potential of the magnesium half-cell.

A table of standard electrode potentials can be produced; this is known as the **electrochemical series** (Table 3.1). In the electrochemical series the most powerful reducing agents appear at the top of the series and the most powerful oxidising agents at the bottom. You will notice that both positive and negative standard electrode potentials appear in the electrochemical series. Negative values are produced if the half-cell being considered releases electrons more easily than the hydrogen half-cell (and is therefore more negative than the hydrogen half-cell). Positive values are produced when the hydrogen half-cell releases electrons more readily than the half-cell being considered.

Table 3.1
The electrochemical series, showing some common half-cells.

Figure 3.13 (right)
A standard half-cell for the process $Fe^{3+}(aq) + e^- \rightleftharpoons Fe^{2+}(aq)$.

Half-cell	Half-equation	$E_{cell}^{\ominus}$ (V)
$Na^+(aq) \parallel Na(s)$	$Na^+(aq) + e^- \rightleftharpoons Na(s)$	-2.71
$Mg^{2+}(aq) \parallel Mg(s)$	$Mg^{2+}(aq) + 2e^- \rightleftharpoons Mg(s)$	-2.37
$Al^{3+}(aq) \parallel Al(s)$	$Al^{3+}(aq) + 3e^- \rightleftharpoons Al(s)$	-1.66
$Zn^{2+}(aq) \parallel Zn(s)$	$Zn^{2+}(aq) + 2e^- \rightleftharpoons Zn(s)$	-0.76
$Fe^{2+}(aq) \parallel Fe(s)$	$Fe^{2+}(aq) + 2e^- \rightleftharpoons Fe(s)$	-0.44
$Pb^{2+}(aq) \parallel Pb(s)$	$Pb^{2+}(aq) + 2e^- \rightleftharpoons Pb(s)$	-0.13
$2H^+(aq) \parallel H_2(g)$	$2H^+(aq) + 2e^- \rightleftharpoons H_2(g)$	0.00
$Cu^{2+}(aq) \parallel Cu^+(aq)$	$Cu^{2+}(aq) + e^- \rightleftharpoons Cu^+(aq)$	$+0.15$
$Cu^{2+}(aq) \parallel Cu(s)$	$Cu^{2+}(aq) + 2e^- \rightleftharpoons Cu(s)$	$+0.34$
$Fe^{3+}(aq) \parallel Fe^{2+}(aq)$	$Fe^{3+}(aq) + e^- \rightleftharpoons Fe^{2+}(aq)$	$+0.77$
$Br_2(aq) \parallel 2Br^-(aq)$	$Br_2(aq) + 2e^- \rightleftharpoons 2Br^-(aq)$	$+1.09$
$Cr_2O_7^{2-}(aq) \parallel 2Cr^{3+}(aq)$	$Cr_2O_7^{2-}(aq) + 14H^+(aq) + 6e^- \rightleftharpoons 2Cr^{3+}(aq) + 7H_2O(l)$	$+1.33$
$Cl_2(aq) \parallel 2Cl^-(aq)$	$Cl_2(aq) + 2e^- \rightleftharpoons 2Cl^-(aq)$	$+1.36$
$MnO_4^-(aq) \parallel Mn^{2+}(aq)$	$MnO_4^-(aq) + 8H^+(aq) + 5e^- \rightleftharpoons Mn^{2+}(aq) + 4H_2O(l)$	$+1.51$

4 Draw a diagram of the electrochemical cell that would be used to determine the standard electrode potential of the $Pb^{2+}(aq) \parallel Pb(s)$ half-cell.

You will notice in Table 3.1 that several of the cells do not contain a solid element which can be used as an electrode; for example, the half-cell containing $Fe^{3+}(aq)$ and $Fe^{2+}(aq)$. To make a half-cell in these cases an inert metal such as platinum is used as the electrode. The solution is made up to have a concentration of $1\,mol\,dm^{-3}$ with respect to each of the aqueous ions being considered (Figure 3.13).

At the top of Table 3.1 is a value for the sodium ion–sodium half-cell. You might wonder how this is actually determined? The problem with this half-cell is that if you place solid sodium (or any reactive metal) into a solution it will react, perhaps violently. To prevent this the sodium is dissolved in mercury and a mercury amalgam is formed. A platinum electrode is then used instead of the solid piece of sodium.

Cell diagrams

A cell diagram is a standard way of writing down the two half-cells in an electrochemical cell, and of showing the changes which are taking place. Consider the electrochemical cell formed between the zinc and copper half-cells. The overall reaction, as we have stated already, is:

$$Zn(s) + Cu^{2+}(aq) \rightarrow Cu(s) + Zn^{2+}(aq)$$

The cell diagram for this particular electrochemical cell would be:

$$Zn(s)\,\big|\,Zn^{2+}(aq)\,\big\|\,Cu^{2+}(aq)\,\big|\,Cu(s) \qquad E_{cell}^{\ominus} = +1.10\,V$$

5 Write cell diagrams for the electrochemical cells formed between:
a the magnesium and copper half-cells
b the magnesium and zinc half-cells.

In the cell diagram the reducing half-cell is shown on the left, showing the change which takes place within it. The two vertical lines represent the salt bridge. On the right-hand side the oxidising half-cell, the one in which reduction occurs, is shown. The negative half-cell is on the left and the positive half-cell on the right.

Uses of standard electrode potentials

Standard electrode potentials have several important uses. The first one we shall consider is their use in determining the $E_{cell}^{\ominus}$ value of a particular electrochemical cell without actually having to set the cell up.

Example What is the $E_{cell}^{\ominus}$ of an electrochemical cell set up from the chlorine and zinc half-cells?

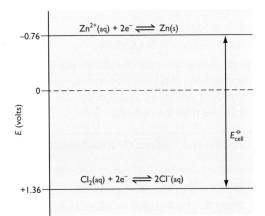

Figure 3.14 (above) An electrode potential chart showing the chlorine and zinc half-cells. Notice it follows the same trend as the electrochemical series with the more negative values at the top.

One way of doing this is to use an electrode potential chart, which is simply a vertical axis onto which the equations for the two half-cells are placed (Figure 3.14).

The $E_{cell}^{\ominus}$ is found by determining the size of 'gap' between the two lines on the chart. In this case, this is equivalent to 2.12 V. Cell voltages are always positive – when was the last time you asked for a -1.5 V battery! The chart can now be used to predict what reaction will occur in each half-cell if the electrochemical cell was set up so as to permit current to flow. The electrons flow from the half-cell with the more negative $E_{cell}^{\ominus}$ to the other half-cell. At the moment the more negative zinc half-cell is written with the electrons on the left-hand side of its half-equation. This half-equation is reversed to give:

$$Zn(s) \rightarrow Zn^{2+}(aq) + 2e^{-}$$

In the chlorine half-cell the aqueous chlorine accepts the electrons and chloride ions are formed:

$$Cl_2(aq) + 2e^{-} \rightarrow 2Cl^{-}(aq)$$

An overall equation can now be found for the reaction that would occur if the two half-cells were not actually separated as they are in the electrochemical cell.

$$Zn(s) \rightarrow Zn^{2+}(aq) + 2e^{-}$$
$$\underline{Cl_2(aq) + 2e^{-} \rightarrow 2Cl^{-}(aq)}$$
$$Zn(s) + Cl_2(aq) \rightarrow Zn^{2+}(aq) + 2Cl^{-}(aq)$$

If zinc metal was placed in an aqueous solution of chlorine, zinc ions and chloride ions would be formed, that is, a solution of zinc chloride.

6 Calculate the cell e.m.f.s formed by connecting the following half-cells, using the information given in Table 3.1 (page 38).
a magnesium and zinc
b manganate(VII) and copper
c aluminium and bromine.

The method of using an electrode potential chart allows us to predict what reaction will occur if the species in the two half-cells under consideration are placed in contact with one another. It does not, however, allow us to say that the reaction will occur; only that if a reaction does occur it will be the one that is predicted. Using standard electrode potentials does not tell us anything about the rate of a particular reaction, so it could be that we predict a particular reaction is feasible only to find that it does not actually occur under standard conditions, due to a large activation energy. The reaction may become feasible if the conditions are changed away from standard conditions, as doing this will change the electrode potential of the half-cells under consideration.

Other examples

In this section we will look at some other examples of predicting reactions which may occur when the species in two half-cells are put together.

Example 1

Earlier in this chapter (pages 31–2) we looked at the use of manganate(VII) ions in the estimation of reducing agents. In that section it was stated that this method could not be used if chloride ions were present in the solution of the reducing agent.

Figure 3.15
The electrode potential chart for manganate(VII) and chlorine half-cells.

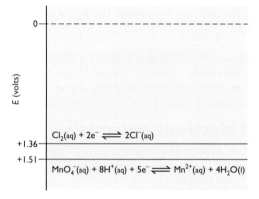

We can now explain this problem, using an electrode potential chart (Figure 3.15). Being the more negative electrode, the chlorine half-cell will release electrons to the manganate(VII) half-cell. So the steps to determine the overall redox equation are as before.

- **Step 1:** Reverse the more negative half-equation:

$$2Cl^-(aq) \rightarrow Cl_2(aq) + 2e^-$$

- **Step 2:** Add it to the more positive half-equation ensuring that electrons cancel out.

To do this in this case, we multiply the more negative process by 5 and the more positive by 2.

$$2Cl^-(aq) \rightarrow Cl_2(aq) + 2e^-$$
$$\underline{MnO_4^-(aq) + 8H^+(aq) + 5e^- \rightarrow Mn^{2+}(aq) + 4H_2O(l)}$$
$$2MnO_4^-(aq) + 16H^+(aq) + 10Cl^- \rightarrow 2Mn^{2+}(aq) + 5Cl_2(aq) + 8H_2O(l)$$

You can now see from this equation that chlorine gas would be produced if chloride ions were present in a solution being titrated against manganate(VII). This does not occur if dichromate(VI) ions are used for the titration as the standard electrode potential for the dichromate(VI) ion half-cell is more negative than that of the chlorine half-cell.

Example 2

Determine the possible reaction between aqueous bromine and aqueous chlorine half-cells.

Figure 3.16
The electrode potential chart for Example 2.

7 Predict the reactions which occur when the following half-cells are connected:
a aluminium and lead(II)
b dichromate(VI) and bromine
c manganate(VII) and iron(II)
d magnesium and iron(II).

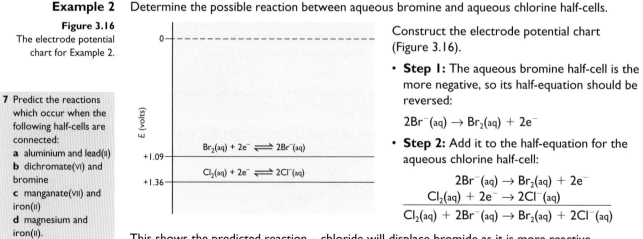

Construct the electrode potential chart (Figure 3.16).

- **Step 1:** The aqueous bromine half-cell is the more negative, so its half-equation should be reversed:

$$2Br^-(aq) \rightarrow Br_2(aq) + 2e^-$$

- **Step 2:** Add it to the half-equation for the aqueous chlorine half-cell:

$$2Br^-(aq) \rightarrow Br_2(aq) + 2e^-$$
$$\underline{Cl_2(aq) + 2e^- \rightarrow 2Cl^-(aq)}$$
$$Cl_2(aq) + 2Br^-(aq) \rightarrow Br_2(aq) + 2Cl^-(aq)$$

This shows the predicted reaction – chloride will displace bromide as it is more reactive.

Rusting

An important, and costly, redox process is that of corrosion or rusting. Rusting only occurs in the presence of water and oxygen. The two half-equations which occur in the rusting of iron are:

$$Fe^{2+}(aq) + 2e^- \rightleftharpoons Fe(s) \qquad E^\ominus = -0.44\,V$$
$$\tfrac{1}{2}O_2(g) + H_2O(l) + 2e^- \rightleftharpoons 2OH^-(aq) \qquad E^\ominus = +0.40\,V$$

Using the method described above, the overall reaction which describes the rusting process can be obtained by reversing the equation for the more negative half-cell and adding it to that of the more positive, giving:

$$Fe(s) + \tfrac{1}{2}O_2(g) + H_2O(l) \rightarrow Fe^{2+}(aq) + 2OH^-(aq)$$

In this process the iron metal is oxidised to iron(II) ions at the centre of the water drop where the oxygen concentration is low, and the electrons released reduce the oxygen at the surface of the drop where oxygen concentration is high (Figure 3.17). The iron(II) ions and hydroxide ions so formed diffuse away from the surface of the iron object. Further oxidation by dissolved oxygen in the water drop forms rust, red hydrated iron(III) oxide, $Fe_2O_3.xH_2O$.

Figure 3.17
The rusting process.

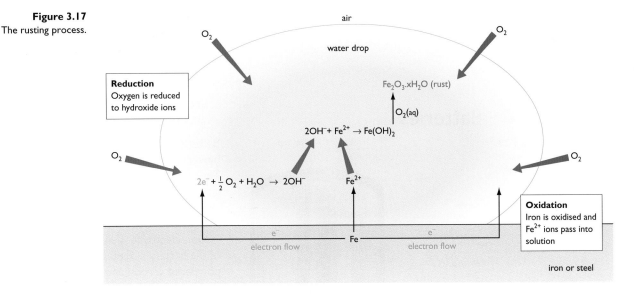

The rusting process can be accelerated by the presence of ionic substances, such as sodium chloride used to help clear iced roads. Ionic substances are the source of ions and when they dissolve in the water it causes an increase in conductivity and hence speeds the rusting process.

Prevention of rusting

There are a variety of methods employed to prevent the rusting of objects made from iron or steel. Painting is a relatively inexpensive method used to protect objects, such as cars, from rusting. The disadvantage with paint is that rusting can occur if the paint becomes scratched. Recently some car manufacturers have employed a different method to prevent rusting. By covering the car body with a thin layer of the metal zinc, a process known as **galvanisation**, better protection is obtained than simply by applying paint. Zinc is a more reactive metal than iron and is therefore oxidised more readily:

$$Zn^{2+}(aq) + 2e^- \rightleftharpoons Zn(s) \qquad E^\ominus = -0.76\,V$$

This ensures that the iron body work of the car is prevented from rusting as the zinc will corrode instead. A similar method is used to prevent ship's hulls from rusting. Large blocks of the even more reactive metal magnesium are attached to the hull. The magnesium corrodes in preference to the steel hull, with new blocks of magnesium being added as appropriate. The use of more reactive metals in the prevention of rusting is known as **sacrificial protection**.

Disproportionation

Disproportionation is the process in which a single species is simultaneously oxidised and reduced. There are many examples of this in Chemistry; some are shown below.

When chlorine is bubbled into a solution of sodium hydroxide the process shown in Figure 3.18 occurs.

Copper(I) ions are very unstable in aqueous solution and they disproportionate to give copper metal and copper(II) ions, as shown in Figure 3.19.

Figure 3.18 (left)
Disproportionation of chlorine.

Figure 3.19 (right)
Disproportionation of copper(I).

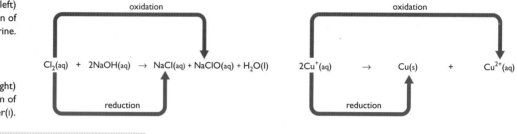

8 In the following processes, which element undergoes disproportionation? Give the oxidation state changes which occur in each case.
a $Cl_2(aq) + H_2O(l) \rightarrow HCl(aq) + HClO(aq)$
b $3IO^-(aq) \rightarrow 2I^-(aq) + IO_3^-(aq)$
c $2H_2O_2(aq) \rightarrow 2H_2O(l) + O_2(g)$

Batteries

Figure 3.20
A standard zinc–carbon battery.

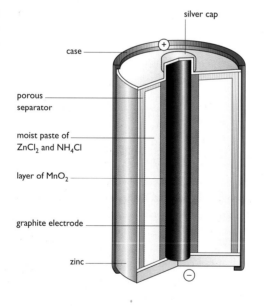

silver cap

case

porous separator

moist paste of ZnCl₂ and NH₄Cl

layer of MnO₂

graphite electrode

zinc

Batteries of all shapes and sizes and of a bewildering number of types are available. The standard battery, or dry **cell**, is based on a system using zinc and carbon electrodes with an electrolyte of ammonium chloride and water in a paste (Figure 3.20). The carbon electrode is situated in the centre of the battery surrounded by a mixture of powdered carbon and manganese(IV) oxide.

The two half-reactions involved can be represented by the half-equations:

$$Zn^{2+}(aq) + 2e^- \rightarrow Zn(s) \qquad E^\ominus = -0.76\,V$$
$$2NH_4^+(aq) + 2e^- \rightarrow 2NH_3(g) + H_2(g)$$
$$E^\ominus = +0.74\,V$$

The overall reaction taking place is, therefore:

$$Zn(s) + 2NH_4^+(aq) \rightarrow Zn^{2+}(aq) + 2NH_3(g) + H_2(g)$$

giving a voltage of 1.5 V.

The ammonia gas dissolves in the water found in the paste and the hydrogen gas is oxidised to water by the manganese(IV) oxide. Both these processes are vital as without them a pressure build-up would occur in the battery.

One of the major problems with this type of battery is as the battery gets older the zinc walls of the battery can become thin as the zinc is oxidised, causing the battery to leak.

The vast majority of batteries used today in everyday items such as CD players, digital cameras and children's toys are alkaline batteries (Figure 3.21), which have long lives when compared with older types. These batteries are also based on the zinc–carbon–manganese(IV) oxide system. They differ from the standard battery in that the zinc is present in the form of a powder and it no longer forms the case of the cell. Alkaline batteries have a steel case which ensures against leakage.

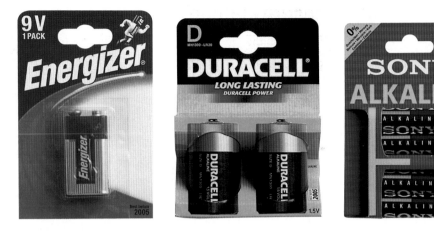

Figure 3.21
Alkaline batteries have longer lives than standard batteries, but are more expensive.

Rechargeable batteries

Many different types of rechargeable batteries are available. One of the most common is the NiCd rechargeable battery which contains a nickel hydroxide cathode and an anode of cadmium. The electrolyte is alkaline potassium hydroxide.

The overall cell reaction is:

$$2NiOOH + 2H_2O + Cd \rightleftharpoons 2Ni(OH)_2 + Cd(OH)_2$$

As the battery is used the forward reaction occurs with the cadmium being oxidised. During the charging process the back reaction occurs. NiCd batteries can be recharged hundreds of times.

● **Key skills** **ICT**
- Construct a spreadsheet which would calculate the percentage of available chlorine in a sample of bleach using the experimental details given on pages 34–5. The spreadsheet should give the percentage if the titration value and concentration of sodium thiosulphate used are entered.

Number
- Calculations associated with the redox titrations mentioned in this chapter.
- Calculations of cell voltages.

● **Skills task** Produce a presentation to compare the advantages and disadvantages of chlorine-based bleaches with hydrogen peroxide-based bleaches.

CHECKLIST After studying Chapter 3 you should know and understand the following terms.

- ● **Electrochemical cell:** An electrochemical cell consists of two half-cells, one for each of the half-equations of the redox process.
- ● **Salt bridge:** The connection between the two half-cells of an electrochemical cell, which allows ions to flow between them without the two solutions mixing.
- ● **Standard hydrogen half-cell:** The half-cell used to measure standard electrode potentials. It sets up an equilibrium between a $1\,mol\,dm^{-3}$ hydrogen ion solution and hydrogen gas, at 1 atmosphere pressure and a temperature of 298 K.
- ● **Standard electrode potential ($E^{\ominus}$):** The potential observed when a half-cell is connected to the standard hydrogen half-cell under standard conditions.
- ● **Cell diagram:** A representation of the two half-cells which make up an electrochemical cell showing the changes which are taking place.
- ● **Disproportionation:** This is the simultaneous oxidation and reduction of the same species in a chemical reaction.
- ● **Cell:** A device in which an electric current is produced by a chemical reaction.
- ● **Battery:** A number of cells connected in series or in parallel to deliver a voltage.

Examination questions

1 The data given below are taken from the electrochemical series.

Reaction at 298 K	$E^{\ominus}$ (V)
$MnO_4^{2-}(aq) + 4H^+(aq) + 2e^- \rightarrow MnO_2(s) + 2H_2O(l)$	+1.55
$MnO_4^-(aq) + 8H^+(aq) + 5e^- \rightarrow Mn^{2+}(aq) + 4H_2O(l)$	+1.51
$MnO_4^-(aq) + e^- \rightarrow MnO_4^{2-}(aq)$	+0.60

Disproportionation is the term used for a reaction in which an element changes from a single oxidation state to two different oxidation states, one being higher and the other lower than the original. This can be illustrated by the reaction

$$Cl_2(aq) + H_2O(l) \rightarrow 2H^+(aq) + Cl^-(aq) + OCl^-(aq)$$

in which Cl(0) becomes Cl(−1) and Cl(1).

Use the information given above, wherever relevant, to answer the questions that follow.

a What is meant by the term **electrochemical series**? (1)

b On warming, OCl^- ions change into Cl^- and ClO_3^-. Write an equation for this disproportionation reaction and determine the oxidation state of chlorine in the ClO_3^- ion. (2)

c How does the overall redox potential show that a reaction is spontaneous? (1)

d The manganate(VI) ion, MnO_4^{2-}, is also unstable and undergoes spontaneous disproportionation to form manganate(VII) ions and solid manganese(IV) oxide. Construct an overall equation for this disproportionation reaction and use values from the electrochemical series to calculate the overall $E^{\ominus}$ value for the reaction. (4)

AQA, A level, Specimen Paper 6421, 2001/2

2 Use the following redox potential data to answer the questions that follow.

Electrode reaction		$E^{\ominus}$ (V)
$Fe(OH)_3$	$+ e^- \rightleftharpoons Fe(OH)_2 + OH^-$	−0.56
Fe^{2+}	$+ 2e^- \rightleftharpoons Fe$	−0.44
Fe^{3+}	$+ 3e^- \rightleftharpoons Fe$	−0.04
H^+	$+ e^- \rightleftharpoons \frac{1}{2}H_2$	0.00
$\frac{1}{2}O_2 + H_2O$	$+ 2e^- \rightleftharpoons 2OH^-$	+0.40
Fe^{3+}	$+ e^- \rightleftharpoons Fe^{2+}$	+0.77
$\frac{1}{2}Cl_2$	$+ e^- \rightleftharpoons Cl^-$	+1.36
Co^{3+}	$+ e^- \rightleftharpoons Co^{2+}$	+1.82

a i Explain why the reaction of iron metal with dilute, aqueous hydrochloric acid gives iron(II) chloride and not iron(III) chloride. (2)

ii Suggest how an aqueous solution of iron(III) chloride could be made from iron metal. Justify your answer. (3)

b Suggest, in outline, the stages in the rusting of iron, using the data given. (3)

c At 25 °C, a cell is constructed as follows

$$Pt\,|\,Co^{2+}(aq, 1\,mol\,dm^{-3}), Co^{3+}(aq, 1\,mol\,dm^{-3})\,\|$$
$$Fe^{3+}(aq, 1\,mol\,dm^{-3}), Fe^{2+}(aq, 1\,mol\,dm^{-3})\,|\,Pt$$

Write the equation for reaction that takes place when a current is allowed to flow. Justify your answer. (2)

Edexcel, A level, Module 3, June 2000

3 a Define the term **standard electrode potential** for a metal/metal ion system. (2)

b The apparatus drawn below was used to measure the standard electrode potential of the $Cu^{2+}(aq)/Cu(s)$ electrode.

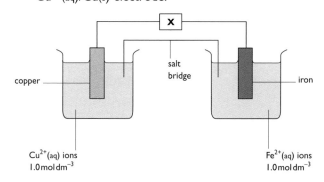

The conventional way of representing this cell is

$$Fe(s)\,|\,Fe^{2+}(aq)\,\|\,Cu^{2+}(aq)\,|\,Cu(s)$$

i What instrument would you use at position **X** to measure the e.m.f. of the cell? (1)

ii The standard electrode potential of the $Fe^{2+}(aq)/Fe(s)$ electrode is −0.44 volts. The e.m.f. of the above cell is +0.78 volts. Calculate the standard electrode potential of the $Cu^{2+}(aq)/Cu(s)$ electrode. (2)

iii Indicate by an arrow, on a copy of the diagram above, the direction in which electrons would flow between the two metals if the external circuit was completed. (1)

iv What would you expect to happen if a small piece of copper was placed in an aqueous solution of iron(II) sulphate? Give your reasoning. (2)

c The standard electrode potentials of the reactions involved in the first stage in the rusting of iron are

$$Fe^{2+}(aq) + 2e^- \rightleftharpoons Fe(s) \qquad -0.44\,volts$$
$$O_2(g) + 2H_2O(l) + 4e^- \rightleftharpoons 4OH^-(aq) \qquad +0.40\,volts$$

i Write an overall equation for the first stage in the rusting of iron. (2)

ii Explain how magnesium metal attached to a sheet of iron prevents it from rusting. (2)

London, A level, Module 3, Jan 2000

4 Redox reactions are an important type of reaction in chemistry.

Explain what is meant by a redox reaction. Illustrate your answer with **two** examples drawn from inorganic chemistry (one of which should involve a transition element) and **two** examples from organic chemistry. (In this question, 1 mark is available for the quality of written communication.) (14) (For details on organic chemistry, see Chapters 9–13.)

OCR, A level, Specimen Paper A7882, Sept 2000

5 A student set up the following electrochemical cell.

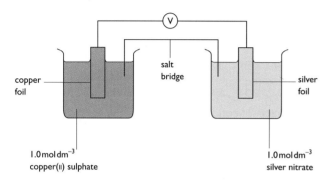

copper foil

salt bridge

silver foil

$1.0 \, mol \, dm^{-3}$ copper(II) sulphate

$1.0 \, mol \, dm^{-3}$ silver nitrate

You are provided with the following standard electrode potentials:

$Cu^{2+}/Cu \quad E^{\ominus} = +0.34 \, V$

$Ag^{+}/Ag \quad E^{\ominus} = +0.80 \, V$

a How could the student have made the salt bridge? (1)

b Write half-equations showing the reactions that occurred in:

i the Cu/Cu^{2+} half-cell

ii the Ag/Ag^{+} half-cell. (2)

c Write an equation for the overall cell reaction. (1)

d i Calculate the standard cell potential for this cell.

ii Identify the electrode at which reduction occurs. Explain your answer. (4)

e The student found that the e.m.f. obtained for this cell was less than the calculated value. Suggest **two** reasons for this. (2)

OCR, A level, Specimen Paper A7882, Sept 2000

4 Further thermodynamics

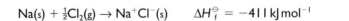

STARTING POINTS
- Hess's law states that the enthalpy change for any chemical reaction is independent of the route taken, provided that the initial and final conditions are identical.
- The standard enthalpy change of formation is the enthalpy change which takes place when 1 mol of a compound is formed from its elements, in their standard states, under standard conditions. Symbol: $\Delta H_f^\ominus$.
- The standard enthalpy change of combustion is the enthalpy change which occurs when 1 mol of a substance is completely burned in oxygen, under standard conditions. Symbol: $\Delta H_c^\ominus$.
- The standard enthalpy change of reaction is the enthalpy change which takes place when the amounts of reactants shown in the balanced chemical equation react together under standard conditions to give the products in their standard states. Symbol: $\Delta H_r^\ominus$.
- The standard enthalpy change of solution is the enthalpy change which takes place when 1 mol of a solute dissolves to form a solution of concentration $1 \, mol \, dm^{-3}$ under standard conditions. Symbol: $\Delta H_{sol}^\ominus$.
- The bond dissociation enthalpy is the enthalpy change which occurs when 1 mol of a particular covalent bond is broken in the gaseous state.
- The bond enthalpy (bond energy) is the mean value of the bond dissociation enthalpies.
- The ionisation energy is the energy required to remove an electron from a gaseous atom or ion.
- The first electron affinity is the enthalpy change which takes place when 1 mol of gaseous atoms of an element gains 1 mol of electrons to form 1 mol of singly negatively charged ions.

Born–Haber cycles

You will have already met Hess's law when dealing with enthalpy changes such as formation, combustion and reaction.

A Born–Haber cycle is an application of Hess's law which is used to look at the enthalpy changes which occur when an ionic compound is formed from its elements. To form an ionic compound, positive and negative ions have to be formed by the movement of electrons from one atom or group of atoms, to another atom or group of atoms.

Consider the reaction between sodium metal and chlorine gas. Sodium reacts with chlorine exothermically to form solid sodium chloride (Figure 4.1a).

$$Na(s) + \tfrac{1}{2}Cl_2(g) \rightarrow Na^+Cl^-(s) \qquad \Delta H_f^\ominus = -411 \, kJ \, mol^{-1}$$

Figure 4.1
a Sodium chloride is an ionic substance formed by the chemical reaction between sodium and chlorine.
b It is an ionic compound and the crystals formed are cubic.

a

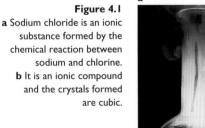

b

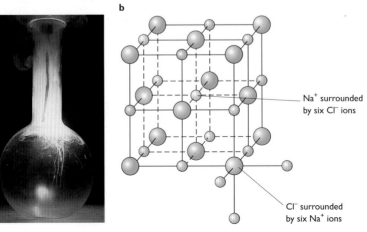

Na^+ surrounded by six Cl^- ions

Cl^- surrounded by six Na^+ ions

It may seem, on the surface, that the reaction involves each of the sodium atoms losing an electron and each of the chlorine atoms gaining an electron to form a chloride ion. The process is, however, made up of many more steps. Each of these steps involves energy changes and can be shown using a Born–Haber cycle. These steps are now detailed below.

- The sodium atom, as stated above, must at some point during the reaction lose an electron to form the Na^+ ion. As you will have learned earlier in your study of Chemistry, the loss of 1 mol of electrons from 1 mol of *gaseous* atoms is known as the first ionisation enthalpy. But first the solid sodium must be converted into gaseous sodium. The change:

$$Na(s) \rightarrow Na(g) \qquad \Delta H_{at}^{\ominus} = +108 \, kJ \, mol^{-1}$$

is an endothermic process needing the input of energy. The term **atomisation enthalpy**, $\Delta H_{at}^{\ominus}$, is used to describe the process in which 1 mol of gaseous atoms is formed from the element in its standard state under standard conditions.

- In order to produce a chloride ion, a chlorine *atom* must first be produced from a chlorine molecule. The energy change involved is equivalent to the **atomisation enthalpy** of chlorine.

$$\tfrac{1}{2}Cl_2(g) \rightarrow Cl(g) \qquad \Delta H_{at}^{\ominus} = +122 \, kJ \, mol^{-1}$$

- Now that the gaseous atoms have been formed, the sodium atom can now lose an electron and donate it to the chlorine atom which gains the electron to form the chloride ion. The energy change which describes the loss of 1 mol of electrons from 1 mol of sodium ions is known as the first ionisation energy of sodium.

$$Na(g) \rightarrow Na^+(g) + e^- \qquad \Delta H_{ie}^{\ominus} = +495 \, kJ \, mol^{-1}$$

- The chlorine atoms now gain an electron to form the chloride ion, $Cl^-(g)$. The enthalpy change which occurs when a gaseous atom gains an electron to form a negatively charged ion is called the **electron affinity**. As with ionisation enthalpies, the gain of a single electron is correctly known as the first electron affinity, while the gain of a second electron by an already negative ion would be called the second electron affinity.

$$Cl(g) + e^- \rightarrow Cl^-(g) \qquad \Delta H_{ea}^{\ominus} = -349 \, kJ \, mol^{-1}$$

- Now that the ions have been formed they bond together to form solid sodium chloride. The enthalpy change which accompanies the formation of solid sodium chloride from its gaseous ions is known as the **lattice enthalpy**, $\Delta H_{LE}^{\ominus}$, of sodium chloride, and it is an exothermic process.

$$Na^+(g) + Cl^-(g) \rightarrow NaCl(s) \qquad \Delta H_{LE}^{\ominus} = -787 \, kJ \, mol^{-1}$$

These steps can now be used to construct a Born–Haber cycle for sodium chloride (Figure 4.2, overleaf). In this example we have given values for all of the individual enthalpy changes which occur during the formation of sodium chloride. The Born–Haber cycle can be used to calculate individual values if all the others are known.

A Born–Haber cycle can be drawn to scale in order to obtain values for specific enthalpy changes, or it can be used as a basis for numeric calculation of a missing value.

The Born–Haber cycle shows that the principle reason that stable ionic compounds form from their elements is because the lattice enthalpy is very exothermic. This more than cancels out the endothermic processes which also occur in the formation of the compound.

In the following we will construct a Born–Haber cycle to calculate the standard enthalpy change of formation of calcium oxide. First we need to determine the individual steps which occur during the formation of calcium oxide, starting from solid calcium and oxygen gas and find, from data sources, enthalpy change values for the steps.

- Calcium must first be atomised to produce gaseous calcium atoms.

$$Ca(s) \rightarrow Ca(g) \qquad \Delta H_{at}^{\ominus} = +178 \, kJ \, mol^{-1}$$

- 1 mol of gaseous oxygen atoms is produced from 0.5 mol of oxygen molecules.

$$\tfrac{1}{2}O_2(g) \rightarrow O(g) \qquad \Delta H_{at}^{\ominus} = +249 \, kJ \, mol^{-1}$$

1 Draw a Born–Haber cycle for potassium chloride using the information given in the text and the following:

$\Delta H_{at}^{\ominus}(K) = +89.2 \, kJ \, mol^{-1}$
$\Delta H_{ie}^{\ominus}(K) = +419 \, kJ \, mol^{-1}$
$\Delta H_{LE}^{\ominus}(KCl) = -711 \, kJ \, mol^{-1}$

Figure 4.2
The Born–Haber cycle for sodium chloride.

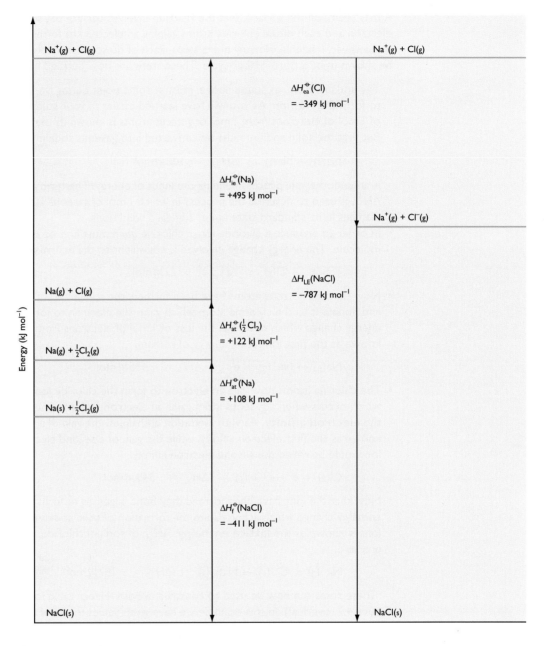

- Each calcium atom now loses two electrons to form gaseous calcium ions, $Ca^{2+}(g)$. This change is the sum of the first and second ionisation enthalpies.

$$Ca(g) \rightarrow Ca^+(g) + e^- \qquad \Delta H_{ie1}^\ominus = +590 \, kJ \, mol^{-1}$$
$$Ca^+(g) \rightarrow Ca^{2+}(g) + e^- \qquad \Delta H_{ie2}^\ominus = +1145 \, kJ \, mol^{-1}$$

- To form the oxide ion the oxygen atom must gain two electrons. The enthalpy change here will be the sum of the first and second electron affinities of oxygen.

$$O(g) + e^- \rightarrow O^-(g) \qquad \Delta H_{ea1}^\ominus = -141 \, kJ \, mol^{-1}$$
$$O^-(g) + e^- \rightarrow O^{2-}(g) \qquad \Delta H_{ea2}^\ominus = +798 \, kJ \, mol^{-1}$$

Notice that the first electron affinity of oxygen is an exothermic process, but the second is endothermic as a result of a second electron being added to an already negative ion.

- Finally the lattice enthalpy of calcium oxide.

$$Ca^{2+}(g) + O^{2-}(g) \rightarrow CaO(s) \qquad \Delta H_{LE}^\ominus = -3454 \, kJ \, mol^{-1}$$

48

The Born–Haber cycle can now be constructed for calcium oxide, and the missing enthalpy change, $\Delta H_f^\ominus$, can be calculated (Figure 4.3). The standard enthalpy of formation of calcium oxide can be found using the expression below:

$$\Delta H_f^\ominus(\text{CaO}) = \Delta H_{at}^\ominus(\text{Ca}) + \Delta H_{at}^\ominus(\text{O}) + \Delta H_{ie1}^\ominus(\text{Ca}) + \Delta H_{ie2}^\ominus(\text{Ca}) + \\ \Delta H_{ea1}^\ominus(\text{O}) + \Delta H_{ea2}^\ominus(\text{O}) + \Delta H_{LE}^\ominus(\text{CaO})$$

$$\Delta H_f^\ominus(\text{CaO}) = 178 + 249 + 590 + 1145 + (-141) + 798 + (-3454) \\ = -635 \text{ kJ mol}^{-1}$$

Figure 4.3
The Born–Haber cycle for calcium oxide.

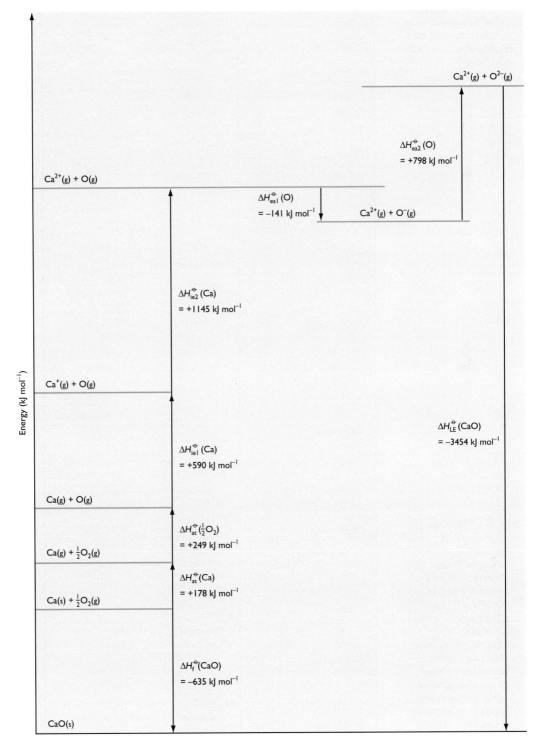

Large negative values of enthalpy of formation are typical of stable ionic compounds. To show this, look at the Born–Haber cycle for the hypothetical ionic compound CaCl (Figure 4.4). To produce this cycle assumptions have to be made about the structure of such a compound. The lattice enthalpy is also hypothetical and is obtained from an equation out of the scope of your course, known as the Born–Mayer equation.

CaCl is not a stable chloride of calcium and this can be seen by the much less exothermic value of $\Delta H_f^\ominus$ compared with that of $CaCl_2$, the stable chloride of calcium, which has a $\Delta H_f^\ominus$ of $-795\,kJ\,mol^{-1}$.

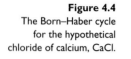

Figure 4.4
The Born–Haber cycle for the hypothetical chloride of calcium, CaCl.

2 Draw a Born–Haber cycle for the hypothetical compound $CaCl_3$. Given that the third ionisation energy of calcium is $4912\,kJ\,mol^{-1}$ and the lattice enthalpy is $-5000\,kJ\,mol^{-1}$, determine the enthalpy of formation of this compound. Comment on the value obtained with reference to the enthalpies of formation of $CaCl_2$ and CaCl.

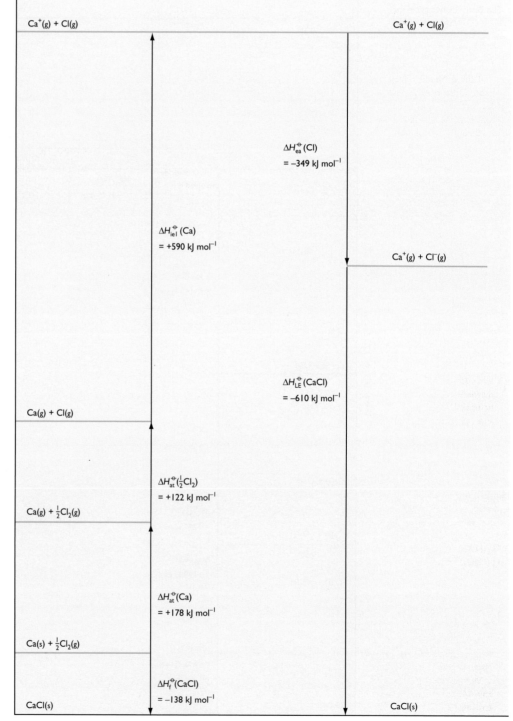

Energy changes in solution

Figure 4.5
Sodium chloride dissolves well in water, whereas silver chloride does not.

Many ionic compounds, such as sodium chloride and potassium fluoride, dissolve well in water to form solutions. Others, such as silver chloride and lead chloride, do not dissolve to a great extent (Figure 4.5, see also Chapter 2). This section looks at the enthalpy changes which occur when an ionic solid is added to water and why some dissolve and some do not.

We saw in the previous section that the enthalpy change which occurs when 1 mol of an ionic solid is formed is known as its lattice enthalpy. When an ionic substance dissolves it is the reverse of this process which needs to occur, that is, the ionic crystal lattice needs to be broken up (Figure 4.6). Lattice enthalpies are always exothermic, so the process to break up an ionic crystal lattice will be endothermic. If the compound is to dissolve the energy needed to do this must be found from other enthalpy changes within the dissolving process.

Table 4.1 shows lattice enthalpies of some common ionic compounds formed by Group 1, 2 and 3 metals.

Figure 4.6
During the dissolving process the ionic crystal lattice is broken up and the ions become spread out through the solution.

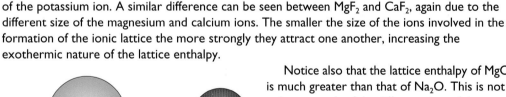

solid sodium chloride
– a regular ionic lattice

sodium chloride dissolved in water

Cl^- $\quad$ Na^+

$Cl^-(aq)$ $\quad$ $Na^+(aq)$

Table 4.1

Group 1 compounds $\Delta H_{LE}^{\ominus}$ (kJ mol^{-1})		Group 2 compounds $\Delta H_{LE}^{\ominus}$ (kJ mol^{-1})		Group 3 compounds $\Delta H_{LE}^{\ominus}$ (kJ mol^{-1})	
NaF	−918	MgF$_2$	−2957		
KF	−817	CaF$_2$	−2630		
Na$_2$O	−2478	MgO	−3791	Al$_2$O$_3$	−15916
K$_2$O	−2232	CaO	−3401		

3 Why do you think that magnesium oxide can be usefully employed as a refractory lining?

There is a pattern in how the size of the lattice enthalpy varies. Notice that the lattice enthalpy of NaF is more exothermic than that of KF. The difference must be due to the larger ionic size of the potassium ion. A similar difference can be seen between MgF$_2$ and CaF$_2$, again due to the different size of the magnesium and calcium ions. The smaller the size of the ions involved in the formation of the ionic lattice the more strongly they attract one another, increasing the exothermic nature of the lattice enthalpy.

Figure 4.7
Smaller, highly charged ions attract one another more strongly than large, singly charged ions.
$\Delta H_{LE}^{\ominus}(NaCl) = -787$ kJ mol^{-1},
$\Delta H_{LE}^{\ominus}(MgO) = -3791$ kJ mol^{-1}.

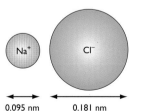

Na$^+$ $\quad$ Cl$^-$

0.095 nm $\quad$ 0.181 nm

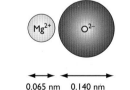

Mg^{2+} $\quad$ O^{2-}

0.065 nm $\quad$ 0.140 nm

Notice also that the lattice enthalpy of MgO is much greater than that of Na$_2$O. This is not only as a result of the different sizes of the metal ions, and hence the different crystal structures, but also to the fact that the sodium ion has a +1 charge and the magnesium has a +2 charge. The larger the ionic charge the stronger the attraction will be between the ions, leading to a more negative lattice enthalpy (Figure 4.7).

In summary, lattice enthalpy values become more exothermic when:

- the ions involved become smaller
- the charge on the ions becomes larger.

4 a Explain why the lattice enthalpy of calcium oxide is greater than that of potassium oxide.
b Would you expect the lattice enthalpy of rubidium fluoride to be more or less exothermic than that of potassium fluoride? Explain your answer.
c Would you expect lithium fluoride or lithium chloride to have the more exothermic lattice enthalpy? Explain your answer.

Since the first step in the dissolving process is to break up the lattice, the amount of energy shown in Table 4.1 must first be put into the process.

Hydration enthalpies

Figure 4.8
The ions become hydrated.

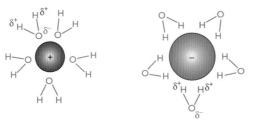

Once the ionic crystal lattice has been broken up, the separated ions become **hydrated** by water molecules. The hydration process involves the negative and positive ions being surrounded by water molecules (Figure 4.8). Polar water molecules are able to form bonds to both negative and positive ions. Because bonds are being formed between water molecules and ions, hydration is an exothermic process. The enthalpy change which occurs is known as the **enthalpy of hydration**, $\Delta H^{\ominus}_{hyd}$, and it is defined as the enthalpy change which occurs when 1 mol of gaseous ions is hydrated to produce a solution which has a concentration of $1\,mol\,dm^{-3}$ with respect to that ion. For example,

$$Mg^{2+}(g) + aq \rightarrow Mg^{2+}(aq) \qquad \Delta H^{\ominus}_{hyd} = -1891\,kJ\,mol^{-1}$$
$$Cl^{-}(g) + aq \rightarrow Cl^{-}(aq) \qquad \Delta H^{\ominus}_{hyd} = -364\,kJ\,mol^{-1}$$

If a solvent other than water is being used the enthalpy change is called the **enthalpy of solvation**, $\Delta H^{\ominus}_{sol}$. Some values for the enthalpy of hydration of some common ions are shown in Table 4.2.

Table 4.2

Group 1 ions $\Delta H^{\ominus}_{hyd}$ (kJ mol^{-1})		Group 2 ions $\Delta H^{\ominus}_{hyd}$ (kJ mol^{-1})		Group 3 ions $\Delta H^{\ominus}_{hyd}$ (kJ mol^{-1})	
Na$^+$	-499	Mg^{2+}	-1891	Al^{3+}	-4613
K$^+$	-390	Ca^{2+}	-1562		
Rb$^+$	-280	Sr^{2+}	-1410		

5 a Would you expect the enthalpy of hydration of the lithium ion to be more or less exothermic than that of the sodium ion? Explain your answer.
b Which hydrated ion would be the larger, sodium or aluminium? Explain your answer.

You will see from Table 4.2 that enthalpies of hydration are more exothermic for smaller ions with higher charges. The aluminium ion, for example, is small and has a 3+ charge. It attracts water molecules strongly to itself, typically about 25 water molecules per aluminium ion, hence the large enthalpy of hydration of the aluminium ion.

Because large numbers of water molecules surround each aluminium ion, the hydrated ion formed is large. The converse is true for the potassium ion, for example. The K$^+$(g) ion is a larger ion than Al^{3+}(g) and has only a 1+ charge. As such it only attracts, on average, four water molecules per ion.

Enthalpies of solution

The dissolving process involves a process which requires the input of energy in order to break up the crystal lattice and another which gives out energy as the ions become hydrated. The **enthalpy of solution**, $\Delta H^{\ominus}_{sol}$, is effectively the difference between these two amounts and can be represented generally by:

$$\Delta H^{\ominus}_{sol} = \Delta H^{\ominus}_{hyd}\,(cation) + \Delta H^{\ominus}_{hyd}\,(anion) - \Delta H^{\ominus}_{LE}$$

Enthalpies of solution can be measured experimentally by adding a known amount of a solid ionic compound to sufficient water to produce a $1\,mol\,dm^{-3}$ solution with respect to both ions in the compound. If the temperature change is measured as dissolving occurs then it can be used to determine the enthalpy of solution of an ionic compound.

Figure 4.9
A Hess cycle to show the dissolving of sodium chloride.

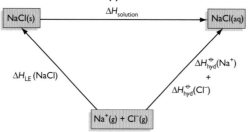

The overall process of dissolving can be shown as a Hess cycle. Figure 4.9 shows a Hess cycle for sodium chloride. It is found that for an ionic solid to dissolve, the enthalpy of solution must be either exothermic, or only slightly endothermic. A large positive value for the enthalpy of solution of an ionic solid will lead to it being insoluble.

The diagram in Figure 4.10 is an energy level diagram showing the processes which occur when silver fluoride dissolves.

The energy level diagram in Figure 4.11 is for silver chloride, an insoluble ionic solid. Notice the larger positive enthalpy of solution than for silver fluoride.

The reason for the difference in the solubilities of silver fluoride and silver chloride, although both of their enthalpies of solution are positive, is related to another energy term involved in the dissolving process which we have yet to discuss – **entropy**.

Figure 4.10
The enthalpy of solution for silver fluoride is just endothermic. Silver fluoride is a soluble ionic solid.

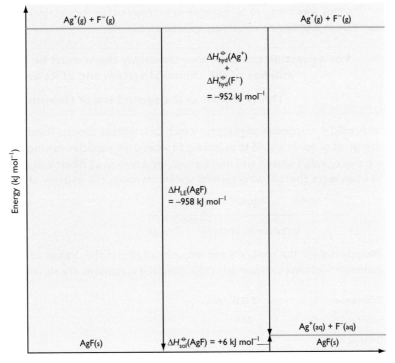

$Ag^+(g) + F^-(g)$ $Ag^+(g) + F^-(g)$

$\Delta H_{hyd}^{\ominus}(Ag^+)$
$+$
$\Delta H_{hyd}^{\ominus}(F^-)$
$= -952$ kJ mol^{-1}

$\Delta H_{LE}(AgF)$
$= -958$ kJ mol^{-1}

Energy (kJ mol^{-1})

$Ag^+(aq) + F^-(aq)$
AgF(s)

AgF(s) $\Delta H_{sol}^{\ominus}(AgF) = +6$ kJ mol^{-1}

Figure 4.11
The enthalpy of solution for silver chloride is more endothermic than for silver fluoride. Silver chloride is an insoluble ionic solid.

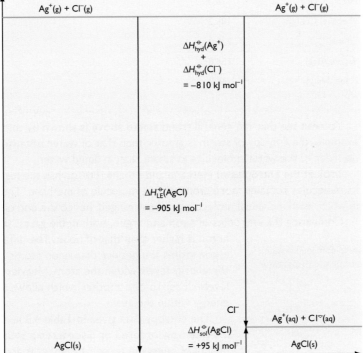
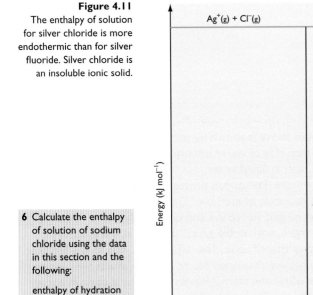

$Ag^+(g) + Cl^-(g)$ $Ag^+(g) + Cl^-(g)$

$\Delta H_{hyd}^{\ominus}(Ag^+)$
$+$
$\Delta H_{hyd}^{\ominus}(Cl^-)$
$= -810$ kJ mol^{-1}

$\Delta H_{LE}^{\ominus}(AgCl)$
$= -905$ kJ mol^{-1}

Energy (kJ mol^{-1})

Cl^-
$Ag^+(aq) + Cl^-(aq)$

$\Delta H_{sol}^{\ominus}(AgCl)$
$= +95$ kJ mol^{-1}

AgCl(s)   AgCl(s)

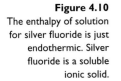

6 Calculate the enthalpy of solution of sodium chloride using the data in this section and the following:

enthalpy of hydration of $Cl^- = -364$ kJ mol^{-1}; lattice enthalpy of sodium chloride $= -787$ kJ mol^{-1}.

● Entropy

Disregarding their rates, why is it that some chemical reactions take place and others do not? To explain this we need to introduce another energy term known as **entropy**. In its simplest form the entropy of a system is a measure of the **degree of disorder**. In a simple everyday context we could say that a freshly tidied bedroom has a lower entropy than the same bedroom several days later when it is messier. In the tidied bedroom there is more order than in the messier bedroom!

Changes which lead to an increase in entropy occur spontaneously. Those which lead to a decrease in entropy do not occur spontaneously.

> **For a reaction to occur spontaneously there must be an overall increase in entropy of the chemical system and of its surroundings.**
>
> **This is known as the second law of thermodynamics.**

A solid is composed of particles which only vibrate about a fixed position. There is a higher degree of order in a solid than in a liquid where the particles can move about. In a gas the particles are more widely spaced and move about randomly in all directions, a high degree of disorder. We can make the following general statement about the entropy of solids, liquids and gases.

solids   liquids   gases
$\xrightarrow{\hspace{5cm}}$
increase in entropy

Entropy is given the symbol S and has units of $JK^{-1}mol^{-1}$. Values for the entropy of some common substances measured under standard conditions are shown in Table 4.3.

Table 4.3

Substance	S ($JK^{-1}mol^{-1}$)
Iron, Fe(s)	27.1
Graphite, C(s)	5.7
Water, H_2O(l)	69.9
Water, H_2O(g)	188.6
Mercury, Hg(l)	76.0
Methane, CH_4(g)	186.2
Ethane, C_2H_6(g)	230.0
Argon, Ar(g)	154.8
Neon, Ne(g)	146.2

You can see that the general trend stated above is shown by the data in Table 4.3. For example, the entropy of steam is greater than that of water because there is a higher level of disorder of the water molecules in steam than in liquid water.

Look at the entropies of methane and ethane. Ethane has the higher entropy because each of its molecules contains more atoms than a molecule of methane. There are, therefore, more ways in which each individual molecule can be arranged, hence the entropy is higher.

Comparing the entropies of neon and argon, both noble gases, shows that the entropy of argon is higher than that of neon. The difference is due to the fact that argon atoms are heavier than neon atoms and this affects the spacing of the energy levels within the atom. Heavier atoms tend to have energy levels closer to one another which allows more ways of arranging the energy within the atom.

The entropy data given in Table 4.3 are values which not only indicate how ordered or disordered a substance is, but also how the quanta of energy in these substances is arranged.

7 For each of the following pairs of compounds state which will have the higher entropy and give a reason for your answer.
a F_2(g) and Br_2(g)
b Cl_2(g) and F_2(g)
c NH_3(g) and H_2O(g)
d propane gas and butane gas
e N_2(g) and CO_2(g)

Let us have a simplified look at the ways in which four quanta of energy can be shared between two molecules.

molecule I	molecule 2
0	4
I	3
2	2
3	I
4	0

8 How many ways are there of sharing:
a six quanta of energy between two molecules?
b four quanta of energy between three molecules?

There are five ways four quanta of energy can be shared between two molecules. But consider the following: in I mol of methane molecules, at a particular temperature, not all the molecules will have the same energy. Collisions between the molecules transfer energy, so some molecules have energies below the average and some above. The energy the molecules have is distributed between its energy levels. There are a number of different types of energy level associated with each molecule, each with different differences in energy (Figures 4.12 and 4.13).

Figure 4.12
Electronic, vibrational and rotational energy levels. Translational energy levels are so close together that they are regarded as being continuous.

Figure 4.13
Vibrational, rotational and translational energy are all different forms of kinetic energy that a molecule can have.

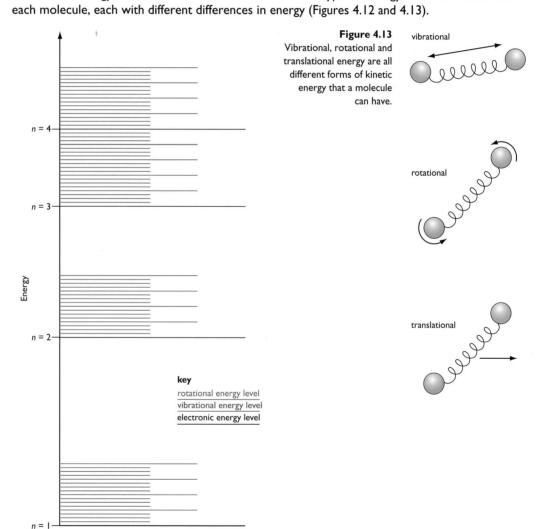

key
rotational energy level
vibrational energy level
electronic energy level

The difference in energy between each type of energy level can be summarised as below:

electronic > vibrational > rotational > translational

So in our mole of methane molecules, not only do the molecules have different amounts of energy but the energy can be distributed in many different ways.

Entropy in practice

Figure 4.14
Is the entropy change for
water condensing
negative or positive?

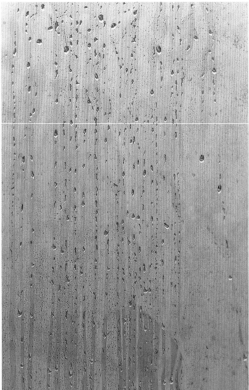

It was stated earlier that if a chemical reaction occurs spontaneously then there must be an increase in entropy associated with it. Consider the everyday phenomena of water condensing on a window pane (Figure 4.14).

During the condensation process water vapour changes into liquid water. A gas into a liquid – surely this process is accompanied by a decrease in entropy? Well this is true, so why does this process occur spontaneously? In our discussion so far on entropy we have only looked at the molecules in the chemical system itself. There are other molecules involved – but where?

When liquid water is formed from its vapour, energy is given out because of the stronger intermolecular bonds which are formed. The energy which is given out is absorbed by molecules in the surrounding materials, for example, the glass which the window is made from, the wooden or uPVC frame and the surrounding air. If these molecules are gaining energy then they will become more disordered, and their particles will move faster. The entropy of the surroundings increases (Figure 4.15).

Figure 4.15
An exothermic process like
the condensing of water
gives out energy to
its surroundings.

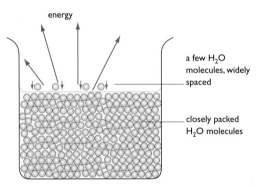

Whenever we calculate the entropy change for any chemical process we must consider both the particles in the reacting materials and the particles in the surroundings.

Entropy changes

A change in entropy is given the symbol ΔS, and has units of $J\,K^{-1}\,mol^{-1}$.

The overall or total entropy change for a chemical process is given by the relationship:

$$\Delta S_{total} = \Delta S_{system} + \Delta S_{surr}$$

where ΔS_{surr} is the entropy change of the surroundings which occurs as a result of the chemical process occurring.

The entropy change of the surroundings can be obtained, knowing the enthalpy change for the process (ΔH), using the relationship:

$$\Delta S_{surr} = \frac{-\Delta H}{T}$$

where T is the temperature at which the process occurs, in kelvin (K).

Let us now use this relationship to quantify the entropy change which occurs when water condenses on a window at 298 K.

$$H_2O(g) \rightarrow H_2O(l) \qquad \Delta H^{\ominus} = -44.0 \, kJ \, mol^{-1}$$
$$\Delta S^{\ominus}_{sys} = -118.0 \, J \, K^{-1} \, mol^{-1}$$

As discussed above, the entropy change of the system is indeed negative because of the more ordered system (liquid water) which is being produced.

$$\Delta S_{surr} = \frac{-\Delta H}{T} = \frac{-(-44000)}{298}$$
$$= 147.7 \, J \, K^{-1} \, mol^{-1}$$

The entropy change of the surroundings is positive. Now the total change in entropy can be calculated:

$$\Delta S_{total} = \Delta S_{sys} + \Delta S_{surr}$$
$$= -118 + 147.7$$
$$= 29.7 \, J \, K^{-1} \, mol^{-1}$$

The overall change in entropy is positive, which is why water condenses spontaneously at room temperature.

> **9** Water shows another change of state when it changes from water to ice at 0 °C. Calculate the entropy change for this process at −15 °C.
>
> $H_2O(l) \rightarrow H_2O(s)$
> $\Delta S_{sys} = -22.0 \, J \, K^{-1} \, mol^{-1}$
> $\Delta H = -6.01 \, kJ \, mol^{-1}$
>
> **10** When calcium carbonate is heated strongly it decomposes to give calcium oxide and carbon dioxide gas.
> **a** Write an equation for this process.
> **b** Given that $\Delta S_{sys} = +160.4 \, J \, K^{-1} \, mol^{-1}$ and that $\Delta H = +118 \, kJ \, mol^{-1}$, find the entropy change for the process at 298 K, 600 K and 1300 K.
> **c** Find the lowest temperature which can be used to thermally decompose calcium carbonate.

Calculations such as Questions 9 and 10 can be used to find out whether a chemical process will occur spontaneously. However, the fact that they may occur spontaneously does not mean they will occur quickly. The kinetics of the reaction would need to be studied to give any information about the rate (see Chapter 1).

Entropy and equilibrium

We all know that water will condense on windows at 298 K, but at what temperature will it not condense. Well, a little thought should perhaps lead you to the temperature of 100 °C (373 K), the boiling point of water. Let us look at the value of ΔS_{total} at this temperature, where an equilibrium has been established:

$$H_2O(g) \rightleftharpoons H_2O(l)$$
$$\Delta S_{total} = -118 + \frac{-(-44000)}{373}$$
$$= 0 \, J \, K^{-1} \, mol^{-1}$$

At 373 K water vapour and liquid water exist together. For any system at equilibrium, ΔS_{total} must be zero.

Dissolving and entropy

In the earlier section on dissolving we ended by saying that substances which had a positive enthalpy of solution still dissolved as a result of entropy. Hopefully you can now see why. Dissolving is a process which proceeds with an increase in entropy. The solution formed has a higher entropy than the pure solute and solvent. It is this increase in entropy which overcomes small endothermic enthalpies of solution to make the dissolving process occur.

● Gibbs free energy

Gibbs free energy is a term named after the American chemist Josiah Willard Gibbs (1839–1903). To obtain an expression for this energy term we need to look back to the relationships we used in relation to entropy changes.

$$\Delta S_{total} = \Delta S_{sys} + \Delta S_{surr}$$

and

$$\Delta S_{surr} = \frac{-\Delta H}{T}$$

Substituting for ΔS_{surr} gives

$$\Delta S_{total} = \Delta S_{sys} + \frac{-\Delta H}{T}$$

Multiplying this expression through by $-T$ gives

$$-T\Delta S_{total} = -T\Delta S_{sys} + \Delta H$$

The term $-T\Delta S_{\textbf{total}}$ is called the **Gibbs free energy change** and is given the symbol **ΔG**. It has units of $J\,mol^{-1}$.

$$\Delta G = -T\Delta S_{total}$$

and

$$\Delta G = \Delta H - T\Delta S_{sys}$$

The beauty of using ΔG to predict whether or not a reaction will proceed spontaneously is that to calculate ΔG we need only have information about the chemical system, its ΔH and its ΔS; we need not worry about the ΔS of the surroundings. A *negative* value of ΔG indicates that a chemical process will proceed *spontaneously* (Figure 4.16). If ΔG is *zero* an *equilibrium* is attained and if it is *positive* then the reverse process to the one being considered is spontaneous.

We can use the above expression to find the Gibbs free energy change for the condensation of water at 298 K.

$$\Delta G = -44\,000\,J\,mol^{-1} - (298\,K) \times (-118\,J\,K^{-1}\,mol^{-1})$$
$$= -8836\,J\,mol^{-1}$$
$$= -8.8\,kJ\,mol^{-1}$$

ΔG has a negative value for this process, indicating that it proceeds spontaneously.

Figure 4.16
The displacement of copper by zinc and the reaction between ammonia and hydrogen chloride both occur spontaneously. What sort of ΔG values do these processes have?

11 Calculate the Gibbs free energy change for the decomposition of calcium carbonate at temperatures of:
a 298 K
b 1300 K.
Use the information given in Question 10 to help you.

$\bullet$ **Key skills** ICT

* Construct a spreadsheet which would allow you to determine the enthalpy of formation of an ionic solid, via a Born–Haber cycle, from given data.

Number

* Calculations of enthalpy and entropy changes.
* Construction of Born–Haber cycles drawn to scale.

$\bullet$ **Skills task** Construct Born–Haber cycles for the hypothetical compound MgCl and for $MgCl_2$. You will need to use an appropriate data book or the Internet to find the relevant data. Comment on your findings.

CHECKLIST After studying Chapter 4 you should know and understand the following terms.

$\bullet$ **Born–Haber cycle:** An application of Hess's law used to look at the enthalpy changes which occur when an ionic compound is formed from its elements.

$\bullet$ **Standard enthalpy of atomisation:** The enthalpy change which takes place when 1 mol of gaseous atoms is formed from the element in its standard state under standard conditions. Symbol: $\Delta H_{at}^{\ominus}$.

$\bullet$ **Lattice enthalpy:** The enthalpy change which occurs when 1 mol of a solid ionic substance is formed from its gaseous ions under standard conditions. Symbol: $\Delta H_{LE}^{\ominus}$.

$\bullet$ **Standard enthalpy of hydration:** The enthalpy change which occurs when 1 mol of a gaseous ion is completely hydrated in aqueous solution, under standard conditions, to form a solution of concentration $1 \, mol \, dm^{-3}$. Symbol: $\Delta H_{hyd}^{\ominus}$.

$\bullet$ **Standard enthalpy of solvation:** The enthalpy change which occurs when 1 mol of gaseous ions is solvated by solvent molecules, under standard conditions, to form a solution of concentration $1 \, mol \, dm^{-3}$.

$\bullet$ **Entropy:** A thermodynamic quantity which is a measure of the degree of disorder within any system. The greater the degree of disorder then the higher the entropy is. Symbol: S.

$\bullet$ **Entropy changes:** The overall entropy change for a chemical process is given by the relationship:

$$\Delta S_{total} = \Delta S_{sys} + \Delta S_{surr}$$

where $\Delta S_{surr} = \dfrac{\Delta H}{T}$

$\bullet$ **Gibbs free energy change:** The thermochemical quantity which allows predictions to be made as to whether a chemical reaction is feasible. A feasible chemical reaction has a negative ΔG value.

● Examination questions

I Sodium bromide is formed from its elements at 298 K according to the equation

$$Na(s) + \tfrac{1}{2}Br_2(l) \rightarrow NaBr(s)$$

The lattice dissociation enthalpy of solid sodium bromide refers to the enthalpy change for the process

$$NaBr(s) \rightarrow Na^+(g) + Br^-(g)$$

The electron addition enthalpy refers to the process

$$Br(g) + e^- \rightarrow Br^-(g)$$

Use this information and the data in the table below to answer the questions which follow.

Standard enthalpies		$\Delta H^\ominus$ (kJ mol^{-1})
$\Delta H_f^\ominus$	Formation of NaBr(s)	-361
$\Delta H_{ea}^\ominus$	Electron addition to Br(g)	-325
$\Delta H_{sub}^\ominus$	Sublimation of Na(s)	$+107$
$\Delta H_{diss}^\ominus$	Bond dissociation of Br$_2$(g)	$+194$
$\Delta H_i^\ominus$	First ionisation of Na(g)	$+498$
$\Delta H_L^\ominus$	Lattice dissociation of NaBr(s)	$+753$

a Construct a Born–Haber cycle for sodium bromide. Label the steps in the cycle with symbols like those used above rather than numerical values. **(6)**

b Use the data above and the Born–Haber cycle in part **a** to calculate the enthalpy of vaporisation, $\Delta H_{vap}^\ominus$, of liquid bromine. **(3)**

AQA, A level, Specimen Paper 6421, 2001/2

2 Use the entropy data in the table below to answer the following questions.

Species	$S^\ominus$ (J K^{-1} mol^{-1})	Species	$S^\ominus$ (J K^{-1} mol^{-1})
C(graphite)	6	H$_2$O(g)	189
C(diamond)	3	H$_2$O(l)	70
H$_2$(g)	131	CH$_4$(g)	186
CO(g)	198	CaO(s)	40
CO$_2$(g)	214	CaCO$_3$(s)	90

Give chemical equations and calculate numerical values of ΔS wherever possible.

a At all temperatures below 100 °C, steam at atmospheric pressure condenses spontaneously to form water. Explain this observation in terms of ΔG and calculate the enthalpy of vaporisation of water at 100 °C. **(4)**

b Explain why the reaction of 1 mol of methane with steam to form carbon monoxide and hydrogen ($\Delta H^\ominus = +210 \, kJ \, mol^{-1}$) is spontaneous only at high temperatures. **(6)**

c Explain why the change of 1 mol of diamond to graphite ($\Delta H^\ominus = -2 \, kJ \, mol^{-1}$) is feasible at all temperatures yet does not occur at room temperature. **(3)**

d The reaction between 1 mol of calcium oxide and carbon dioxide to form calcium carbonate ($\Delta H^\ominus = -178 \, kJ \, mol^{-1}$) ceases to be feasible above a certain temperature, T_s. Determine the value of T_s. **(2)**

AQA, A level, Specimen Paper 6421, 2001/2

3 It has recently been reported that an incident occurred at the Dounreay fast-reactor nuclear plant in Scotland in May 1977 in which about 2 kg of sodium were dumped down a shaft which had earlier been used for the disposal of radioactive waste. (Liquid sodium is used as a coolant in this type of reactor.) The shaft was partially flooded with seawater, and the violent reaction between sodium and water led to an explosion which scattered radioactive material over the nearby area.

a A number of highly exothermic reactions occur when sodium comes into contact with water. The principal reaction is

$$2Na(s) + 2H_2O(l) \rightarrow 2NaOH(aq) + H_2(g) \quad \text{(equation 1)}$$

i Suggest why the reaction of sodium with water, in the restricted situation of the old mine shaft, gave rise to an explosion. **(3)**
ii Calculate the standard enthalpy change for the reaction in equation 1 using the standard enthalpy change of formation values which follow. **(2)**

$$\Delta H_f^\ominus \, (kJ \, mol^{-1}): H_2O(l) = -286, \, NaOH(aq) = -470$$

b Sodium and magnesium are neighbours in the periodic table. Sodium hydroxide is considerably more soluble in water than magnesium hydroxide, Mg(OH)$_2$. In part, solubility is controlled by the enthalpy change of solution (ΔH_{soln}). This is itself determined by other enthalpy changes: for example, the enthalpies of hydration (ΔH_{hyd}) of the cations and anions. Some data for the cations, Na$^+$ and Mg^{2+}, are given in the following table.

	Na$^+$	Mg^{2+}
Ionic radius (nm)	0.102	0.072
Enthalpy of hydration (kJ mol^{-1})	-390	-1891
Extent of hydration (average number of attached water molecules)	5	15

i Explain why water molecules are able to interact with both cations and anions. **(2)**
ii Explain, in terms of the three quantities: charge, ionic radius and extent of hydration, why ΔH_{hyd} for Mg^{2+} is more negative than ΔH_{hyd} for Na$^+$. (In this question 1 mark is available for the quality of written communication.) **(6)**
iii Name another enthalpy change which contributes to the enthalpy change of solution. **(1)**

c Solubility is also controlled by the entropy change (ΔS) which accompanies solution.
i In terms of the number of ions per mole of each compound, explain why this entropy change would be expected to be more positive for $Mg(OH)_2$ than for NaOH. (3)
ii In terms of the arrangement of water molecules, explain why this entropy change would be expected to be more negative for $Mg(OH)_2$ than for NaOH. (2)
iii Name another entropy change which contributes to the total entropy change accompanying solution, and explain how it arises. (2)

OCR, A level, Specimen Paper A7887, Sept 2000

4 The lattice enthalpy of rubidium chloride, RbCl, can be determined indirectly using a Born–Haber cycle.
a Use the data in the table below to construct the cycle and to determine a value for the lattice enthalpy of rubidium chloride.

Enthalpy change	Energy (kJ mol^{-1})
Formation of rubidium chloride	−435
Atomisation of rubidium	+81
Atomisation of chlorine	+122
1st ionisation energy of rubidium	+403
1st electron affinity of chlorine	−349

(6)

b Explain why the lattice enthalpy of lithium chloride, LiCl, is more exothermic than that of rubidium chloride. (2)

OCR, A level, Specimen Paper A7882, Sept 2000

5 The Born–Haber cycle for the formation of sodium chloride from sodium and chlorine may be represented by a series of stages labelled **A** to **F** as shown.

a i Copy and complete the table below writing the letters **A** to **F** next to the corresponding definition.

Definition	Letter	ΔH (kJ mol^{-1})
1st ionisation energy of sodium		+494
1st electron affinity of chlorine		−364
The enthalpy of atomisation of sodium		+109
The enthalpy of atomisation of chlorine		+121
The lattice enthalpy of sodium chloride		−770
The enthalpy of formation of sodium chloride		

(3)

ii Calculate the enthalpy of formation of sodium chloride from the data given. (2)

b The lattice enthalpies can be calculated from theory as well as determined experimentally.

	Experimental ΔH (kJ mol^{-1})	Theoretical ΔH (kJ mol^{-1})
Sodium chloride	−770	−766
Silver iodide	−889	−778

Why is the experimental value of the lattice enthalpy of silver iodide (-889 kJ mol^{-1}) so different from the value calculated theoretically? (2)

c Explain the trend in first ionisation energies of the elements of Group 1 in the periodic table. (3)

Edexcel, A level, Module 3, June 2000

6 Define the term **lattice enthalpy**. (2)

London, A level, Module 3, Jan 2000

5 Acids and bases

STARTING POINTS
● Bases are chemical opposites of acids.
● Neutralisation is the process by which the acidity or alkalinity of a substance is destroyed.
● Volumetric analysis is a basic chemical technique, using solutions as reactants, that allows us to find out the quantities of substances reacting if the concentration of one of the substances is known.
● Indicators are substances which change colour when a reaction is complete.

Figure 5.1
a Acids are all around us. They are present in the food we eat as well as being important industrially in the manufacture of detergents and paints.
b Bases are also all around us. They are found in bleach, washing powder and toothpaste as well as being used in the manufacture of fertilisers (along with acids such as nitric acid).

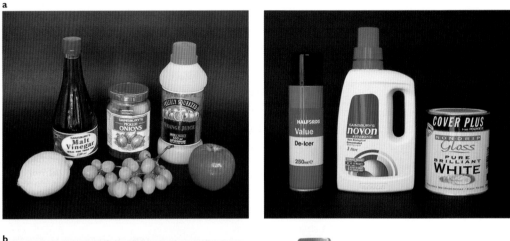

Acids and bases are important in everyday life as well as economically in the chemical industry. The word acid is derived from the Latin word meaning 'sour' related to taste. **Acids:**

- are soluble in water
- are corrosive
- react with bases to form a salt and water, for example

sodium hydroxide + hydrochloric acid → sodium chloride + water
$$NaOH(aq) \ + \ HCl(aq) \ \rightarrow \ NaCl(aq) \ + H_2O(l)$$
(soluble base)

copper(II) oxide + sulphuric acid → copper(II) sulphate + water
$$CuO(s) \ + \ H_2SO_4(aq) \rightarrow \ CuSO_4(aq) \ + H_2O(l)$$
(insoluble base)

- react with the 'MAZIT' metals (**m**agnesium, **a**luminium, **z**inc, **i**ron and **t**in) to form a salt and hydrogen gas, for example

$$\text{zinc} \quad + \quad \text{hydrochloric acid} \rightarrow \text{zinc chloride} + \text{hydrogen}$$
$$Zn(s) + \quad 2HCl(aq) \quad \rightarrow ZnCl_2(aq) + \quad H_2(g)$$

- react with carbonates to form salts, water and carbon dioxide, for example

$$\text{copper(II) carbonate} + \quad \text{nitric acid} \quad \rightarrow \text{copper(II) nitrate} + \quad \text{water} + \text{carbon dioxide}$$
$$CuCO_3(s) \quad + 2HNO_3(aq) \rightarrow Cu(NO_3)_2(aq) + H_2O(l) + \quad CO_2(g)$$

Bases are also corrosive but are very different from acids; they are said to be the chemical opposites of acids. Bases:

- will remove the sharp taste from an acid
- have a soapy feel when rubbed between a damp forefinger and thumb.

Alkalis are bases which dissolve in water to produce hydroxide ions, $OH^-(aq)$. Common alkalis include potassium, sodium and ammonium hydroxides. The term 'alkali' is also used in relation to other substances which, when added to water, react to produce hydroxide ions; an example of such a substance is sodium carbonate. The carbonate ion reacts with water to produce hydrogencarbonate ions and hydroxide ions.

$$CO_3{}^{2-}(aq) + H_2O(l) \rightarrow HCO_3{}^-(aq) + OH^-(aq)$$

It would be too dangerous to use the 'taste' or 'feel' test to find out if a substance was an acid or base. Instead we use substances called **indicators** which change colour when they are added to acids or bases. Many indicators are dyes which have been extracted from natural sources. For example, litmus is a purple dye which has been extracted from lichens. Litmus turns red when it is added to an acid and turns blue when added to a base that is soluble. Some other indicators are shown in Table 5.1, along with the colours they turn in acidic and basic solutions.

Table 5.1
Some indicators and their colours in acid or base solution.

Indicator	Colour in acid solution	Colour in base solution
Methyl red	Red	Yellow
Methyl orange	Pink	Yellow
Phenolphthalein	Colourless	Pink

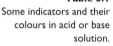

Theories of acids and bases

Figure 5.2
Antoine Lavoisier (1743–94) at work in his laboratory.

The first real attempt to define the difference between acids and bases took place in 1777, when that originator of so much of our early chemistry, Antoine Lavoisier (Figure 5.2), suggested that acids were substances that contained oxygen. However, it was not long after this that the 'hydro-halic' acids (HCl, HBr, and others) were discovered and they had no oxygen present in them.

This produced a modified theory (1810) when the English chemist Sir Humphrey Davy suggested that all acids contain hydrogen as the important element. However, it was pointed out that there were many hydrogen-containing substances that were not acids.

Figure 5.3
Justus von Liebig (1803–73)
defined acids more closely.

The German chemist Justus von Liebig (Figure 5.3) made the next useful proposal about acids (1838), when he suggested that acids were substances that can react with metals to produce hydrogen gas.

This theory was followed in 1884 by the first really comprehensive theory of acids and bases. This was produced by the Swedish chemist **Svante Arrhenius** (1859–1927). He suggested that since these acid solutions were electrolytes, their solutions contained many ions. According to Arrhenius' theory acids produce hydrogen ions (H^+) when they dissolve in water, whereas bases produce hydroxide ions (OH^-). It was thus recognised that water plays an important part in the acidity of a substance. This led to the suggestion that the hydrogen ion cannot exist alone in aqueous solution. In fact it is always associated in aqueous solution with a water molecule and produces the **oxonium ion**, H_3O^+:

$$H^+(aq) + H_2O(l) \rightarrow H_3O^+(aq)$$

It should be noted that gaseous hydrogen chloride, $HCl(g)$, is not acidic but when it dissolves in water an acidic solution is produced because the oxonium ion is formed.

Neutralisation is the following reaction:

$$H_3O^+(aq) + OH^-(aq) \rightarrow 2H_2O(l)$$

These ideas were rather limiting since they only applied to aqueous solutions. There were situations where acid–base reactions were taking place in solvents other than water, or even in no solvent at all. This problem was addressed in 1923 by the Danish chemist Johannes Brønsted (1879–1947) and the English chemist Thomas Lowry (1874–1936) when they independently proposed a more general definition of acids and bases. As a result, the study of acids and bases took a great step forward. This theory became known as the **Brønsted–Lowry** theory of acids and bases.

The Brønsted–Lowry theory of acids and bases

This theory defined:

- an acid as a proton (or H^+ ion) donor
- a base as a proton (or H^+ ion) acceptor.

The theory explains why a pure acid behaves differently from its aqueous solution, since for an acid to behave as a proton donor it must have another substance present to accept the proton. So the water, in the aqueous solution, is behaving as a **Brønsted–Lowry base** and accepting a proton (H^+). Generally:

$$HA(aq) + H_2O(l) \rightarrow H_3O^+(aq) + A^-(aq)$$
$$\text{acid} \qquad \text{base} \qquad \text{conjugate acid} \qquad \text{conjugate base}$$

The base ($H_2O(l)$) is linked to the oxonium ion, $H_3O^+(aq)$, by the transfer of a proton. $H_3O^+(aq)$ is the **conjugate acid** of the base $H_2O(l)$, and $A^-(aq)$ is the **conjugate base** of the acid HA. Every acid has a conjugate base and every base a conjugate acid. They are called conjugate acid–base pairs.

If a substance can behave both as a **Brønsted–Lowry acid** and a Brønsted–Lowry base then it is **amphoteric**. Water has this ability. As well as reacting with acids (see above), it can also react with Brønsted–Lowry bases, such as ammonia, in the following way to form the conjugate base OH^-:

$$NH_3(aq) + H_2O(l) \rightarrow NH_4^+(aq) + OH^-(aq)$$
$$\text{base} \qquad \text{acid} \qquad \text{conjugate acid} \qquad \text{conjugate base}$$

The reaction between hydrogen chloride gas and ammonia gas (Figure 5.4) can be described as an acid–base reaction under this theory. The hydrogen chloride molecule acts as a proton donor and the ammonia molecule acts as the proton acceptor:

$$HCl(g) + NH_3(g) \rightarrow NH_4^+Cl^-(s)$$
$$\text{acid} \qquad \text{base}$$

1 Complete the following equations and write down the conjugate acid–base pairs for the following systems:
a $HBr + H_2O \rightarrow$
b $CH_3COOH + OH^- \rightarrow$
c $H_2SO_4 + H_2O \rightarrow$

Figure 5.4
The hydrogen chloride molecule (from concentrated hydrochloric acid) acts as a proton donor and the ammonia molecule (from concentrated ammonia) acts as the proton acceptor.

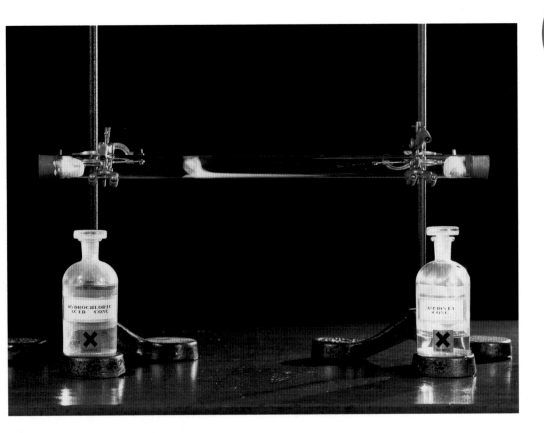

2 How would you describe the NH_4^+ ion and the Cl^- ions using the Brønsted–Lowry theory?

The relative strengths of acids and bases

The relative strength of an acid is found by comparing one acid with another. The strength of any acid depends upon how many molecules dissociate (split up) when the acid is dissolved in water. The relative strength of a base is found by comparing one base with another and is again dependent upon the dissociation of the base in aqueous solution.

Strong and weak acids

A typical strong acid is hydrochloric acid. It is formed by dissolving hydrogen chloride gas in water. In hydrochloric acid the ions formed separate completely:

hydrogen chloride + water → oxonium ions + chloride ions
$$HCl(g) + H_2O(l) \rightarrow H_3O^+(aq) + Cl^-(aq)$$

For hydrochloric acid *all* the hydrogen chloride molecules break up to form H^+ ions and Cl^- ions. Any acid that behaves in this way is termed a **strong acid**. Sulphuric and nitric acids also behave in this way and are therefore also called strong acids. All these acids have a high concentration of hydrogen ions in solution ($H^+(aq)$).

When strong acids are neutralised by strong alkalis the following reaction takes place between oxonium ions and hydroxide ions:

$$H_3O^+(aq) + OH^-(aq) \rightarrow 2H_2O(l)$$

This is an exothermic reaction. In the case of strong acids such as hydrochloric acid and nitric acid 57.1 kJ are released to the surroundings for every mole of oxonium ions neutralised. This is known as the **standard enthalpy of neutralisation**, $\Delta H_n^{\ominus}$.

A **weak acid** such as ethanoic acid, found in vinegar, produces fewer oxonium ions when it dissolves in water compared to a strong acid of the same concentration.

3 Why is the ΔH_n for sulphuric acid twice the value for the strong acids quoted in the text?

ethanoic acid + water ⇌ oxonium ions + ethanoate ions
$$CH_3COOH(l) + H_2O(l) \rightleftharpoons H_3O^+(aq) + CH_3COO^-(aq)$$

The $\rightleftharpoons$ sign means that the reaction is *reversible*. This means that if the ethanoic acid molecule breaks down to give hydrogen ions and ethanoate ions then they will react together to re-form ethanoic acid molecules. The fact that fewer ethanoic acid molecules dissociate, compared to a strong acid, and that the reaction is reversible means that fewer oxonium ions are present in the solution. Other examples of weak acids are citric acid, found in oranges and lemons, carbonic acid, found in soft drinks, sulphurous acid (acid rain), and ascorbic acid (vitamin C) (Figure 5.5).

Figure 5.5
Weak acids are found in all the everyday items shown here.

In a similar manner to that discussed above for strong acids, when weak acids are neutralised by strong alkalis then the reaction taking place is

$$H_3O^+(aq) + OH^-(aq) \rightarrow 2H_2O(l)$$

The neutralisation reaction pulls the dissociation equilibrium to the right, and so all of a weak acid will be neutralised. However, the $\Delta H_n^\ominus$ will be less than -57.1 kJ mol.

All acids when in aqueous solution produce oxonium ions, $H_3O^+(aq)$. To say an acid is a strong acid does not mean it is concentrated. The strength of an acid tells you how easily it dissociates or ionises to produce oxonium ions. The concentration of an acid indicates the proportions of water and acid present in aqueous solution.

It is important to emphasise that a strong acid is still a strong acid even when it is in dilute solution and a weak acid is still a weak acid even when it is concentrated.

Strong and weak alkalis

An alkali is a substance which produces hydroxide ions, $OH^-(aq)$, when it is dissolved in water. Sodium hydroxide is a **strong alkali** because when it dissolves in water its lattice breaks up completely to produce ions:

sodium hydroxide $\xrightarrow{\text{water}}$ sodium ions $+$ hydroxide ions
$$NaOH(aq) \longrightarrow Na^+(aq) + OH^-(aq)$$

Substances that are strong alkalis produce large quantities of hydroxide ions. Common, strong alkalis include sodium hydroxide and potassium hydroxide.

A **weak alkali**, such as ammonia, produces fewer hydroxide ions when it dissolves in water than a strong alkali of the same concentration.

ammonia $+$ water $\rightleftharpoons$ ammonium ions $+$ hydroxide ions
$$NH_3(g) + H_2O(l) \rightleftharpoons NH_4^+(aq) + OH^-(aq)$$

The ammonia molecules react with the water molecules to form ammonium ions and hydroxide ions. However, fewer ammonia molecules do this, so only a low concentration of hydroxide ions is produced.

Acid dissociation constants

We have seen already that weak acids and weak bases only partly dissociate (ionise) in water. This means that owing to the reversibility of the process, a dynamic equilibrium is established. In the case of weak acids we can represent the process by the general equation:

weak acid + water ⇌ oxonium ion + weak acid anion

$$HA(aq) + H_2O(l) \rightleftharpoons H_3O^+(aq) + A^-(aq)$$

We can write an equilibrium constant for this process (Chapter 2, page 17):

$$K_c = \frac{[H_3O^+(aq)][A^-(aq)]}{[HA(aq)][H_2O(l)]}$$

Very few molecules of a weak acid will dissociate when it is dissolved in water. This means that only a very small number of water molecules will be used to combine with the $H^+(aq)$ ions produced. The concentration of the water will remain virtually constant. We can therefore combine this constant with the equilibrium constant K_c to give a new constant K_a:

$$K_a = \frac{[H_3O^+(aq)][A^-(aq)]}{[HA(aq)]}$$

where K_a is the **acid dissociation constant**.

The acid dissociation constant is a measure of the strength of a weak acid since it tells you the extent of the ionisation taking place. A large value for the acid dissociation constant will indicate that there is a high degree of ionisation taking place and it is a stronger acid compared to another with a smaller value for K_a which indicates that there is a small amount of ionisation taking place.

However, because the K_a values for different acids vary over a wide range it is common to refer more conveniently to the negative logarithm to base ten of the K_a value. This is called **pK_a**.

$$pK_a = -\log_{10} K_a$$

Table 5.2 shows values of pK_a alongside the K_a values for various weak acids. The higher the acid dissociation constant, the lower is the pK_a value.

Table 5.2
K_a and pK_a values for some weak acids at 298 K.

Weak acid	Formula	K_a (mol dm^{-3})	pK_a
Methanoic acid	HCOOH	1.78×10^{-4}	3.75
Ethanoic acid	CH_3COOH	1.74×10^{-5}	4.76
Chloroethanoic acid	$CH_2ClCOOH$	1.38×10^{-3}	2.86
Bromoethanoic acid	$CH_2BrCOOH$	1.26×10^{-3}	2.90
Benzoic acid	C_6H_5COOH	6.31×10^{-5}	4.20
Hydrocyanic acid	HCN	3.98×10^{-10}	9.40
Phenol	C_6H_5OH	1.0×10^{-10}	10.00

4 Place the weak acids from Table 5.2 in order of increasing acidity.

5 Write an expression for K_a for benzoic acid and for hydrocyanic acid.

A base dissociation constant for weak bases, similar to that shown for a weak acid, can be produced using the general equation:

weak base + water ⇌ conjugate acid ion + hydroxide ion

$$B(aq) + H_2O(l) \rightleftharpoons BH^+(aq) + OH^-(aq)$$

We can write an equivalent base dissociation constant for this process:

$$K_b = \frac{[BH^+(aq)][OH^-(aq)]}{[B(aq)]}$$

6 Write the expression for K_b for:
a ammonia (NH_3)
b methylamine (CH_3NH_2)
c phenylamine ($C_6H_5NH_2$).

7 The K_b values for methylamine and phenylamine are 4.37×10^{-4} mol dm^{-3} and 4.17×10^{-10} mol dm^{-3}, respectively. Which is the weaker base? Explain your answer.

where K_b is the **base dissociation constant**. K_b values vary over a wide range, as with K_a. A high value for the base dissociation constant will indicate that there is a high degree of ionisation taking place and it is a strong base, whilst a small value indicates that there is a small amount of ionisation taking place and it is a weak base. For example, ammonia has a K_b value of 1.78×10^{-5} mol dm^{-3} and is a weaker base than methylamine, which has a K_b value of 4.37×10^{-4} mol dm^{-3}.

Again it is more convenient to show the pK_b value for a weak base, where p$K_b = -\log K_b$. The lower the pK_b value then the stronger the base.

pH scale

Figure 5.6
Universal indicator in
solution, showing the
colour range related to
the pH scale shown
opposite in Figure 5.9.

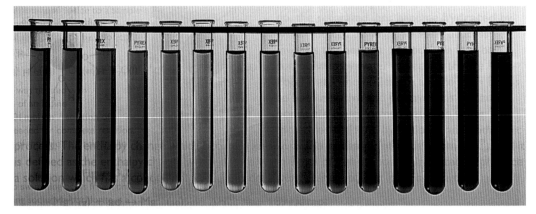

We can determine how acidic a solution is by measuring the quantity, or concentration, of oxonium ions present in aqueous solution. However, the values for these concentrations vary very greatly. For example, a very weak acid may have a value for the $[H_3O^+(aq)]$ of $3 \times 10^{-7}\,mol\,dm^{-3}$, whilst a very strong acid may have a value as high as $10\,mol\,dm^{-3}$. The figures for weak acids are often hardly convenient to use and so it is common to use the pH scale, which was introduced in 1909 by the Danish biochemist SPL Sorenson (1868–1939). He defined the **pH** of a solution as the negative logarithm to base ten of the concentration of the oxonium ions:

$$pH = -\log_{10}[H_3O^+(aq)]$$

This is often abbreviated to

$$pH = -\log[H^+(aq)]$$

$[H_3O^+(aq)]$ is measured in $mol\,dm^{-3}$. Figure 5.7 shows the pH scale for a full range of $[H_3O^+(aq)]$. Because pH is equal to *minus* $\log[H_3O^+(aq)]$ you will notice that as $[H_3O^+(aq)]$ increases the pH decreases, so that low pH values correspond to a high concentration of $[H_3O^+(aq)]$ and vice versa.

Figure 5.7
The variation of pH with
$[H_3O^+(aq)]$.

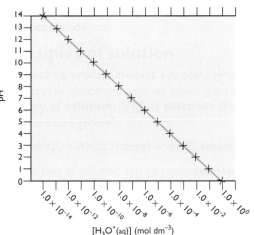

$[H_3O^+(aq)]$ (mol dm^{-3})

The most accurate method of obtaining the pH of a solution is to use a pH meter (Figure 5.8). This consists of a combination glass and reference electrode attached to a meter which measures the potential difference created by the solution. The potential difference measured depends upon and is proportional to the concentration of oxonium ions in solution. The pH reading is given on a digital display.

Another method of measuring pH is to use full range universal indicator paper. The colour produced by dipping this indicator paper into the solution to be tested is then compared with a colour chart (Figure 5.9).

Figure 5.8 (left)
This modern pH meter will allow you to get an accurate measure of pH to 0.01 pH units even when the solution is coloured.

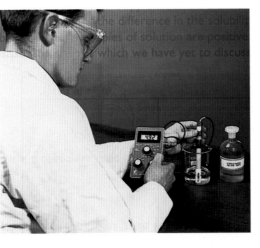

Figure 5.9 (right)
The colours appropriate to the pH scale for universal indicator.

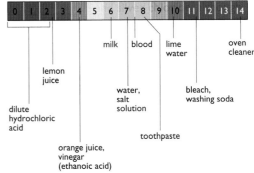

Calculating pH values

For strong acids, such as hydrochloric acid, calculating pH is relatively straightforward since the reaction of a strong acid with water goes effectively to completion. Therefore, in the case of hydrochloric acid, the amount, in moles, of the oxonium ion is equal to the amount in moles of the acid in solution. So it is then possible to substitute into the equation for pH:

$$pH = -\log[H_3O^+(aq)]$$

Example What is the pH of a $0.1\,mol\,dm^{-3}$ solution of hydrochloric acid?

Hydrochloric acid is a strong acid and therefore:

$$[H_3O^+(aq)] = [HCl(aq)]$$
$$= 0.1\,mol\,dm^{-3}$$
$$pH = -\log[H_3O^+(aq)]$$

Therefore,

$$pH = -\log(0.1)$$
$$= -(-1)$$
$$= 1$$

8 Calculate the pH of hydrochloric acid solutions of the following concentrations:
  a $0.01\,mol\,dm^{-3}$
  b $0.001\,mol\,dm^{-3}$
  c $0.0001\,mol\,dm^{-3}$.
9 What do you notice happening to the pH of the solutions as they are diluted? Explain your answer.

Most scientific calculators will allow you to enter the concentration, press the 'log' button and then change the sign to arrive at the pH value. Alternatively, you may need to do the following: press 'log', enter the concentration, press 'equals' and then change the sign to arrive at the pH value.

Problems do arise, however, when you have to decide whether the solution you are dealing with is that of a strong acid or a weak acid. For example, both a dilute strong acid and a concentrated weak acid could give solutions with a pH of 4 ($[H_3O^+(aq)] = 1 \times 10^{-4}\,mol\,dm^{-3}$), but the two solutions would have different concentrations of acid present. It is important to distinguish between concentration, which is the amount of substance present in a given volume, and the strength of the acid, which is the measure of the extent or ability of an acid to donate H^+ ions.

pH and weak acids

For a weak acid, more information is required before its pH can be calculated; we need to know the acid dissociation constant, K_a, as well as the concentration of the solution.

Example   Calculate the pH of a $0.1\,mol\,dm^{-3}$ solution of ethanoic acid at $298\,K$ given that the K_a for ethanoic acid is $1.7 \times 10^{-5}\,mol\,dm^{-3}$.

The equilibrium for this weak acid is:

$$CH_3COOH(aq) + H_2O(l) \rightleftharpoons CH_3COO^-(aq) + H_3O^+(aq)$$

$$K_a = \frac{[CH_3COO^-(aq)]\,[H_3O^+(aq)]}{[CH_3COOH(aq)]}$$

Since the ethanoate ions and oxonium ions must be produced in equal proportions

$$K_a = \frac{[H_3O^+(aq)]^2}{[CH_3COOH(aq)]}$$

Also, since the value of K_a is small for weak acids, suggesting that the acid is only weakly ionised, we can approximate $[CH_3COOH(aq)] = 0.1\,mol\,dm^{-3}$.
   Substituting for K_a:

$$1.7 \times 10^{-5}\,mol\,dm^{-3} = \frac{[H_3O^+(aq)]^2}{0.1\,mol\,dm^{-3}}$$

Rearranging this gives:

$$1.7 \times 10^{-5}\,mol\,dm^{-3} \times 0.1\,mol\,dm^{-3} = [H_3O^+(aq)]^2$$

So

$$[H_3O^+(aq)] = \sqrt{(1.7 \times 10^{-5})\,mol\,dm^{-3} \times 0.1\,mol\,dm^{-3}}$$
$$= 1.3 \times 10^{-3}\,mol\,dm^{-3}$$

Therefore:

$$pH = -\log[1.3 \times 10^{-3}] = -(-2.88) = 2.88$$

> **10** Calculate the pH of a $0.50\,mol\,dm^{-3}$ solution of the weak acid methanoic acid at $298\,K$, given that K_a for this acid is $1.6 \times 10^{-4}\,mol\,dm^{-3}$.
>
> **11** Calculate the concentration of a solution of ethanoic acid which has a pH of 3.5. K_a for ethanoic acid at $298\,K$ is $1.7 \times 10^{-5}\,mol\,dm^{-3}$.

Calculation of the pH of a strong base

To calculate the pH of a strong base it is necessary to know the $[H_3O^+]$ in that solution. This tiny concentration of oxonium ions can be calculated using the **ionic product** for water.

The ionic product for water

Pure water always ionises to a small extent:

$$H_2O(l) \rightleftharpoons H^+(aq) + OH^-(aq)$$

or this may be written:

$$H_2O(l) + H_2O(l) \rightleftharpoons H_3O^+(aq) + OH^-(aq)$$

We can write an equilibrium constant for this water dissociation:

$$K_a = \frac{[H_3O^+(aq)]\,[OH^-(aq)]}{[H_2O(l)]}$$

Since water only slightly dissociates into its ions, the equilibrium lies well over to the left. The concentration of water can be regarded as constant, so we may write:

$$K_w = K_a[H_2O(l)] = [H_3O^+(aq)]\,[OH^-(aq)]$$

K_w is called the **ionic product of water** and it has a value of $1.0 \times 10^{-14}\,mol^2\,dm^{-6}$ at $298\,K$.

Like all equilibrium constants, K_w varies with temperature, as shown in Figure 5.10. This rise in K_w with increasing temperature is not difficult to understand since the enthalpy change associated with the forward reaction in this equilibrium:

$$H_2O(l) \rightleftharpoons H^+(aq) + OH^-(aq)$$

is $+57.5\,kJ\,mol^{-1}$ (endothermic). Le Chatelier's principle (see Chapter 2, page 19) would lead us to expect that an endothermic reaction will move further to the right with an increase in temperature.

We can now use this expression for K_w to find the concentration of oxonium ions in the solution of a strong base.

Figure 5.10
As the temperature is increased for a sample of pure water so does K_w.

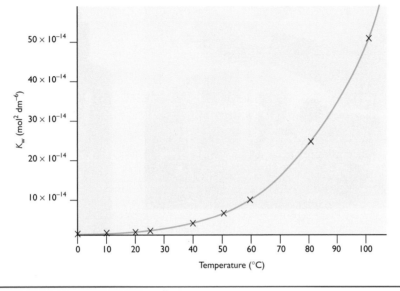

Example   Find the pH of a $0.2\,mol\,dm^{-3}$ solution of potassium hydroxide.

We can assume that this strong base is completely ionised in aqueous solution. Therefore in this case:

$$[KOH(aq)] = [OH^-(aq)]$$
$$= 0.2\,mol\,dm^{-3}$$

Using the relationship for K_w:

$$K_w = [H_3O^+(aq)]\,[OH^-(aq)]$$
$$= 1.0 \times 10^{-14}\,mol^2\,dm^{-6}$$

We can now substitute for $[OH^-(aq)]$:

$$[H_3O^+(aq)] \times 0.2\,mol\,dm^{-3} = 1 \times 10^{-14}\,mol^2\,dm^{-6}$$

So

$$[H_3O^+(aq)] = \frac{1.0 \times 10^{-14}\,mol^2\,dm^{-6}}{0.2\,mol\,dm^{-3}}$$

$$= 5.0 \times 10^{-14}\,mol\,dm^{-3}$$

12 Calculate the pH of a $0.50\,mol\,dm^{-3}$ solution of sodium hydroxide. K_w has a value of $1.0 \times 10^{-14}\,mol^2\,dm^{-6}$ at 298 K.

The pH of this solution is found using

$$pH = -\log[H_3O^+(aq)]$$
$$= -\log[5.0 \times 10^{-14}]$$
$$= -(-13.3)$$
$$= 13.3$$

To calculate the pH of a solution of a weak base requires the value of the base dissociation constant, K_b, in addition to the concentration of the solution of the weak base.

Buffers

A **buffer solution** is one which resists changes in its pH when small amounts of either acid or alkali are added to it. This results in buffer solutions being used to control the pH environment for chemical and biochemical processes. For example, blood has to be maintained at a constant pH of 7.4 to ensure efficient enzyme activity (Figure 5.11). This is done via the buffering action in blood plasma of the equilibrium between carbonic acid and the hydrogencarbonate and oxonium ions:

$$H_2CO_3(aq) + H_2O(l) \rightleftharpoons HCO_3^-(aq) + H_3O^+(aq)$$

Figure 5.11
Blood has a pH of 7.4 due to the buffering action of dissolved carbon dioxide.

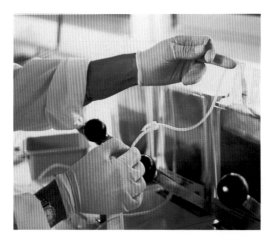

Figure 5.12
To adjust the pH of shampoos, both citric acid and sodium hydroxide can be added to the shampoo. Sodium hydroxide is used as a processing aid in the formulation to maintain an alkaline environment in one of the side processes. As a result of this there is a need to adjust the pH of the shampoo back to its desired pH, which is slightly acidic, using citric acid. Sodium citrate is the salt formed. This creates a slightly acid buffer involving sodium citrate and citric acid.

Acid buffers

A typical acid buffer, one which maintains a pH of less than 7, can be made by mixing a weak acid with a salt of that acid (which contains the conjugate base). For example, a mixture of ethanoic acid and sodium or potassium ethanoate would act as an acid buffer. An acid buffer of this type acts in the following way:

$$CH_3COOH(aq) + H_2O(l) \rightleftharpoons CH_3COO^-(aq) + H_3O^+(aq) \qquad \text{(equation 1)}$$

Since ethanoic acid is a weak acid the above equilibrium lies over to the left, giving a large reservoir of H^+ ions (in CH_3COOH) which can dissociate if the $[H_3O^+(aq)]$ falls.

Sodium ethanoate is a strong electrolyte. Strong electrolytes are completely ionised in water, in this case forming a large reservoir of $CH_3COO^-(aq)$ ions:

$$CH_3COONa(s) \xrightarrow{\text{(aq)}} CH_3COO^-(aq) + Na^+(aq) \qquad \text{(equation 2)}$$

The large quantity of these ions ensures that the equilibrium in equation 1 is moved further to the left.

The buffering action on addition of an acid

When an acid is added to a buffer solution such as the one above, the concentration of the oxonium ions is increased. These combine with the ethanoate ions and equation 1 moves to the left, according to Le Chatelier's principle, 'soaking up' the additional oxonium ions. This ensures that the pH of the solution varies very little from the original value.

The buffering action on addition of an alkali

When an alkali is added to an acid buffer solution the added $OH^-(aq)$ ions react with the oxonium ions formed by equation 1 to produce water:

$$H_3O^+(aq) + OH^-(aq) \rightarrow 2H_2O(l)$$

To replace the oxonium ions that have reacted the equilibrium (equation 1) moves to the right, generating more oxonium ions. Again this ensures that the pH of the solution varies only slightly from the original value.

Alkaline buffers

If a buffer solution is required to maintain a pH of greater than 7, it is made by mixing together a weak base, such as aqueous ammonia, and a soluble salt of that base, such as ammonium chloride. An alkaline buffer of this type acts as shown by equations 3 and 4:

$$NH_3(aq) + H_2O(l) \rightleftharpoons NH_4^+(aq) + OH^-(aq) \qquad \text{(equation 3)}$$

$$NH_4Cl(s) \xrightarrow{\text{(aq)}} NH_4^+(aq) + Cl^-(aq) \qquad \text{(equation 4)}$$

Because aqueous ammonia is a weak base it contains a high concentration of ammonia molecules, $NH_3(aq)$. The ammonium chloride completely ionises, ensuring a reservoir of $NH_4^+(aq)$. This causes the equilibrium in equation 3 to move further to the left.

The buffering action on addition of an alkali

When an alkali is added to an alkaline buffer of this type, the added hydroxide ions are absorbed by reacting with the $NH_4^+(aq)$ ions in equation 3, forcing the equilibrium to the left. This ensures that the pH of the solution varies only very slightly from the original value.

The buffering action on addition of an acid

When an acid is added to an alkaline buffer, the added oxonium ions react with hydroxide ions to produce water:

$$H_3O^+(aq) + OH^-(aq) \rightarrow 2H_2O(l)$$

To replace the hydroxide ions which have reacted the equilibrium (equation 3) moves to the right, generating more hydroxide ions. Again this ensures that the pH of the solution varies only slightly from the original value.

If large amounts of a strong acid or a strong alkali are added to either type of buffer solution the pH will change due to saturation of the solution with either oxonium or hydroxide ions, respectively.

Calculations involving buffer solutions

Consider an acid buffer, for example that formed by the mixing of ethanoic acid and sodium ethanoate. The expression for K_a of ethanoic acid is:

$$K_a = \frac{[H_3O^+(aq)][CH_3COO^-(aq)]}{[CH_3COOH(aq)]}$$

This relationship can be written as:

$$[H_3O^+(aq)] = K_a \times \frac{[CH_3COOH(aq)]}{[CH_3COO^-(aq)]}$$

As a buffer contains a weak acid not all the acid molecules dissociate and the equilibrium

$$CH_3COOH(aq) + H_2O(l) \rightleftharpoons CH_3COO^-(aq) + H_3O^+(aq)$$

lies to the left. This means that at equilibrium the concentration of the ethanoic acid molecules will be almost the same as the initial concentration of the acid shown in the equation as $[CH_3COOH(aq)]$. Therefore, the concentration of the ethanoate ions in the buffer will effectively be the initial concentration of the sodium ethanoate (salt) as this dissociates completely. Taking account of these assumptions, the expression for a general buffer becomes:

$$[H_3O^+(aq)] = K_a \times \frac{[acid]}{[salt]}$$

Taking $-\log$ of each side gives:

$$-\log[H_3O^+(aq)] = -\log K_a - \log\left(\frac{[salt]}{[acid]}\right)$$

or

$$pH = pK_a - \log\left(\frac{[salt]}{[acid]}\right)$$

It can be seen from this relationship that the pH of a buffer solution depends on the ratio of [salt] : [acid]. Note that if [salt] = [acid] then the relationship simplifies to $pH = pK_a$ (or $[H_3O^+] = K_a$).

Example

Stirring 4.10 g of sodium ethanoate into 1000 cm³ of 0.1 mol dm⁻³ ethanoic acid makes a buffer solution. What is the pH of this buffer solution? (K_a for ethanoic acid is 1.7×10^{-5} mol dm⁻³ at 298 K.)

We know that:

$$[H_3O^+(aq)] = K_a \times \frac{[acid]}{[salt]}$$

M_r of sodium ethanoate is 82 g mol⁻¹, so:

$$[salt] = \frac{4.10\,g}{82\,g\,mol^{-1}} \times \frac{1}{1.0\,dm^3}$$

$$= 0.05\,mol\,dm^{-3}$$
$$[acid] = 0.10\,mol\,dm^{-3}$$

Substituting:

$$[H_3O^+(aq)] = 1.7 \times 10^{-5}\,mol\,dm^{-3} \times \frac{0.10\,mol\,dm^{-3}}{0.05\,mol\,dm^{-3}}$$

$$= 3.4 \times 10^{-5}\,mol\,dm^{-3}$$

13 Calculate the pH of the buffer solution produced when 2.05 g of sodium ethanoate are stirred into 1.0 dm³ of 0.05 mol dm⁻³ ethanoic acid.

Using $pH = -\log[H_3O^+(aq)]$ we can find the pH of the buffer solution:

$$pH = -\log[H_3O^+(aq)]$$
$$= -\log[3.4 \times 10^{-5}]$$
$$= -(-4.47)$$
$$= 4.47$$

Salt hydrolysis

14 Copy out the two equations in the 'Salt hydrolysis' section and underneath each of the ions or molecules indicate the acid–base conjugate pairs.

15 Predict whether the solutions of the following salts will be acidic or alkaline.
  a CH_3COONa
  b $FeCl_3$
  c CH_3COONH_4

Some salts, for example sodium chloride, dissolve in water to produce neutral solutions where the pH is 7. However, some other salts react with water to form alkaline or acidic solutions. This particular type of reaction is called **salt hydrolysis**.

For example, if ammonium chloride (the salt of a weak base and a strong acid) is added to water then the dissolved ammonium ions react:

$$NH_4^+(aq) + H_2O(l) \rightarrow NH_3(aq) + H_3O^+(aq)$$

This results in the formation of an acidic solution.

When sodium carbonate (the salt of a strong base and a weak acid) is added to water, the aqueous carbonate ions react with the water:

$$CO_3^{2-}(aq) + H_2O(l) \rightarrow HCO_3^-(aq) + OH^-(aq)$$

This results in the formation of an alkaline solution.

Indicators

There are different types of indicators. The most familiar will be pH indicators, which show when an acid has just been neutralised by an alkali. Other indicators include those that detect the completion of oxidation, precipitation and complex formation reactions.

pH indicators are weak acids which have a conjugate base with a different colour to the acid itself. Indicators are given the general formula of HIn. Since they are weak acids they do not dissociate completely:

Figure 5.13
Phenolphthalein is colourless in acid but pink in alkaline conditions.

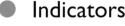

$$\underset{\text{colour 1}}{HIn(aq)} + H_2O(l) \rightleftharpoons H_3O^+(aq) + \underset{\text{colour 2}}{In^-(aq)}$$

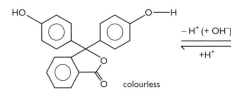

colourless   pink

If an acid is added to an indicator solution there will be an increase in the concentration of $H_3O^+(aq)$ and, according to Le Chatelier's principle, this will move the position of the equilibrium to the left. So colour 1 will be observed. When an alkali is added to the indicator solution, the added hydroxide ions, $OH^-(aq)$, react with the oxonium ions, using them up to produce water. This causes the position of equilibrium to move to the right and colour 2 will be observed.

A common indicator used for acid–alkali titration is phenolphthalein. The formula of phenolphthalein (Figure 5.13) shows the complex chemical nature of indicators and the structures of the weak acid molecule and its conjugate base.

It is a misconception to believe that all acid–alkali indicators change their colour at a pH of 7. In fact, very few do. The pH at which indicators do change their colour can be found by looking again at equilibrium constants.

Consider again the general equilibrium process present in an indicator solution:

$$\underset{\text{colour 1}}{HIn(aq)} + H_2O(l) \rightleftharpoons H_3O^+(aq) + \underset{\text{colour 2}}{In^-(aq)}$$

For indicators the equilibrium constant K_{In} is used. For the above process K_{In} is given by:

$$K_{In} = \frac{[H_3O^+(aq)][In^-(aq)]}{[HIn(aq)]}$$

At the changeover from one colour to the other there will be the same concentration of $In^-(aq)$ as $HIn(aq)$. They can then be cancelled from the above expression, leaving

$$K_{In} = [H_3O^+(aq)]$$

It follows that

$$pK_{In} = pH$$

This expression tells us that an indicator will change its colour when the pH of the solution is equal to the pK_{In} value for that indicator. Table 5.3 lists pK_{In} values for some common acid–alkali indicators.

Table 5.3
pK_{In} values of some common acid–alkali indicators.

Indicator	pK_{In}	pH range	Colour change
Methyl orange	3.7	3.2–4.4	Red to yellow
Methyl red	5.1	4.2–6.3	Red to yellow
Bromothymol blue	7.0	6.0–7.8	Yellow to blue
Phenolphthalein	9.3	8.2–10.0	Colourless to pink

Figure 5.14
Change in pH can be monitored during an acid–alkali titration using a pH meter.

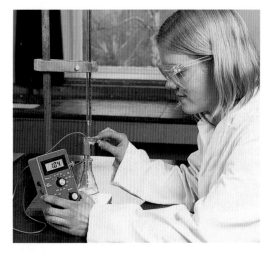

The point at which a sudden change of colour is observed in an acid–alkali titration is called the **end-point**. Most people can, however, start to observe the gradual change in colour at an earlier point. It can be shown that the pH range of an indicator is about 2 pH units, roughly 1 either side of the pK_{In}.

The choice of indicator to use during an acid–alkali titration depends on whether the acid and alkali are strong or weak, as this affects the way in which the pH changes during the titration. The change in pH can be monitored using a pH meter as the titration is carried out (Figure 5.14). A graph of pH against the volume of a alkali added is called a **titration curve**.

Titration curves

There are four types of acid–alkali titration curve:

- strong acid against strong alkali (for example, HCl(aq) and NaOH(aq))
- strong acid against weak alkali (for example, HCl(aq) and NH₃(aq))
- weak acid against strong alkali (for example, CH₃COOH(aq) and NaOH(aq))
- weak acid against weak alkali (for example, CH₃COOH(aq) and NH₃(aq)).

These four types will now be discussed in turn.

Strong acid against a strong alkali

Figure 5.15
The titration curve for the titration of a strong acid (25 cm³ of 0.1 mol dm⁻³) with a strong alkali (0.1 mol dm⁻³).

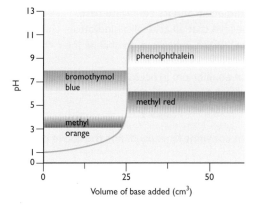

If 0.1 mol dm⁻³ HCl(aq) is titrated against 0.1 mol dm⁻³ NaOH(aq) the pH changes only very slightly at first. Near the point of neutralisation, known as the equivalence point, the pH changes dramatically on the addition of only a small amount of alkali. The pH change then slows down and eventually levels off as alkali is added to excess (Figure 5.15). The actual equivalence point, the point at which the amounts of acid and alkali balance, can be taken as the mid-point of the vertical section of the titration curve, which is at pH 7 in this case. The end-point of the indicator chosen and the equivalence point should be close together for an accurate titration to be carried out. Using Table 5.3 it can be seen that because of the large vertical section covering pH 3–11 associated with this titration then any of the indicators would be suitable.

Strong acid against a weak alkali

If $0.1\,mol\,dm^{-3}$ HCl(aq) is titrated against $0.1\,mol\,dm^{-3}$ NH$_3$(aq) then once again the pH changes very little until near the equivalence point, when the pH dramatically changes. The pH levels off again but this time at a lower pH due to the alkali being weak (Figure 5.16). The vertical section of the graph and the equivalence point at a pH of 5.8 would allow methyl orange or bromothymol blue to be used as indicators, but phenolphthalein would not be suitable as its range does not coincide with this vertical section.

Figure 5.16
The titration curve for the titration of a strong acid ($25\,cm^3$ of $0.1\,mol\,dm^{-3}$) with a weak alkali ($0.1\,mol\,dm^{-3}$).

Figure 5.17
The titration curve for the reaction of sodium carbonate ($50\,cm^3$ of $0.1\,mol\,dm^{-3}$) with hydrochloric acid ($0.1\,mol\,dm^{-3}$).

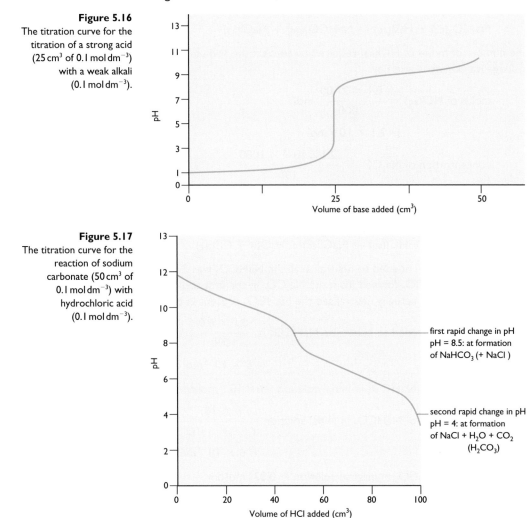

Another interesting chemical system to consider in this section is the titration curve produced by the reaction of hydrochloric acid with sodium carbonate (Figure 5.17). This is a two-step process. In the first step sodium hydrogencarbonate is produced:

$$Na_2CO_3(aq) + HCl(aq) \rightarrow NaHCO_3(aq) + NaCl(aq)$$

In the second step the sodium hydrogencarbonate reacts further with the hydrochloric acid:

$$NaHCO_3(aq) + HCl(aq) \rightarrow NaCl(aq) + H_2O(l) + CO_2(g)$$

From Figure 5.17 it can be seen that there are two equivalence points during the whole process. An indicator that could be used at the first equivalence point, pH 8.5, is phenolphthalein and at the second equivalence point, pH 4, methyl orange could be used.

Use can be made of this two-step process in the estimation of sodium carbonate in a mixture of sodium carbonate and sodium hydrogencarbonate or sodium carbonate in a mixture of sodium carbonate and sodium hydroxide.

Example   A student had a 100 cm³ solution which contained a mixture of sodium carbonate and sodium hydrogencarbonate to analyse. A 25.00 cm³ portion of this solution required 21.00 cm³ of a 0.1 mol dm⁻³ solution of hydrochloric acid to change the colour of the phenolphthalein indicator from pink to colourless. Methyl orange was then added and a further 27.00 cm³ of the acid were needed to turn this indicator from yellow to red. Find the concentrations of sodium carbonate and sodium hydrogencarbonate in the original solution using the data provided.

In the first step the following reaction takes place:

$$Na_2CO_3(aq) + HCl(aq) \rightarrow NaHCO_3(aq) + NaCl(aq)$$

The number of moles of HCl(aq) required to convert the sodium carbonate to sodium hydrogencarbonate is given by:

$$\text{moles of HCl(aq)} = \frac{0.1 \times 21.00}{1000} \text{ mol}$$

$$= 2.1 \times 10^{-3} \text{ mol}$$

$$\text{concentration of } Na_2CO_3 = \frac{2.1 \times 10^{-3} \times 1000}{100}$$

$$= 0.021 \text{ mol dm}^{-3}$$

In the second step NaHCO₃(aq) reacts with the HCl(aq) as shown:

$$NaHCO_3(aq) + HCl(aq) \rightarrow NaCl(aq) + H_2O(l) + CO_2(g)$$

The volume of HCl needed to neutralise all the NaHCO₃ was 27 cm³. Of this, 21.0 cm³ neutralised the NaHCO₃ formed from the Na₂CO₃ in the first step. The difference between these volumes, 6 cm³, actually neutralised the NaHCO₃ present in the initial solution.

$$\text{mol of HCl needed to neutralise } NaHCO_3 = \frac{0.1 \times 6.0}{1000}$$

$$= 6 \times 10^{-4} \text{ mol dm}^{-3}$$

Hence the amount of NaHCO₃ in initial solution $= 6 \times 10^{-4} \text{ mol dm}^{-3}$

$$\text{concentration of } NaHCO_3 \text{ in initial solution} = \frac{6 \times 10^{-4} \times 1000}{100}$$

$$= 6 \times 10^{-3} \text{ mol dm}^{-3}$$

Concentration of Na₂CO₃ in original solution $= 0.021 \text{ mol dm}^{-3}$
Concentration of NaHCO₃ in original solution $= 6 \times 10^{-3} \text{ mol dm}^{-3}$

Weak acid against a strong alkali

If a 0.1 mol dm⁻³ ethanoic acid solution is titrated with a 0.1 mol dm⁻³ NaOH(aq) solution then the titration curve shown in Figure 5.18 is produced. Again there is a slow increase in pH at the beginning but a dramatic increase around the neutralisation point. This time an equivalence point at about pH 8.7 is obtained, making phenolphthalein a suitable indicator for this type of titration.

Weak acid against a weak alkali

16 Draw titration curves for the following reactions, as well as choosing suitable indicators:
a 0.1 mol dm⁻³ methanoic acid titrated with 0.1 mol dm⁻³ potassium hydroxide solution
b 0.1 mol dm⁻³ nitric acid titrated with 0.1 mol dm⁻³ ammonia solution
c 0.1 mol dm⁻³ nitric acid titrated with 0.1 mol dm⁻³ potassium hydroxide solution.

If 0.1 mol dm⁻³ CH₃COOH(aq) is titrated against 0.1 mol dm⁻³ NH₃(aq) a very different titration curve results (Figure 5.19). This time no dramatic change in pH occurs and hence no vertical section is seen in the titration curve; the pH changes gradually throughout the titration process. This makes it very difficult to use any indicator to follow this reaction and a pH meter is often used to identify the end-point of this type of titration.

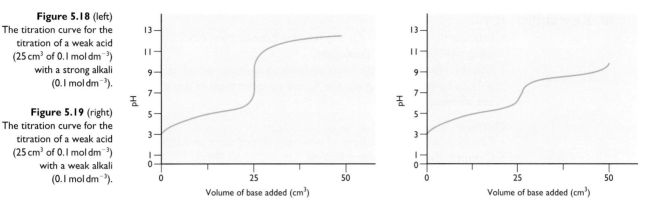

Figure 5.18 (left)
The titration curve for the titration of a weak acid (25 cm³ of 0.1 mol dm⁻³) with a strong alkali (0.1 mol dm⁻³).

Figure 5.19 (right)
The titration curve for the titration of a weak acid (25 cm³ of 0.1 mol dm⁻³) with a weak alkali (0.1 mol dm⁻³).

pK_a from titration curves

Titration curves can be used to determine the pK_a of a weak acid, experimentally (Figure 5.20).

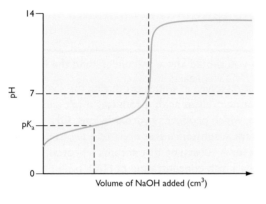

Figure 5.20
The pK_a of ethanoic acid can be determined from its titration curve.

The pK_a is found by determining the pH value at a point when half of the volume of alkali needed to neutralise the acid has been used. At the point where half of the alkali has been added, the concentrations of $CH_3COOH(aq)$ and its conjugate base $CH_3COO^-(aq)$ are equal. So from the expression for K_a of the weak acid:

$$K_a = \frac{[H^+(aq)][CH_3COO^-(aq)]}{[CH_3COOH(aq)]}$$

we obtain

$$K_a = [H^+(aq)]$$

or

$$pK_a = pH$$

Titration curves for di- and tri-protic acids

Diprotic acids, those which can donate two H^+ ions, and triprotic acids, those which can donate three H^+ ions, produce different titration curves to the monoprotic acids mentioned above.

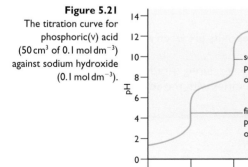

Figure 5.21
The titration curve for phosphoric(v) acid (50 cm³ of 0.1 mol dm⁻³) against sodium hydroxide (0.1 mol dm⁻³).

If the triprotic acid phosphoric(v) acid (H_3PO_4) is titrated against a strong alkali the titration curve shown in Figure 5.21 is produced. Phosphoric(v) acid undergoes its complete dissociation in three steps:

$$H_3PO_4(aq) \rightarrow H^+(aq) + H_2PO_4^-(aq)$$
$$H_2PO_4^-(aq) \rightarrow H^+(aq) + HPO_4^{2-}(aq)$$
$$HPO_4^{2-}(aq) \rightarrow H^+(aq) + PO_4^{3-}(aq)$$

Each of these dissociations is characterised by a different vertical section of the titration curve. The K_a for each of these dissociations can be determined, as shown in the previous section.

● **Key skills** **ICT**

- Produce titration curves for the reaction between phosphoric(v) acid and sodium hydroxide using a pH probe and a datalogger.
- Use the equation on page 74 to construct a spreadsheet which will allow you to calculate the pH of a buffer solution given the K_a, the concentration of the acid and the concentration of the salt solution.

Number

- Calculate the pH of solutions of weak and strong acids and alkalis.
- Calculate the pH of buffer solutions.
- Construct graphs to produce titration curves.

● **Skills task** Some oven cleaners contain sodium hydroxide. $10.0\,cm^3$ of such an oven cleaner was dissolved in water and made up to $250\,cm^3$. $25.0\,cm^3$ of the oven cleaner solution was titrated with $0.25\,mol\,dm^{-3}$ hydrochloric acid solution using phenolphthalein as the indicator. $20.50\,cm^3$ of the acid was required to neutralise the oven cleaner solution. Calculate the mass of sodium hydroxide in a $500\,cm^3$ bottle of oven cleaner.

CHECKLIST   After studying Chapter 5 you should know and understand the following terms.

- **Acid:** A substance that dissolves in water, producing the oxonium ion, H_3O^+(aq).
- **Base:** A substance that neutralises acids, producing a salt and water as the only products.
- **Alkali:** A soluble base, which produces hydroxide ions, OH^-(aq), in water.
- **Brønsted–Lowry acid:** A substance that donates a proton, H^+(aq).
- **Brønsted–Lowry base:** A substance that accepts a proton, H^+(aq).
- **Strong acid:** Produces a high concentration of H_3O^+(aq) in aqueous solution.
- **Weak acid:** Produces a low concentration of H_3O^+(aq) in aqueous solution.
- **Strong alkali:** Produces a high concentration of OH^-(aq) in aqueous solution.
- **Weak alkali:** Produces a low concentration of OH^-(aq) in aqueous solution.
- **Acid dissociation constant:** A measure of the extent of dissociation of a weak acid in aqueous solution. Symbol: K_a.
- **pK_a:** $pK_a = -\log K_a$.
- **Base dissociation constant:** A measure of the extent of dissociation of a weak base in aqueous solution. Symbol: K_b.
- **pH:** $pH = -\log[H_3O^+$(aq)$]$.
- **Ionic product of water:** A constant derived from the equilibrium produced by the ionisation of water.
  $$K_w = [H_3O^+(aq)]\,[OH^-(aq)]$$
  $$= 1.0 \times 10^{-14}\,mol^2\,dm^{-6} \text{ at } 298\,K$$
- **Buffer solution:** A solution that resists changes in its pH when a small amount of either an acid or an alkali is added to it.
- **Indicator:** Substances used to detect the end-point during titrations.
- **Titration curve:** A graph that shows how pH varies with the addition of acid or base to a base or acid.

Examination questions

1 a i Write an expression for the dissociation constant K_a of propanoic acid, CH_3CH_2COOH.
ii Write an expression for pK_a in terms of K_a.
iii Calculate the pH of a $0.10\,mol\,dm^{-3}$ solution of propanoic acid, given that $K_a = 1.35 \times 10^{-5}\,mol\,dm^{-3}$ for this acid at $25\,°C$. (6)
b Explain why an aqueous solution containing propanoic acid and its sodium salt constitutes a buffer system able to minimise the effect of added hydrogen ions. (3)

AQA, A level, Specimen Paper 6421, 2001/2

2 The pH of blood is maintained in healthy individuals by various buffering systems. One of the most important systems contains carbon dioxide, CO_2, and hydrogencarbonate ions, HCO_3^-, linked by the reaction:

$$CO_2(aq) + H_2O(l) \rightleftharpoons H^+(aq) + HCO_3^-(aq) \quad \text{(equation I)}$$

K_a for this reaction has the value of $4.5 \times 10^{-7}\,mol\,dm^{-3}$ at the temperature in the body.

a i Give the mathematical definition of pH. (1)
ii Write the expression for K_a for the equilibrium shown in equation I in terms of the concentrations of $CO_2(aq)$, $H^+(aq)$ and $HCO_3^-(aq)$. (2)
b In the blood of a healthy person:

$$[HCO_3^-(aq)] = 2.5 \times 10^{-2}\,mol\,dm^{-3}$$

(This comes from other processes as well as equation I.)

$$[CO_2(aq)] = 1.25 \times 10^{-3}\,mol\,dm^{-3}$$

i Use this data, your answer to part **a ii** and the value of K_a to calculate the concentration of H^+ ions in the blood of a healthy person. (2)
ii Calculate the pH of the blood of a healthy person. (1)
c Use equation I and the information in part **b** to explain how blood acts as a buffer solution. (In this question I mark is available for the quality of written communication.) (5)
d The equilibrium reaction in equation I plays a dual role in controlling the acidity of the oceans and controlling the concentration of carbon dioxide in the atmosphere.
Write two further equations, with state symbols, to show how dissolved carbon dioxide in the oceans can be converted to calcium carbonate in seashells. (2)

OCR, A level, Specimen Paper A7887, Sept 2000

3 This question refers to different aspects of acid–base chemistry:

a Hydrochloric acid, HCl, is classed as a **strong** acid but it can have both **concentrated** and **dilute** solutions. Explain why this is so. (3)
b Sodium phosphate, Na_3PO_4, a water-softening agent, can be prepared in the laboratory by neutralising phosphoric acid.
A student prepared this compound in the laboratory from $20.0\,cm^3$ of $0.100\,mol\,dm^{-3}$ phosphoric acid and $0.250\,mol\,dm^{-3}$ sodium hydroxide:

$$H_3PO_4(aq) + 3NaOH(aq) \rightarrow Na_3PO_4(aq) + 3H_2O(l)$$

i Deduce the oxidation state of phosphorus in sodium phosphate, Na_3PO_4.
ii Calculate the volume of $NaOH(aq)$ that the student would need to use to just neutralise the phosphoric acid using the quantities above. (4)
c Calculate the pH of the $NaOH(aq)$ used in part **b**. ($K_w = 1.00 \times 10^{-14}\,mol^2\,dm^{-6}$) (4)

OCR, A level, Specimen Paper A7882, Sept 2000

4 The values of K_a for a series of organic acids at $25\,°C$ are shown in the table below:

Acid	Formula	K_a $(mol\,dm^{-3})$
Ethanoic acid	CH_3COOH	1.75×10^{-5}
Chloroethanoic acid	$CH_2(Cl)COOH$	1.38×10^{-3}
Bromoethanoic acid	$CH_2(Br)COOH$	1.26×10^{-3}
Iodoethanoic acid	$CH_2(I)COOH$	6.76×10^{-4}

a i Give the expression for K_a for ethanoic acid. (1)
ii Calculate the pH of $0.100\,mol\,dm^{-3}$ ethanoic acid. (3)
iii Calculate the pH of a mixture which is $0.100\,mol\,dm^{-3}$ with respect to ethanoic acid and $0.125\,mol\,dm^{-3}$ with respect to sodium ethanoate. (2)
iv What property is shown by such a mixture? (2)
b i By consideration of bond polarity, suggest why 1-chloroethanoic acid is a stronger acid than ethanoic acid. (2)
ii Suggest, with an explanation, whether 1-fluoroethanoic acid is likely to be a stronger or a weaker acid than 1-chloroethanoic acid. (2)

London, AS level, Synoptic Paper (CH5), Jan 2000

5 a Use Brønsted–Lowry theory of acid/base behaviour and acid/base conjugate pairs to explain how sulphuric acid shows its acidic nature when added to water. (3)

b If the enthalpy of neutralisation ΔH_{neut} is measured for hydrobromic acid, HBr, or for nitric acid, HNO_3, both with sodium hydroxide solution, the value is $-57.6\,kJ\,mol^{-1}$; for the reaction of ethanoic acid, CH_3COOH, with sodium hydroxide solution it is $-55.2\,kJ\,mol^{-1}$.

i Why is ΔH_{neut} the same for both HBr and HNO_3 with NaOH? (2)

ii Give one reason why ΔH_{neut} for the reaction of ethanoic acid with sodium hydroxide solution is less exothermic than for hydrobromic acid with sodium hydroxide solution. (2)

c i What is the function of a buffer solution? (2)

ii When sodium hydrogensulphate, $Na^+HSO_4^-$, is added to a solution of sodium sulphate the mixture behaves as a buffer solution. Show how it functions by writing equations for the reactions which occur on the addition of acid or base. (2)

London, AS/A level, Module 2, June 1999

6 Citral can be oxidised to an acid $C_9H_{15}COOH$ ($M_r = 168$) which ionises in water

$$C_9H_{15}COOH + H_2O \rightleftharpoons H_3O^+ + C_9H_{15}COO^-$$

4.62 g of this acid was dissolved in water to give a solution of volume $250\,cm^3$. This solution had a pH of 2.91.

a Write the expression for K_a for this acid. (1)

b Calculate the concentration of the acid in $mol\,dm^{-3}$. (1)

c Calculate the value of K_a of the acid. (3)

London, AS/A level, Module 2, Jan 2000

7 a Water partially dissociates into ions.

i Write an equation for this dissociation.

ii Write an expression for the ionic product, K_w, of water.

iii At 303 K the ionic product of water, K_w, has a value of $1.47 \times 10^{-14}\,mol^2\,dm^{-6}$. Use this value, together with your knowledge of the value of K_w at 298 K, to deduce the sign of the enthalpy change when water dissociates into ions. (5)

b i Calculate the pH of a $0.300\,mol\,dm^{-3}$ solution of NaOH at 298 K.

ii Calculate the pH of the solution formed when $25.0\,cm^3$ of $0.300\,mol\,dm^{-3}$ NaOH are added to $225\,cm^3$ of water at 298 K.

iii Calculate the pH of the solution formed when $25.0\,cm^3$ of $0.300\,mol\,dm^{-3}$ NaOH are added to $75.0\,cm^3$ of $0.200\,mol\,dm^{-3}$ HCl at 298 K. (9)

NEAB, AS/A level, CH02, Mar 1998

8 The pH of solution **A**, a $0.15\,mol\,dm^{-3}$ solution of a weak monoprotic acid HX, is 2.69.

a Calculate $[H^+]$ in solution **A** and hence determine the value of the acid dissociation constant, K_a, of HX. (3)

b i A $25\,cm^3$ sample of **A** is titrated with $0.25\,mol\,dm^{-3}$ sodium hydroxide. Calculate the volume of sodium hydroxide needed to reach equivalence in the titration. Give the best estimate you can of the pH of the neutralised solution, stating a reason.

ii Calculate the pH of the titration solution when HX is exactly half-neutralised and $[HX] = [X^-]$.

iii Calculate the pH of the titration solution when a total of $25\,cm^3$ of $0.25\,mol\,dm^{-3}$ sodium hydroxide has been added. (8)

NEAB, A level, CH04, June 1998

9 a When the molar concentration of hydrogen ions and the molar concentration of hydroxide ions in a sample of water are multiplied together a value is obtained which is constant at a fixed temperature.

i Name this constant.

ii State the value of this constant at 298 K.

iii State qualitatively how the value of this constant changes when the temperature of the water is increased. Explain your answer. (5)

b A $4.00\,dm^3$ sample of an aqueous solution of a strong acid was neutralised exactly by the addition of 13.4 g of anhydrous sodium carbonate. The neutralisation reaction is given by the equation

$$2H^+ + CO_3^{2-} \rightarrow H_2O + CO_2$$

i Calculate the number of moles of Na_2CO_3 in the 13.4 g sample.

ii Calculate the number of moles of H^+ ions present in $4.00\,dm^3$ of the acid solution. Use this result to calculate the hydrogen ion concentration in the solution and hence its pH.

iii Calculate the pH of the resulting solution formed if 13.4 g of NaOH, rather than 13.4 g of Na_2CO_3, had been added to $4.00\,dm^3$ of the original acid solution. (11)

NEAB, AS/A level, CH02, June 1998

II
INORGANIC CHEMISTRY

Alchemical symbol for iron. Abundant in the Earth's crust, iron is widely used in industry, and is a component of haemoglobin responsible for transporting oxygen in the blood.

6 Period 3

STARTING POINTS ● The modern periodic table is based upon the work of Dmitri Mendeleev, Lord Rutherford and Henry Moseley.
● All the elements, with the exception of hydrogen, are arranged into one of four blocks in the periodic table: the s block, the p block, the d block and the f block.
● The horizontal rows are called periods and are numbered according to the principle quantum number '*n*'.
● Periodic trends of physical properties are observed across the periodic table.

Figure 6.1
Period 3, Na through to Ar (shaded in green), can be used to highlight trends associated with other periods in the periodic table.

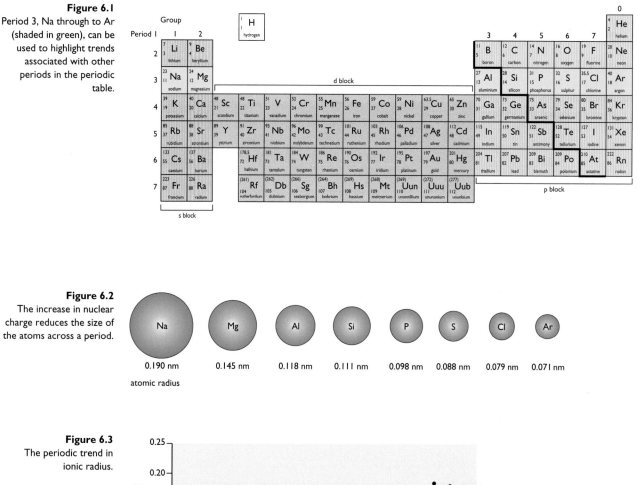

Figure 6.2
The increase in nuclear charge reduces the size of the atoms across a period.

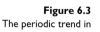

Na	Mg	Al	Si	P	S	Cl	Ar
0.190 nm	0.145 nm	0.118 nm	0.111 nm	0.098 nm	0.088 nm	0.079 nm	0.071 nm

atomic radius

Figure 6.3
The periodic trend in ionic radius.

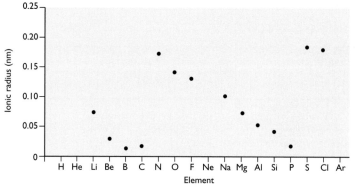

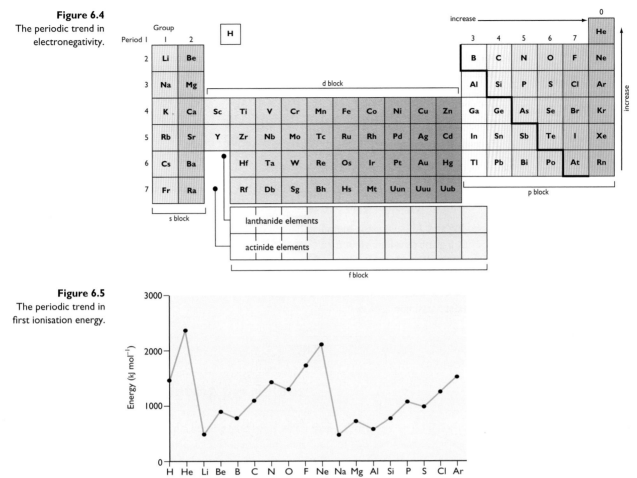

Figure 6.4
The periodic trend in electronegativity.

Figure 6.5
The periodic trend in first ionisation energy.

Earlier in your study of Chemistry at advanced level (see *Introduction to Advanced Chemistry*, Chapter 12) you covered the development of the periodic table. You also looked at regular, or periodic, trends related to:

- atomic radius (Figure 6.2)
- ionic radius (Figure 6.3)
- electronegativity (Figure 6.4) and
- first ionisation energy (Figure 6.5).

The reactions that are related to these properties can be used to illustrate trends for other periods in the periodic table.

From your previous study of the patterns within the periodic table, try to answer the following questions.

1 Which element in the periodic table do you think will have the largest atomic radius?

2 What is the explanation for:
 a the decrease in first ionisation energy from beryllium to boron?
 b the decrease in first ionisation energy from nitrogen to oxygen?

3 Which group of elements has the highest electronegativities?

4 Which element will have the lowest electronegativity?

Trends in structure of the elements of Period 3

Before examining the trends in reactivity of the elements of Period 3 it is useful to remind ourselves of the trends in structure that these elements possess. Table 6.1 shows the trends in structure of the elements in Period 3, that is, Na (metal) to Ar (non-metal).

Table 6.1
Trends in structure in Period 3.

Element	Na	Mg	Al	Si	P	S	Cl	Ar
Group	1	2	3	4	5	6	7	0
Element structure	Giant metallic			Giant covalent	Simple molecular			

Looking across Period 3 there is a change from giant metallic structure, through giant covalent structure to simple molecular elements:

- Na, Mg and Al have giant metallic structures (Figure 6.6)
- Si has a giant covalent structure (Figure 6.7)
- P exists as P_4 molecules (Figure 6.8)
- S exists as S_8 molecules (Figure 6.9)
- Cl exists as diatomic molecules, Cl_2 (Figure 6.10)
- Ar exists as single atoms.

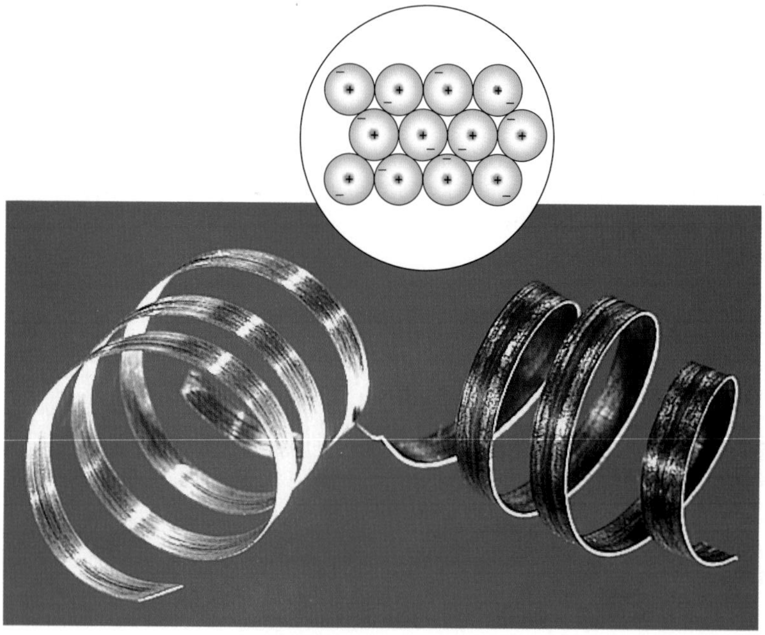

5 Why has the element Ar not been included in the illustrations of simple molecular elements which contain covalent bonding (Figures 6.8 to 6.10)?

Figure 6.6
In the metallic bond, for example in magnesium, the negatively charged delocalised electrons attract all the positive metal ions in the giant structure and bond them together with strong electrostatic forces of attraction.

Figure 6.7
A small part of the structure of silicon.

Trends in the reactivity of the elements of Period 3

Sodium and magnesium have low ionisation energies and are therefore quite electropositive metals. For example, sodium has a first ionisation energy of $496\,kJ\,mol^{-1}$ whilst magnesium has a first ionisation energy of $738\,kJ\,mol^{-1}$ and a second ionisation energy of $1451\,kJ\,mol^{-1}$. In their reactions, therefore, they are more likely to produce ionic substances. These substances contain quite large metal cations, with the ionic radius of Na^+ being $0.098\,nm$ and that of Mg^{2+} $0.065\,nm$.

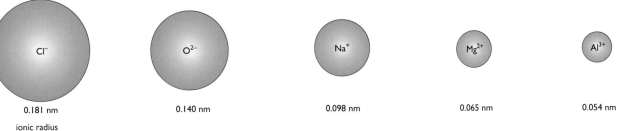

Cl^-	O^{2-}	Na^+	Mg^{2+}	Al^{3+}
0.181 nm	0.140 nm	0.098 nm	0.065 nm	0.054 nm

ionic radius

Aluminium forms highly charged ions with a $3+$ charge. They are also relatively small, with an ionic radius of $0.054\,nm$. It is a highly polarising ion. This makes aluminium oxide ionic but aluminium chloride tends to be covalent. This is because the chloride ion is larger than the oxide ion, $0.181\,nm$ compared to $0.140\,nm$, and so the chloride ion is more polarisable.

The non-metals, Si to Cl, have high ionisation energies and are small in size compared to the metal elements of the period. For example, Si has an atomic radius of $0.111\,nm$ and Cl has an atomic radius of $0.079\,nm$ (see Figure 6.2). They therefore react with varying vigour as the period is traversed, producing covalent compounds.

The reactions of the elements in Period 3 with oxygen, chlorine and water are summarised in Tables 6.2–6.7 (pages 88–94).

Figure 6.8
Phosphorus forms P_4 molecules.

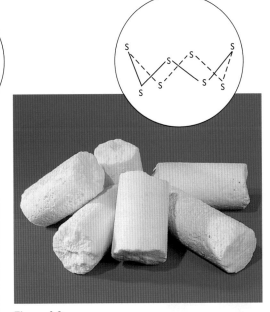

Figure 6.9
Sulphur forms S_8 molecules.

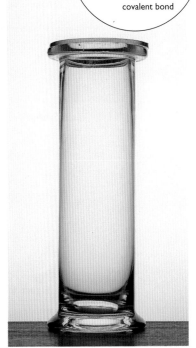

Figure 6.10
Chlorine is a simple diatomic molecule. Its atoms are held together by a single covalent bond.

Reactions of the Period 3 elements with oxygen

Table 6.2

Element	Reaction	Equation for reaction
Sodium	Sodium is a soft metal, of low density which tarnishes very rapidly in air. When heated in air or oxygen it burns with a yellow flame to give a mixture of sodium oxide (white) and sodium peroxide (yellow).	Oxide formation: $4Na(s) + O_2(g) \rightarrow 2Na_2O(s)$ Peroxide formation: $2Na(s) + O_2(g) \rightarrow Na_2O_2(s)$
Magnesium	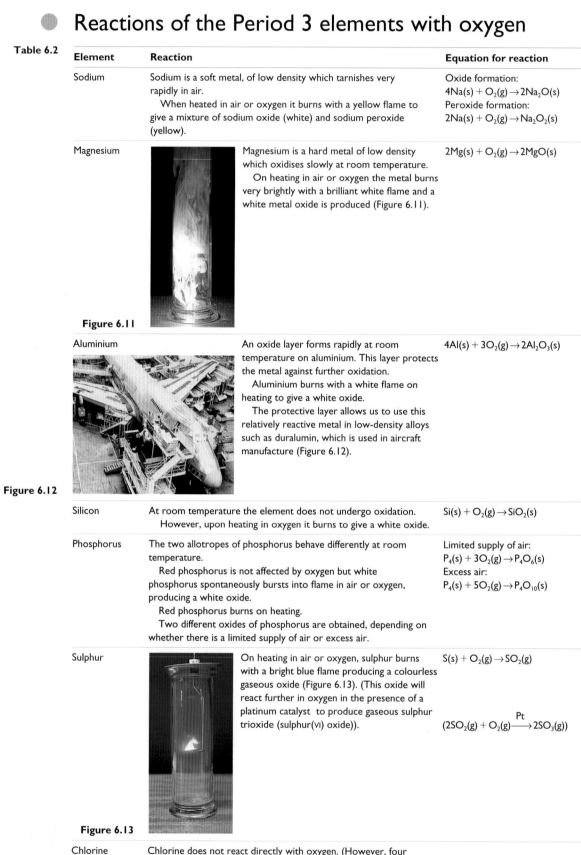 **Figure 6.11** Magnesium is a hard metal of low density which oxidises slowly at room temperature. On heating in air or oxygen the metal burns very brightly with a brilliant white flame and a white metal oxide is produced (Figure 6.11).	$2Mg(s) + O_2(g) \rightarrow 2MgO(s)$
Aluminium	An oxide layer forms rapidly at room temperature on aluminium. This layer protects the metal against further oxidation. Aluminium burns with a white flame on heating to give a white oxide. The protective layer allows us to use this relatively reactive metal in low-density alloys such as duralumin, which is used in aircraft manufacture (Figure 6.12). **Figure 6.12**	$4Al(s) + 3O_2(g) \rightarrow 2Al_2O_3(s)$
Silicon	At room temperature the element does not undergo oxidation. However, upon heating in oxygen it burns to give a white oxide.	$Si(s) + O_2(g) \rightarrow SiO_2(s)$
Phosphorus	The two allotropes of phosphorus behave differently at room temperature. Red phosphorus is not affected by oxygen but white phosphorus spontaneously bursts into flame in air or oxygen, producing a white oxide. Red phosphorus burns on heating. Two different oxides of phosphorus are obtained, depending on whether there is a limited supply of air or excess air.	Limited supply of air: $P_4(s) + 3O_2(g) \rightarrow P_4O_6(s)$ Excess air: $P_4(s) + 5O_2(g) \rightarrow P_4O_{10}(s)$
Sulphur	On heating in air or oxygen, sulphur burns with a bright blue flame producing a colourless gaseous oxide (Figure 6.13). (This oxide will react further in oxygen in the presence of a platinum catalyst to produce gaseous sulphur trioxide (sulphur(VI) oxide)). **Figure 6.13**	$S(s) + O_2(g) \rightarrow SO_2(g)$ $(2SO_2(g) + O_2(g) \xrightarrow{Pt} 2SO_3(g))$
Chlorine	Chlorine does not react directly with oxygen. (However, four oxides of chlorine, Cl_2O, ClO_2, Cl_2O_6, and Cl_2O_7, can be produced indirectly. Apart from Cl_2O they are all unstable and explosive.)	

The oxides of elements in Period 3 change gradually from being ionic and basic on the left-hand side of Period 3 to being covalent and acidic on the right-hand side of Period 3 (Table 6.3). This observed effect is in keeping with a change in structure of the elements from giant ionic, through giant covalent to simple molecular as we move from the metallic elements to the non-metallic elements of Period 3.

Table 6.3
Oxides of Period 3.

Formula of the oxide	Oxide structure	Type of oxide
Na_2O	Giant ionic	Basic
MgO	Giant ionic	Basic
Al_2O_3	Giant ionic	Amphoteric
SiO_2	Giant covalent	Acidic
P_4O_6 P_4O_{10}	Simple molecular	Acidic
SO_2 SO_3	Simple molecular	Acidic
Cl_2O ClO_2 Cl_2O_6 Cl_2O_7	Simple molecular	Acidic

6 Draw dot-cross diagrams to represent the bonding in:
a Na_2O b MgO
c SO_2 d Cl_2O.

7 Study the equations shown in Table 6.2. From your knowledge of oxidation number, decide which element has been oxidised and which has been reduced. Explain your answers in terms of oxidation number.

Acid–base properties of the Period 3 oxides and hydroxides

The covalent character of the oxides increases from left to right with increasing atomic number of the elements in Period 3. This reflects the gradual change from metallic to non-metallic character of the elements in Period 3.

Figure 6.14 (right)
In the manufacture of soap, sodium hydroxide is used to neutralise the fatty acids found in the oils used to make soap.

The ionic oxides of sodium and magnesium produce alkaline solutions when they are dissolved in water. They are classified as **basic oxides**. For example:

$$Na_2O(s) + H_2O(l) \rightarrow 2NaOH(aq)$$

Sodium hydroxide is a strong and relatively cheap alkali and is produced in large quantities by the electrolysis of brine (saturated sodium chloride solution). It exists completely dissociated in aqueous solution as the hydrated ions, $Na^+(aq)$ and $OH^-(aq)$. It has a vast range of uses, ranging from soap and paper manufacture (Figure 6.14) to production of textiles such as Rayon.

Aluminium oxide is insoluble in water. However, it is amphoteric; it will react with both acids and alkalis.

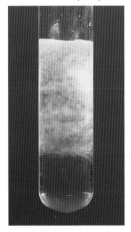

Figure 6.15 (below)
Aluminium hydroxide is a white, gelatinous precipitate.

With acid the reaction is:

$$Al_2O_3(s) + 3H_2SO_4(aq) \rightarrow Al_2(SO_4)_3(aq) + 3H_2O(l)$$

With alkali the reaction is:

$$Al_2O_3(s) + 2OH^-(aq) + 3H_2O(l) \rightarrow 2Al(OH)_4^-(aq)$$
aluminate ion

It should be noted, however, that aluminium oxide has a very strong lattice and so it does not react too readily.

Aluminium hydroxide is formed as a gelatinous precipitate when an alkali is added to a solution containing aluminium ions (Figure 6.15):

$$Al^{3+}(aq) + 3OH^-(aq) \rightarrow Al(OH)_3(s)$$

8 Aluminium hydroxide is amphoteric. Explain what this means in terms of its reactions with acids and bases.

Silicon(IV) oxide (silicon dioxide) is insoluble in water. However, it is an acidic oxide and reacts with hot concentrated sodium hydroxide solution to form the silicate(IV) ion. The need for the 'hot and concentrated' solution of sodium hydroxide is again due to the strength of the giant covalent lattice of SiO_2.

$$SiO_2(s) + 2OH^-(aq) \rightarrow \underset{\text{silicate(IV) ion}}{SiO_3^{2-}(aq)} + H_2O(l)$$

Phosphorus(III) oxide and phosphorus(V) oxide are both acidic and react with water to form acidic solutions of phosphonic acid (phosphorus acid) and phosphoric(V) acid (orthophosphoric acid), respectively.

Figure 6.16

Sulphur dioxide emissions are one of the major causes of acid rain. Acid rain has a major effect on trees.

$$\underset{\text{phosphorus(III) oxide}}{P_4O_6(s)} + 6H_2O(l) \rightarrow \underset{\text{phosphonic acid}}{4H_3PO_3(aq)}$$

$$\underset{\text{phosphorus(V) oxide}}{P_4O_{10}(s)} + 6H_2O(l) \rightarrow \underset{\text{phosphoric(V) acid}}{4H_3PO_4(aq)}$$

The oxides of sulphur, SO_2 and SO_3, are both acidic and react with water to form acidic solutions of sulphurous acid (sulphuric(IV) acid) and sulphuric acid, respectively. Sulphurous acid is a weak acid.

$$SO_2(g) + H_2O(l) \rightleftharpoons \underset{\text{sulphurous acid}}{H_2SO_3(aq)}$$

$$SO_3(g) + H_2O(l) \rightarrow \underset{\text{sulphuric acid}}{H_2SO_4(aq)}$$

Dichloride oxide, Cl_2O, is an acidic oxide and reacts with water, producing chloric(I) acid:

$$Cl_2O(g) + H_2O(l) \rightarrow 2HOCl(aq)$$

All the other oxides of chlorine dissolve in water to form acidic solutions.

Patterns in formulae of the oxides of the elements of Period 3

There is a pattern in the formulae of the oxides. The formulae are:

| Na_2O | MgO | Al_2O_3 | SiO_2 | P_4O_6
P_4O_{10} | SO_2
SO_3 | Cl_2O
Cl_2O_7 |

There is a trend in that for each of the oxides there is an increase of 0.5 mole of oxygen per mole of the particular element in Period 3.

Na_2O	→	MgO	→	Al_2O_3	→	SiO_2	→	P_4O_{10}	→	SO_3	→	Cl_2O_7
0.5		1.0		1.5		2.0		2.5		3.0		3.5

9 Explain how the pattern shown in the formulae of the oxides in Period 3 is due to the number of electrons that are involved in the bonding within the particular oxide.

10 Write down the oxidation states for the Period 3 elements in the compounds shown in Table 6.3.

11 Using your knowledge of the oxides of Period 3, write a brief summary of the similar reactions for the elements of Period 4 (exclude the d block elements, Sc–Zn).

The pattern shown in the formulae is due to the number of electrons that are involved in the bonding within each oxide.

Reactions of the Period 3 elements with chlorine

Table 6.4 shows a summary of the reactions of the elements of Period 3 with chlorine.

Table 6.4

Element	Reaction	Equation for reaction
Sodium	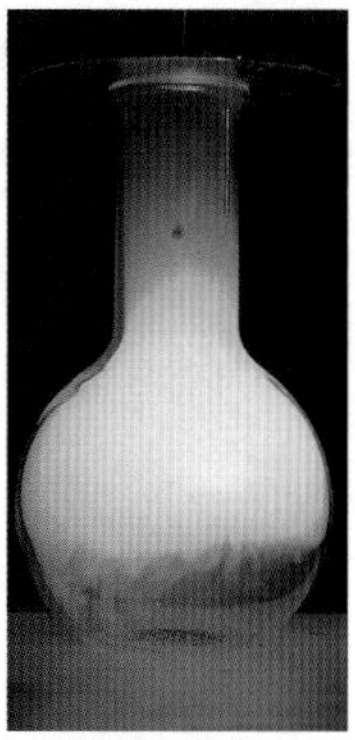 Reacts quite vigorously when heated in chlorine, producing a white powder – sodium chloride (Figure 6.17).	$2Na(s) + Cl_2(g) \rightarrow 2NaCl(s)$
	Figure 6.17	
Magnesium	Magnesium burns brightly, producing a white powder – magnesium chloride.	$Mg(s) + Cl_2(g) \rightarrow MgCl_2(s)$
Aluminium	If chlorine is passed over heated aluminium under anhydrous conditions, then a white powder is produced – aluminium chloride (Figure 6.18).	$2Al(s) + 3Cl_2(g) \rightarrow Al_2Cl_6(s)$
Silicon	Silicon forms a colourless covalent liquid when chlorine is passed over heated silicon under anhydrous conditions.	$Si(s) + 2Cl_2(g) \rightarrow SiCl_4(l)$
Phosphorus	When chlorine is passed over heated white phosphorus the phosphorus burns with a pale green flame and phosphorus trichloride (phosphorus(III) chloride) distils over. It is a colourless covalent liquid.	$P_4(s) + 6Cl_2(g) \rightarrow 4PCl_3(l)$
	If cold PCl_3 liquid is dripped into a flask into which chlorine is passing, then phosphorus pentachloride (phosphorus(V) chloride) is produced. It is a white solid that dissociates quite easily.*	$PCl_3(l) + Cl_2(g) \rightleftharpoons PCl_5(s)$
Sulphur	When chlorine is passed over heated sulphur under anhydrous conditions a red, covalent liquid, disulphur dichloride, is produced.	$2S(s) + Cl_2(g) \rightarrow S_2Cl_2(l)$
	A further chloride of sulphur is SCl_2. It is formed by the reaction of more chlorine with S_2Cl_2 at $0\,°C$.	$S_2Cl_2(l) + Cl_2(g) \rightarrow 2SCl_2(l)$

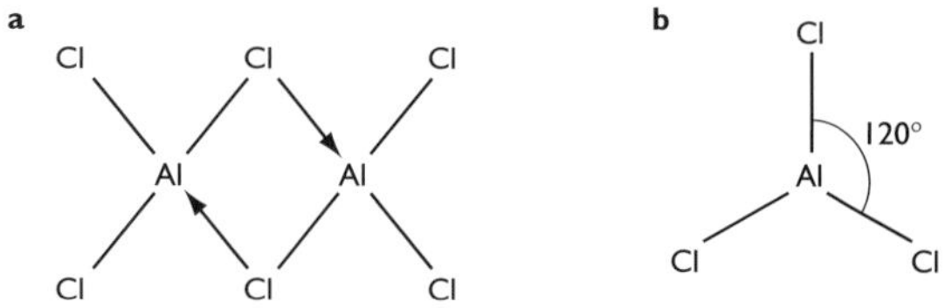

Figure 6.18
a In the structure of the Al_2Cl_6 dimer the arrangement of the chlorine atoms about each of the aluminium atoms is roughly tetrahedral.[†]
b At high temperatures aluminium chloride exists as the monomer, $AlCl_3$, in which the molecule is trigonal planar.

[†]This sort of electron deficient behaviour is also shown by the Group 3 element boron in, for example, BF_3. Another substance that shows this sort of behaviour is beryllium, a group 2 element, in $BeCl_2$. Due to the small size of the atom and the Be^{2+} ion, this element has a marked tendency to form covalent compounds such as $BeCl_2$.

*In the vapour state PCl_5 exists as a trigonal bipyramid structure (Figure 6.19a). In the solid state it is ionic, having a structure containing both $[PCl_4^+]$ and $[PCl_6^-]$ (Figure 6.19b).

Figure 6.19
a PCl_5 has a trigonal bipyramid structure with bond angles of 90° and 120°.
b $[PCl_4^+]$ has a tetrahedral structure with bond angles of 109.5°, whilst $[PCl_6^-]$ has an octahedral structure with bond angles of 90°.

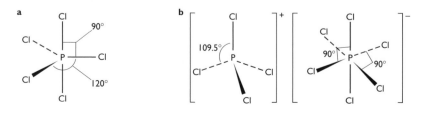

The experimental set-up used to prepare aluminium chloride under anhydrous conditions is shown in Figure 6.20. The chlorides of silicon, phosphorus and sulphur are prepared in a similar way.

Figure 6.20
The preparation of aluminium chloride – also the method for preparation of the chlorides of silicon, phosphorus and sulphur.

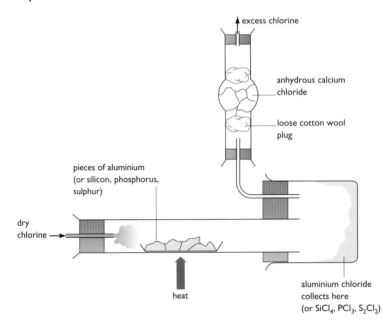

The chlorides change gradually from being ionic on the left-hand side of Period 3 to being covalent on the right-hand side of Period 3. There is also a change in structure from giant ionic lattices through to simple covalent molecules (Table 6.5).

Table 6.5

Formula of the chloride	Chloride structure
NaCl	Giant ionic
$MgCl_2$	Giant ionic
$AlCl_3$	Molecular
$SiCl_4$	Molecular
PCl_3	Molecular
PCl_5	*Ionic*
S_2Cl_2	Molecular
SCl_2	

It should be noted that due to the larger size generally of the atoms in Period 3 versus Period 2, the maximum number of electron pairs which can surround a central atom is greater in Period 3; for example, PCl_5 versus NCl_3.

Reactions of Period 3 chlorides with water

It is found that the tendency towards hydrolysis increases as the chlorides of Period 3 become more covalent.

Sodium chloride dissolves in water without any hydrolysis to produce the hydrated ions:

$$NaCl(s) \xrightarrow{\text{(aq)}} Na^+(aq) + Cl^-(aq)$$

Magnesium chloride, however, undergoes mild hydrolysis to a small extent when dissolved in water:

$$MgCl_2(s) + H_2O(l) \rightarrow MgOHCl(aq) + HCl(aq)$$
$$MgOHCl(aq) + H_2O(l) \rightarrow Mg(OH)_2(aq) + HCl(aq)$$

Aluminium chloride has a much higher degree of covalent character in its bonds. The dimer first of all follows the reaction below to form hexaaquaaluminate(III) ions, $[Al(H_2O)_6]^{3+}$:

$$Al_2Cl_6(s) + 12H_2O(l) \rightarrow 2[Al(H_2O)_6]^{3+}(aq) + 6Cl^-(aq)$$

Al^{3+} is a highly charged ion and so the hexaaquaaluminate(III) ions, $[Al(H_2O)_6]^{3+}$, then undergo the following reactions, with the water molecules acting as base:

$$[Al(H_2O)_6]^{3+}(aq) + H_2O(l) \rightleftharpoons [Al(H_2O)_5OH]^{2+}(aq) + H_3O^+(aq)$$
$$[Al(H_2O)_5OH]^{2+}(aq) + H_2O(l) \rightleftharpoons [Al(H_2O)_4(OH)_2]^+(aq) + H_3O^+(aq)$$
and so on.

Figure 6.21
$SiCl_4$ fumes in moist air, producing acidic fumes of HCl that turn blue litmus red.

The above occurs because the electric field associated with the highly charged Al^{3+} ion is so great that the electrons in the O—H bonds of the water ligands are drawn to it. These water molecules thus become good proton donors. Solutions of aluminium salts are acidic (as acidic as vinegar!).

The remaining covalent halides of Period 3 all produce very acidic solutions.

The covalent liquid, $SiCl_4$, fumes in moist air and is hydrolysed by water (Figure 6.21):

$$SiCl_4(l) + 2H_2O(l) \rightarrow SiO_2(s) + 4HCl(aq)$$

The chlorides of phosphorus are hydrolysed according to the following equations:

$$PCl_3(l) + 3H_2O(l) \rightarrow H_3PO_3(aq) + 3HCl(aq)$$
$$PCl_5(l) + 4H_2O(l) \rightarrow H_3PO_4(aq) + 5HCl(aq)$$

Both S_2Cl_2 and SCl_2 are hydrolysed, giving a precipitate of sulphur:

$$2S_2Cl_2(l) + 2H_2O(l) \rightarrow SO_2(aq) + 3S(s) + 4HCl(aq)$$
$$2SCl_2(l) + 2H_2O(l) \rightarrow SO_2(aq) + S(s) + 4HCl(aq)$$

Table 6.6 provides a summary of these reactions.

Table 6.6

Formula of the chloride	Reaction of chloride with water	Product of reaction
NaCl	None – it dissolves	$Na^+(aq)Cl^-(aq)$
$MgCl_2$	Dissolves	$Mg(OH)_2(aq) + HCl(aq)$
$AlCl_3$	Vigorous reaction	$[Al(H_2O)_6]^{3+}(aq)$
$SiCl_4$	Vigorous reaction	$SiO_2(s) + HCl(aq)$
PCl_3 PCl_5	Vigorous reaction	$H_3PO_3(aq)$ $H_3PO_4(aq) + HCl(aq)$
S_2Cl_2 SCl_2	Vigorous reaction	$SO_2(aq) + S(s) + HCl(aq)$

Patterns in formulae of the chlorides of Period 3

There is a pattern in the formulae of the chlorides, up to phosphorus. The formulae are:

NaCl $MgCl_2$ $AlCl_3$ $SiCl_4$ PCl_3
                                                  PCl_5

For each of the elements there is an increase of 1 mole of chlorine per mole of the particular element in Period 3.

NaCl $\rightarrow$ $MgCl_2$ $\rightarrow$ $AlCl_3$ $\rightarrow$ $SiCl_4$ $\rightarrow$ PCl_5

12 Using your knowledge of the chlorides of Period 3, write a brief summary of the similar reactions for Period 4 (exclude the d block elements, Sc–Zn).

Reactions of the Period 3 elements with water

Table 6.7 shows a summary of the reactions of the elements of Period 3 with water.

Table 6.7

Element	Reaction		Equation for reaction
Sodium	**Figure 6.22** 	Sodium reacts violently at room temperature with water. The reaction is very exothermic and the metal whizzes around the surface of the water and melts (Figure 6.22). Hydrogen is evolved, which when ignited burns with a yellow flame. An alkaline solution of sodium hydroxide is also formed.	$2Na(s) + 2H_2O(l) \rightarrow 2NaOH(aq) + H_2(g)$
Magnesium	**Figure 6.23** 	Magnesium reacts extremely slowly with cold water. However, if steam is passed over the heated metal then an strongly exothermic reaction is produced. The metal glows white and gives the oxide and hydrogen (Figure 6.23).	$Mg(s) + 2H_2O(l) \rightarrow Mg(OH)_2(s) + H_2(g)$ $Mg(s) + H_2O(g) \rightarrow MgO(s) + H_2(g)$
Aluminium		There is no observable reaction between cold water and aluminium. However, if steam is passed over heated aluminium powder then the metal oxide is produced, together with hydrogen.	$2Al(s) + 3H_2O(g) \rightarrow Al_2O_3(s) + 3H_2(g)$
Silicon		No reaction with cold or hot water.	
Phosphorus		No reaction with cold or hot water.	
Sulphur		No reaction with cold or hot water.	
Chlorine		Chlorine dissolves in water forming an acidic solution. This solution contains chloric(I) acid (hypochlorous acid) and hydrochloric acid, as well as hydrated chlorine molecules (Figure 6.24).	$Cl_2(g) + H_2O(l) \rightarrow HOCl(aq) + HCl(aq)$

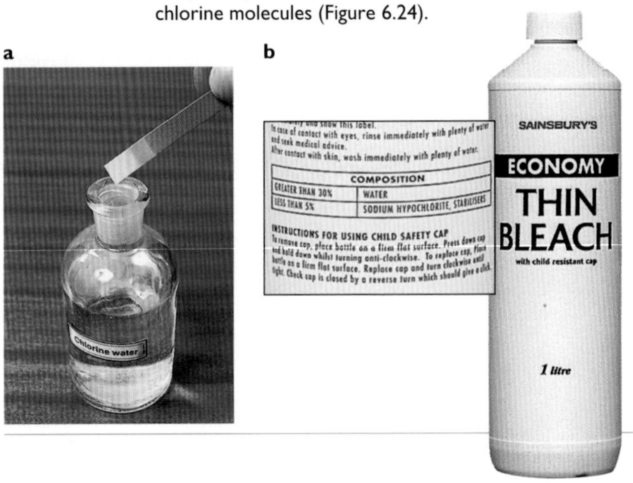

Figure 6.24
a Chlorine in solution is acidic. Note the red colour just above the bleached area. The bleaching agent is the chlorate(I) ion formed from chloric(I) acid.
b Bleaches contain sodium chlorate(I) (sodium hypochlorite) which contains the chlorate(I) ion.

In summary:

- sodium and magnesium, electropositive metals, react with water to produce the metal oxide and hydrogen
- aluminium, a less electropositive metal, reacts initially with water to produce hydrogen but the surface coating of the oxide that is formed stops the reaction
- silicon, phosphorus and sulphur, non-metals, do not react with water
- chlorine, a halogen, reacts with water to form an acidic solution.

Key skills

ICT

- Use a spreadsheet and graphical display to identify trends and variations of properties across Period 3.
- Use a CD-ROM, a database or the Internet to retrieve chemical or physical data related to the elements of Period 3.

Number

- Selection of relevant numeric data for the elements of Period 3 from a database ensuring correct units are applied.

Skills task

Use the Internet and other sources of information to produce a booklet highlighting the physical and chemical properties of the elements of Period 3, along with their uses.

CHECKLIST   After studying Chapter 6 you should know and understand the following terms.

- **Period 3:** The row of elements Na to Ar in the periodic table. Within this period the atoms of all the elements have the same number of occupied electron shells but have an increasing number of electrons in the outer shell, $n = 3$.
- **Periodic trends:** Regular trends in both physical and chemical properties found when moving across a period.

Examination questions

1 a Write equations to show what happens when the following oxides are added to water and predict approximate values for the pH of the resulting solutions.
i sodium oxide
ii sulphur dioxide (4)

b What is the relationship between bond type in the oxides of the Period 3 elements and the pH of the solutions which result from addition of the oxides to water? (2)

c Write equations to show what happens when the following chlorides are added to water and predict approximate values for the pH of the resulting solutions.
i magnesium chloride
ii silicon tetrachloride (4)

AQA, A level, Specimen Paper 6421, 2001/2

2 The table below relates to oxides of Period 3 in the periodic table.

Oxide	Na_2O	MgO	Al_2O_3	SiO_2	P_4O_{10}	SO_3
Melting point (°C)	1275	2827	2017	1607	580	33
Bonding						
Structure						

a Complete the table using the following guidelines.
i Complete the 'bonding' row using **only** the words: *ionic* or *covalent*.
ii Complete the 'structure' row using **only** the words: *simple molecular* or *giant*.
iii Explain, in terms of forces, the difference between the melting points of MgO and SO_3. (5)

b The oxides Na_2O and SO_3 were each added separately to water.
For each oxide, construct a balanced equation for its reaction with water.
i SO_3 reaction with water
ii Na_2O reaction with water (2)

OCR, A level, Specimen Paper A7882, Sept 2000

3 Aluminium chloride occurs in the anhydrous state and in the hydrated state. You may regard the structure of the anhydrous state as having the formula $AlCl_3$. When water is added to solid anhydrous aluminium chloride, steamy acidic fumes are seen.
a What are the steamy fumes? (1)
b Write an equation for the reaction occurring. (1)
c Explain by reference to the structure of $AlCl_3$ how the first step of this reaction occurs. (2)

London, A level, Module Test 3, June 1998

4 The table below shows the melting temperatures, T_m, and the atomic radii, r, of the Period 3 elements.

Element	Na	Mg	Al	Si	P	S	Cl	Ar
T_m (K)	371	923	933	1680	317	392	172	84
r (nm)	0.191	0.160	0.130	0.118	0.110	0.102	0.099	0.095

a Explain the variation in atomic radius.
b In terms of structure and bonding, explain the irregular variation in melting temperature.
c Predict the variation in T_m across Period 2. Explain your answer. (15)

NEAB, AS/A level, Paper 1 Section C, June 1998

5 a Sulphur dioxide, SO_2, and carbon dioxide, CO_2 both give acidic solutions when dissolved in water. Sulphur dioxide solutions gradually oxidise in air to sulphuric acid, but solutions of carbon dioxide do not oxidise in air.
i The bonds in both CO_2 and SO_2 are polar, but only the SO_2 molecule has an overall dipole whereas the CO_2 molecule has no dipole. Explain this in terms of the structures of the two molecules. (3)
ii Explain in terms of oxidation states why only SO_2 oxidises in air. (2)
iii Solutions of carbon dioxide contain the following equilibrium:

$$2H_2O(l) + CO_2(aq) \rightleftharpoons H_3O^+(aq) + HCO_3^-(aq)$$

If a solution of sodium hydroxide containing phenolphthalein indicator, which is purple, is added drop by drop to a solution of carbon dioxide, the purple colour takes several seconds to disappear. What does this suggest about the position of the equilibrium and the kinetics of the reaction between carbon dioxide and water? (2)
iv A saturated solution of sulphur dioxide in water was left in air until it had all been converted to sulphuric acid. $25.0\,cm^3$ portions of this (sulphuric acid) solution required $20.8\,cm^3$ of $0.400\,mol\,dm^{-3}$ sodium hydroxide solution. What is the solubility of SO_2 in $mol\,dm^{-3}$ at the temperature and pressure of the experiment? (3)
b i Magnesium and sulphur form oxides which have different types of bonds. Explain why this is so. (3)
ii Explain why MgO and SO_2 react differently with water. (2)
c Explain why aluminium chloride is covalent and readily dimerises to Al_2Cl_6 and draw the structure of the dimer. (You are not required to show the shape of the molecule.) (3)

London, A level, CH5, Jan 1998

6 a Describe the nature of the attractive forces which hold the particles together in magnesium metal and in magnesium chloride. (4)

b Name the type of bond between aluminium and chlorine in aluminium chloride and explain why the bonding in aluminium chloride differs from that in magnesium chloride. (3)

c Write an equation, including state symbols, to show what happens when magnesium chloride dissolves in water. Explain, in terms of bonding, the nature of the interaction between water and magnesium in this solution. (3)

NEAB, AS/A level, CH01, Mar 1998

7 a Explain why the first ionisation energy of aluminium is less than the first ionisation energy of magnesium. (3)

b Explain why the first ionisation energy of aluminium is less than the first ionisation energy of silicon. (2)

c Explain why the second ionisation energy of aluminium is greater than the first ionisation energy of aluminium. (2)

d Write an equation to illustrate the third ionisation energy of aluminium. (1)

e Explain why the third ionisation energy of aluminium is much less than the third ionisation energy of magnesium. (2)

NEAB, AS/A level, CH01, Mar 1998

8 a What is meant by the term **amphoteric** as applied to a metal hydroxide? (1)

b Describe what you would see if aqueous sodium hydroxide solution was added dropwise, until in excess, to an aqueous solution of aluminium sulphate. Give the formula of the final aluminium-containing species produced in this reaction and write an equation for its formation. (4)

c A solution of aluminium ions was contaminated with iron(III) ions. Describe, giving essential practical details, how you would remove the iron(III) ions so as to obtain a solution containing aluminium ions free from this impurity. (4)

NEAB, A level, CH05, Mar 1998

9 a Name the shape of a molecule of $AlCl_3$ and give its bond angle. (2)

b Explain why a molecule of $AlCl_3$ is able to form a bond with a chloride ion and name the type of bond formed. (3)

c Write equations for the reactions of $AlCl_3$ and $SiCl_4$ with water. (2)

d Sketch the arrangement of oxygen atoms around silicon in the silicon-containing species formed by reaction of $SiCl_4$ with water. Indicate a value for one of the bond angles on your diagram.

NEAB, AS/A level, CH01, June 1998

10 a Explain the meaning of the term **periodic trend** when applied to trends in the periodic table. (2)

b Explain why atomic radius decreases across Period 2 from lithium to fluorine. (2)

c The table below shows the melting temperatures, T_m, of the Period 3 elements.

Element	Na	Mg	Al	Si	P	S	Cl	Ar
T_m (K)	371	923	933	1680	317	392	172	84

Explain the following in terms of structure and bonding.

i Magnesium has a higher melting temperature than sodium.

ii Silicon has a very high melting temperature.

iii Sulphur has a higher melting temperature than phosphorus.

iv Argon has the lowest melting temperature in Period 3. (8)

NEAB, AS/A Level, CH01, June 1998

11 a Each of the elements sodium to chlorine in Period 3 will react with oxygen given suitable conditions.

i Choose an element from this period which gives a basic oxide, and write equations both for its reaction with oxygen and to illustrate the basic nature of the oxide. (2)

ii Choose an element which forms an amphoteric oxide and write equations which illustrate this amphoteric nature. (2)

iii Carbon dioxide reacts readily with dilute aqueous sodium hydroxide whereas silicon dioxide does not. Explain this difference and suggest conditions under which silicon dioxide would react. (2)

London, AS/A level, Module Test 1, Jan 1998

12 a Suggest why the molar mass of aluminium(III) chloride appears to vary between 133.5 and 267 depending upon the temperature. (3)

b Explain why aqueous solutions of aluminium(III) chloride are acidic. (2)

London, A level, Module Test 3, Jan 2000

7 Group 4

STARTING POINTS ● Elements that have similar chemical properties are found in the same group of the periodic table.
● Elements within the same group have the same outer electron configuration.
● Group 4 elements occur within the p block of the periodic table.
● Group 4 contains non-metallic, metalloid and metallic elements.

Figure 7.1

a Carbon as natural diamonds are used in jewellery while manufactured diamonds are used industrially on saw blade tips or drill bits for drilling for oil. The 'diamond' structure is also found in silicon and germanium.

b Graphite is a much softer allotrope of carbon which, because of its properties, is used in pencils and as a lubricant.

c This is a computer graphic of C_{60}, known as buckminsterfullerene, a further allotrope of carbon. Uses of it are being investigated.

d This tennis racket has had its structure reinforced with carbon fibre.

a

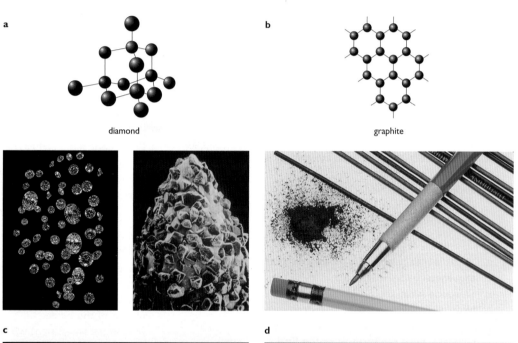

diamond

b

graphite

c

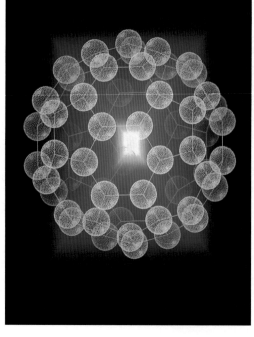

d

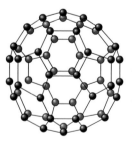

buckminsterfullerene

Figure 7.2
Silicon has revolutionised our lifestyles over the past 30 years. Integrated circuits on silicon chips are key components in mobile phones, computers and so on.

Figure 7.3 (left)
Molten tin is used to produce flat glass.

Figure 7.4 (right)
Both tin and lead are used in alloying. For example, solder contains 50% tin and 50% lead. Other alloys include pewter (80% Sn, 20% Pb) and 'type metal' (10% Sn, 75% Pb, 15% Sb).

Figure 7.5
a Lead-based paints were used extensively by painters in the eighteenth century. **b** Lead metal has the ability to absorb radioactive emissions, so it is used as a shield against them.

a

b

Trends in Group 4

Figure 7.6
The Group 4 elements
(shaded in red) are found
within the p block of the
periodic table.

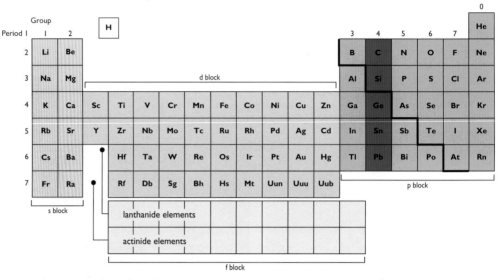

Group 4 consists of the five elements carbon, silicon, germanium, tin and lead and is situated towards the centre of the p-block elements of the periodic table (Figure 7.6). The general outer electron structure for this group of elements is ns^2np^2. These elements show a clear trend from non-metal (carbon) through metalloid or semi-metals (silicon and germanium) to more metallic elements at the bottom of the group (tin and lead). This trend is seen in an increase in the electropositive character of the elements. This in turn is due to a decrease in the ionisation energies as the group is descended. Some data about the elements, their full electron configurations and some physical properties are shown in Table 7.1.

Table 7.1
The Group 4 elements.

Element	Type of structure	Atomic number	Atomic radius (nm)	Density $(g\,cm^{-3})$	Melting point (K)	Boiling point (K)	Electronegativity	Electronic configuration	First ionisation energy $(kJ\,mol^{-1})$
Carbon, C	Giant molecular	6	0.092	3.51*	3823*	5100	2.5	$1s^22s^22p^2$	1086
Silicon, Si	Giant molecular (similar to diamond)	14	0.117	2.33	1683	2628	1.8	$1s^22s^22p^63s^23p^2$	787
Germanium, Ge	Giant molecular (similar to diamond)	32	0.123	5.35	1210	3103	1.8	$[Ar]3d^{10}4s^24p^2$	762
Tin, Sn	Giant metallic	50	0.140	7.28	505	2543	1.8	$[Kr]4d^{10}5s^25p^2$	709
Lead, Pb	Giant metallic	82	0.175	11.34	601	2013	1.0	$[Xe]4f^{14}5d^{10}6s^26p^2$	716

* diamond

From Table 7.1 it can be seen that there are trends in the physical properties of the elements as the group is descended. These are:

- an increase in the size of the atoms and the density of the elements
- a decrease in the melting point and boiling point, electronegativity and first ionisation energy
- a change in the structure of the element from a giant molecular lattice in carbon, silicon and germanium to giant metallic lattice in tin and lead.

The following points should be noted.

1 As the atoms of these elements increase in size the bonds between the atoms get progressively weaker (C—C $347\,kJ\,mol^{-1}$; Si—Si $226\,kJ\,mol^{-1}$; Ge—Ge $167\,kJ\,mol^{-1}$). This makes the structures less hard and gives rise to lower melting points and boiling points.

2 In the case of the first ionisation energy there is little fall in the values from Si to Pb after the relatively large fall which takes place from C to Si. This is due to the larger increase in nuclear charge that can be associated with the filling of d and f sub-shells.

Carbon, silicon and germanium

The lighter elements, carbon and silicon, are non-metallic.

Carbon is found naturally as diamond and graphite. It has a third allotrope, buckminsterfullerene, which was discovered in 1985. The properties of diamond and graphite are shown in Table 7.2. Uses of these two allotropes are shown in Table 7.3 and their structures are shown in Figure 7.1. It is the different structures that give rise to the different physical properties and lead to the allotropes being used in different ways.

Table 7.2
Properties of diamond and graphite.

	Graphite	Diamond
Appearance	Dark grey shiny, solid	Colourless, transparent crystal which sparkles in light
Electrical conductivity	Conducts electricity	Does not conduct electricity
Hardness	Soft material with a slippery feel	Very hard substance
Density ($g\,cm^{-3}$)	2.25	3.51

Table 7.3
Uses of diamond and graphite.

Graphite	Diamond
In pencils	Jewellery
Electrodes	Glass cutters
Lubricant	Diamond studded saws
	Tips of drill bits
	Surface of polishers

The new allotrope of carbon was formed by the action of a laser beam on a sample of graphite. It was named 'buckminsterfullerene' after the American architect Robert Buckminster Fuller, who built complex geometric structures that resemble the structure of this new allotrope of carbon (see Figure 7.1c and Figure 7.7). After investigation it was found that the structure was made up of 60 carbon atoms covalently bonded together. Since this discovery, other spheroidal forms of carbon containing 70, 72 and 84 carbon atoms have been identified. Collectively these are known as 'bucky balls' or fullerenes.

Uses of these allotropes of carbon have still to be developed fully, but scientists have suggested that because of their large surface area they may have uses as catalysts. A further development has seen the creation of 'bucky tubes'. These are based on elongated pipes formed of carbon hexagons (Figure 7.8). The ends of these tubes are closed with 12-carbon pentagons. Bucky tubes are believed to be stronger than diamond and may have many applications. One of the most important is their possible use in microelectronics where they allow carbon-based structures to be built on the nanoscale.

Figure 7.7
A dome designed by Robert Buckminster Fuller (1895–1983).

Figure 7.8
Bucky tubes may have many applications.

Figure 7.9
Window frames, drain pipes and weather boarding are examples of objects constructed using the polymer PVC. The polymer consists of very long chains of molecules made up from the monomer chloroethene (vinyl chloride). These long polymer chains each contain many thousands of carbon atoms.

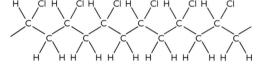

chloroethene

PVC, poly(chloroethene)

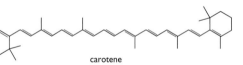

carotene

From your previous study of organic chemistry you will be aware that carbon has the unusual ability of forming strong covalent bonds between its atoms. This gives rise to the unique property amongst all the elements of being able to form very large chains of atoms (Figure 7.9). The carbon atoms may be joined by single, double or triple covalent bonds. It is also capable of forming ring structures (Figure 7.10). The carotene molecule shown in Figure 7.10 shows within the same molecule carbon's ability to form both chains and rings. It is carotene that gives rise to the orange colour of carrots. All living things are made up of organic compounds, some of which contain long carbon chains. The ability to form long molecules with large numbers of atoms bonded together is called **catenation**.

Silicon is the second most abundant element on this planet, making up almost 30% of the crust. It occurs as silicates in clays and rocks as well as silicon(IV) oxide in sand and sandstone. Silicon is the basis of the semiconductor industry worldwide (see Figure 7.2). The silicon has to be ultra-pure for its use in the semiconductor industry. It is purified by zone refining impure silicon extracted from silicon(IV) oxide.

Germanium, as germanium oxide, is found in trace amounts in coal as well as in flue dust of power stations. After extraction and purification it also is used in the semiconductor industry.

Figure 7.10
The orange pigment in carrots is carotene. It is the alternating double and single bonds in the structure of the molecule which give rise to the orange colour.

1 Which is the most abundant element on this planet?

Tin and lead

The heavier elements of Group 4, tin and lead, are predominantly metallic. Tin occurs mainly as cassiterite (tin(IV) oxide) whilst lead occurs as the sulphide ore, galena (PbS). Both metals are ultimately obtained by reduction of their ore with carbon. However, in the case of lead it is necessary to roast the sulphide ore in air first to convert it to the oxide, PbO.

Both tin and lead are heavily used as alloy metals (see Figure 7.4). Tin is used to coat the surface of steel to make tinplate. One use of this material is to make 'tins' for food. Tin is also used in the production of 'flat glass' (see Figure 7.3). In this process molten glass is passed over molten tin.

Lead is used as a screen to protect against radioactivity when radioactive materials are used in industry (see Figure 7.5b). Lead is also used in storage batteries of the lead–acid accumulator type.

2 Write an equation for the conversion of galena to lead(II) oxide by roasting in air. What major pollution problem has to be catered for in this process?

3 Write equations for the reduction using carbon of:
a lead(II) oxide
b tin(IV) oxide.

Compounds of Group 4 elements

The elements of Group 4 are either weakly electronegative non-metals or weakly electropositive metals (see Table 7.1). Therefore, they tend to be relatively unreactive at room temperature.

Reaction with oxygen

Carbon, silicon, germanium and tin are all unaffected by air at room temperature. However, they burn when strongly heated in air or oxygen to form the dioxide, MO_2, in all cases. For example:

$$C(s) + O_2(g) \rightarrow CO_2(g)$$
$$Sn(s) + O_2(g) \rightarrow SnO_2(s)$$
$$\text{tin(IV) oxide}$$

Figure 7.11
a Car exhaust fumes contain carbon monoxide. Carbon monoxide is poisonous since it can replace oxygen in haemoglobin. It is formed from the incomplete combustion of the hydrocarbons in petrol. The amount of carbon monoxide put into the atmosphere has been vastly reduced by the use of catalytic exhaust systems.
b It is important to ensure that any room which has a gas fire in it is well ventilated. If there is not enough air getting into the room then carbon monoxide will be produced!

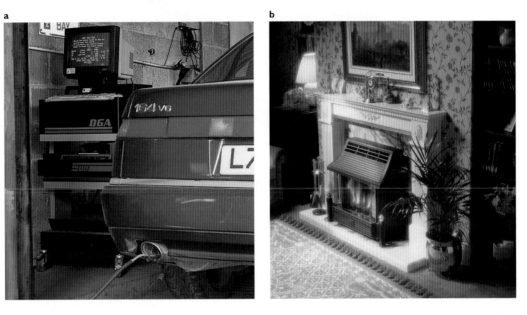

It should be noted that another oxide of carbon, carbon monoxide, which has carbon in the $+2$ oxidation state, also exists (Figure 7.11). It reacts rapidly with oxygen to form CO_2 with carbon in the $+4$ oxidation state. This can be seen in the extraction of iron from haematite; in this process the reducing agent is CO:

$$3CO(g) + Fe_2O_3(s) \rightarrow 2Fe(l) + 3CO_2(g)$$

Lead, on the other hand, tends to form an oxide in the $+2$ oxidation state. If lead is heated gently in air it will form a yellow coating of lead(II) oxide:

$$2Pb(s) + O_2(g) \rightarrow 2PbO(s)$$

The formation of the $+2$ state in PbO is due, in part, to the $6s^2$ electrons of lead being rather inert and difficult to remove during chemical reactions. This is called the **inert pair effect**. This effect gives rise, in part, to the increased stability of the $+2$ state for lead in Group 4 and also the stability of the $+1$ state for the heavier elements of Group 3.

If PbO is heated very strongly in air or oxygen to $450\,°C$ then the lead oxidises further to trilead tetraoxide (red lead), Pb_3O_4 (Figure 7.12).

Figure 7.12
Lead(II) oxide, PbO, is oxidised further to red lead, Pb_3O_4, on heating to $450\,°C$.

The oxides formed by carbon, silicon and germanium are acidic. For example, when carbon dioxide is bubbled through calcium hydroxide (limewater) then a white precipitate of calcium carbonate, a salt of carbonic acid, is produced:

$$CO_2(g) + Ca(OH)_2(aq) \rightarrow CaCO_3(s) + H_2O(l)$$

Figure 7.13
Impurities such as SiO_2 are removed as slag from iron ore by an acid–base reaction.

Figure 7.14
The disordered structure of silicates in soda glass.

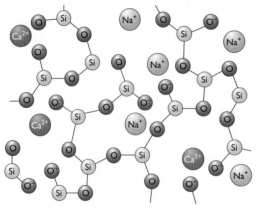

Silicon(IV) oxide is present as impurities in the iron ore and limestone that is put into the blast furnace (Figure 7.13). Calcium oxide, a base produced by the thermal decomposition of calcium carbonate in the blast furnace, reacts with this acidic non-metal oxide, forming calcium silicate (slag):

$$SiO_2(s) + CaO(s) \rightarrow CaSiO_3(l)$$

Similarly germinates are formed by an acid–base reaction. For example, sodium germinate(IV) is made by the reaction of the base Na_2O with the acidic germanium(IV) oxide.

$$GeO_2(s) + Na_2O(s) \overset{heat}{\rightarrow} Na_2GeO_3(s)$$
sodium germinate(IV)

The various types of glass are made from mixtures of silicates:

• soda glass, the common glass used for most bottles and jars, is an irregular mixture of calcium and sodium silicates (Figure 7.14)
• adding small quantities of transition metal ions can give the glass colour (Figure 7.15a)
• adding boron oxide gives borosilicate glass, which can withstand high temperatures (Figure 7.15b).

Figure 7.15
Various uses of glass.

a

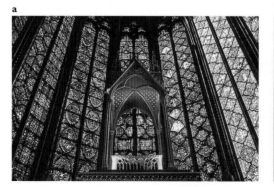

b

Figure 7.16
Uses of glass fibres.
a Structurally in the whole of this yacht.
b As an insulator in roof spaces.
c In communications as optical fibre.

Glass can be made into fibres:

- glass fibre embedded in resin can be used to produce very strong and light structures, for example, boats (Figure 7.16a)
- glass fibre can be used for insulation (Figure 7.16b)
- glass optical fibres can carry more information than copper wires; optical links are now being used for many communication connections (Figure 7.16c).

a

b

c

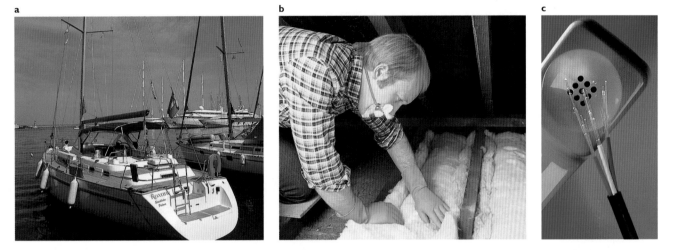

Silicates can also form ordered two-dimensional structures, based on the SiO_4^{4-} tetrahedron (Figure 7.17).

In an acid–base reaction the base sodium oxide will react with germanium(IV) oxide:

$$GeO_2(s) + Na_2O(s) \rightarrow Na_2GeO_3(s)$$
$$\text{sodium germanate(IV)}$$

Figure 7.17
Talc and muscovite (a type of mica) are sheet silicates, made of giant two-dimensional lattices of silicate units, based on SiO_4^{4-} tetrahedra. Clays are formed when some silicon atoms are replaced by aluminium.

talc

muscovite

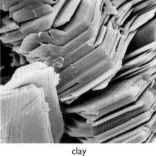

clay

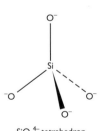

SiO_4^{4-} tetrahedron

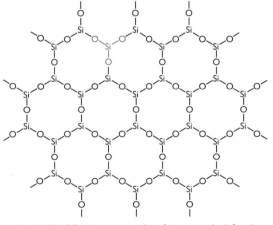

basis of the SiO_2 sheet structure (here Si represents both Si and the O^- above the plane) a basic SiO_4 unit is shown in red

The oxides of tin and lead are both amphoteric and therefore can react with both acid and alkalis.

With acid:

$$SnO_2(s) + 6HCl(aq) \rightarrow \underset{\text{hexachlorostannate(IV) ion}}{[SnCl_6^{2-}](aq)} + 2H_3O^+(aq)$$
$$\text{conc.}$$

$$PbO(s) + 2HCl(aq) \rightarrow PbCl_2(s) + H_2O(l)$$

With fused sodium hydroxide SnO_2 forms the stannate(IV) ion, SnO_3^{2-}, in Na_2SnO_3, whilst with concentrated aqueous sodium hydroxide PbO forms the plumbate(II) ion, $Pb(OH)_4^{2-}$, in $Na_2Pb(OH)_4$.

Reaction with chlorine

Carbon, silicon, germanium and tin all react if heated with chlorine, to form the chloride, MCl_4. These are all covalent liquids with a tetrahedral structure (Figure 7.18). For example:

$$Si(s) + 2Cl_2(g) \rightarrow SiCl_4(l)$$

$$Ge(s) + 2Cl_2(g) \rightarrow GeCl_4(l)$$

The reaction between carbon and chlorine is quite slow. Hence, CCl_4 is usually manufactured by the reaction of chlorine with carbon disulphide:

$$CS_2(l) + 3Cl_2(g) \rightarrow CCl_4(l) + S_2Cl_2(l)$$

This shows the tendency of these Group 4 elements to form compounds in the +4 oxidation state.

However, the +2 oxidation state in tin(II) chloride is quite stable due to the increasing stability of the +2 state as the group is descended; the inert pair effect. Tin(II) chloride is prepared by passing hydrogen chloride over heated tin:

$$Sn(s) + 2HCl(g) \rightarrow SnCl_2(s)$$

Because of the inert pair effect, the most stable oxidation state of lead as a chloride is +2. $PbCl_2$ is formed as the most stable chloride. Adding chloride ions to a soluble lead salt, such as the nitrate, forms this chloride:

$$Pb^{2+}(aq) + 2Cl^-(aq) \rightarrow PbCl_2(s)$$

Lead(II) chloride is soluble in hot water but insoluble in cold water (Figure 7.19a).

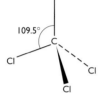

Figure 7.18 (above) CCl_4 is tetrahedral, with bond angles of 109.5°. This shape is due to equal repulsion of the four bonding pairs of electrons present.

4 Write the equation for the reaction of tin with chlorine.

Figure 7.19
a Lead(II) chloride is insoluble in cold water but soluble in hot water.
b Lead(II) iodide is formed as a yellow precipitate when lead(II) ions and iodide ions react together.

5 Write ionic equations to show the formation of the following precipitates:
a lead(II) iodide (Figure 7.19b)
b lead(II) hydroxide
c lead(II) sulphate.

a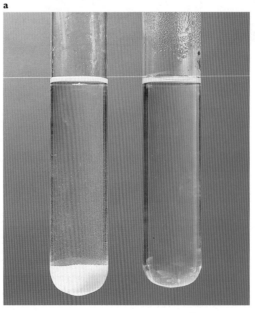

b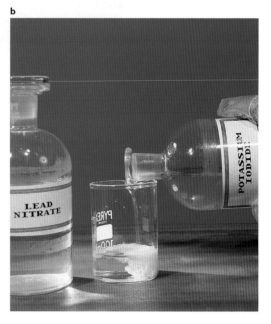

All the chlorides with a +4 oxidation state, except CCl_4, are hydrolysed in water to give the oxide MO_2 and HCl. For example:

$$SiCl_4(l) + 2H_2O(l) \rightarrow SiO_2(s) + 4HCl(aq)$$

Hydrolysis takes place in the case of the covalent chlorides of silicon, germanium and tin because during the mechanism for this reaction the Group 4 atom briefly has five pairs of electrons around it (Figure 7.20). To do this, there must be available d sub-shells for the extra electrons from the water molecule to occupy. This is the addition stage of the addition–elimination reaction that is taking place.

Figure 7.20
The hydrolysis of $SiCl_4$ involves the addition of water and the elimination of HCl.

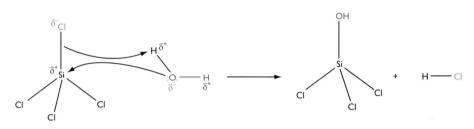

6 a Write an equation to show the hydrolysis of $GeCl_4$.
b Write out the addition–elimination mechanism for the hydrolysis reaction in part **a**.
7 Using standard enthalpy of formation data taken from a data book, confirm that the calculated enthalpy of reaction for the theoretical hydrolysis of CCl_4 is approximately $-60\,kJ\,mol^{-1}$.

enthalpy change of formation ($\Delta H_f^\ominus$) of HCl(aq) = $-165\,kJ\,mol^{-1}$

Tetrachloromethane does not hydrolyse because there are no d sub-shells available in carbon. Thermodynamically this reaction should take place; it has a calculated enthalpy change of $-60\,kJ\,mol^{-1}$ and is therefore energetically feasible. However, the process is kinetically hindered; at room temperature there is no reasonable mechanism for it to occur by.

● **Key skills** **ICT**
• Use a CD-ROM or database to retrieve chemical and physical data related to the Group 4 elements.
• Use a spreadsheet and graphical display to investigate trends and variations in the properties of Group 4 elements.

Number
• Select relevant numeric data related to the Group 4 elements from a database ensuring correct units.

● **Skills task** Use the Internet, as well as other sources, to produce a Powerpoint presentation to describe the discovery, properties and uses of buckminsterfullerene (C_{60}).

CHECKLIST After studying Chapter 7 you should know and understand the following terms.

● **Group 4:** The elements carbon, silicon, germanium, tin and lead.
● **Catenation:** The ability to form long molecules with large numbers of atoms of the same element bonded together.
● **Inert pair effect:** Gives rise to the increased stability of the +2 state towards the lower end of Group 4. It is due to the 'inert' nature of the ns^2 electrons in the heavier elements of the group.

Examination questions

I Consider the following data for the hydrides of the Group 4 elements.

Element	Formula of hydride	Boiling temperature (K)	Relative molecular mass
Carbon	CH_4	109	16
Silicon	SiH_4	161	32
Germanium	GeH_4	185	77
Tin	SnH_4	221	123
Lead	PbH_4	–	–

a Represent the trend in the boiling temperature with relative molecular mass for these hydrides. (2)

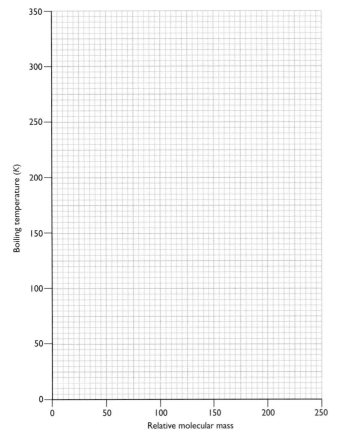

b i Account for the increase in boiling temperature as the relative molecular mass increases. (2)
ii Use your diagram in part **a** to suggest a likely value for the boiling temperature of PbH_4. (1)
iii PbH_4 is actually thermally unstable and unlikely to have the boiling temperature you have suggested. Suggest why PbH_4 is thermally unstable. (2)
c Predict the shape of a molecule of SiH_4 and justify your prediction. (3)

d i Define the term **first ionisation energy**. (2)
ii The successive ionisation energies of silicon are given in the table below.

Number of ionisation	Ionisation energy (kJ mol^{-1})
1	789
2	1580
3	3230
4	4360
5	16 100
6	19 800
7	23 800
8	29 200
9	33 900
10	38 800
11	45 900
12	50 500
13	235 000
14	258 000

Use these data to deduce the electronic structure of silicon. (2)

Edexcel, AS/A level, Module 1, June 2000

2 a In carbon dioxide the carbon atom is joined to each oxygen atom by a double covalent bond, $O=C=O$. Each of the double bonds is made up of one σ bond and one π bond.
Explain, either in words or in clear diagrams, what is meant by a
i σ bond
ii π bond. (2)
b i Write an equation to illustrate the acidic character of carbon dioxide. (1)
ii Give the name of the most basic Group 4 oxide and write an equation for a reaction which illustrates its basic character. (2)
c Lead(IV) oxide can be used to prepare chlorine gas from concentrated hydrochloric acid. The equation for the reaction is:

$$PbO_2(s) + 4HCl(aq) \rightarrow PbCl_2(s) + 2H_2O(l) + Cl_2(g)$$

i Give the function of the lead(IV) oxide in the reaction. (1)
ii Calculate the minimum volume of 12 mol dm^{-3} hydrochloric acid required to react with 5.0 g lead(IV) oxide, and calculate the volume of chlorine gas that would be produced. (The molar volume of a gas at the temperature of the experiment is 24 dm^3.) (4)

London, AS/A level, Module 1, June 1999

8 The d block elements: the transition elements

STARTING POINTS ● The transition elements are known as the d block elements because their final electrons are fed into a d sub-shell.
● The transition elements are some of our most useful metals.
● Many transition elements are used in industry as catalysts.

Figure 8.1
The inner portion of the £2 coin is an alloy of copper (75%) and nickel (25%). The outer portion is an alloy of copper (76%), nickel (4%) and zinc (20%).

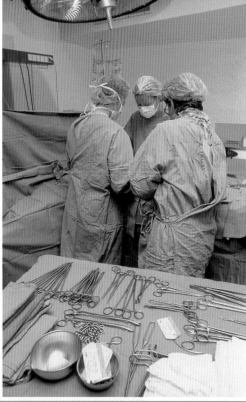

b

Figure 8.2
a (below) Iron is a widely used engineering metal. It is the second most abundant metal in the Earth's crust.
b (right) As well as iron, stainless steel contains a quite large percentage of chromium which prevents the steel from rusting. Nickel in this alloy makes it harder.

a

Figure 8.3
This titanium roof has been used for its high strength and low density properties.

Figure 8.4 (below)
This exhaust system contains a catalyst of platinum or rhodium which speeds up the reactions in the exhaust system. This cuts down the pollutants produced by a car exhaust.

The majority of the metals we use on a day-to-day basis (Figures 8.1–8.4) are found as members of the **d block** or **transition elements**. This block of elements is found between Groups 2 and 3 of the periodic table (Figure 8.5).

The chemistry of the transition elements is quite different from the chemistry of the s block and p block elements. This results from the fact that the electron configurations of the transition elements are associated with the filling of the d sub-shells. The differences between the elements within this block of elements are less marked than within a group of elements in the s block or p block. Moving across the first transition series of elements (those in Period 4), that is the ten elements scandium (atomic number 21) to zinc (atomic number 30), shows many similarities. This means that we can discuss this first transition series as a collection of elements with many features of their chemistry in common. Periods 5 and 6 also contain transition elements.

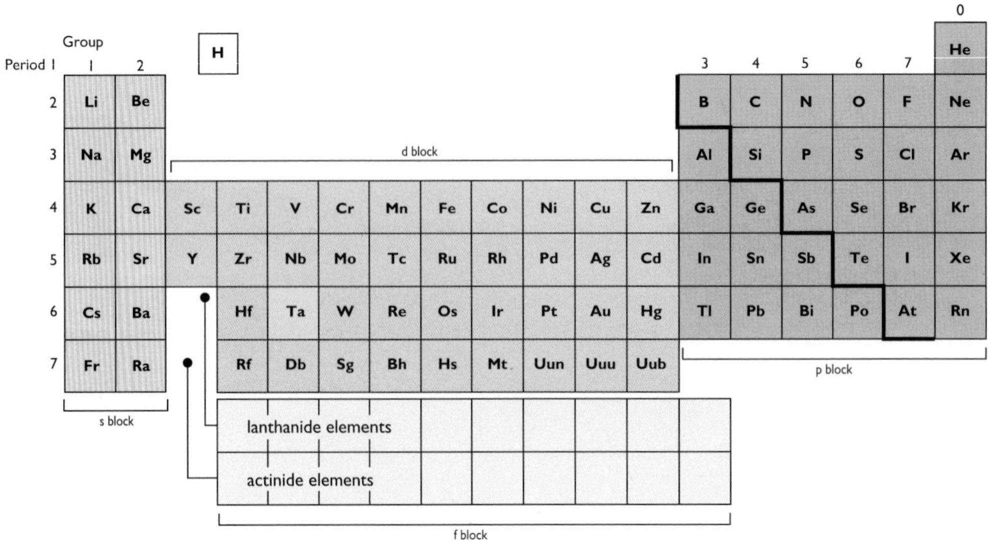

Figure 8.5
The d block elements lie between the s block and p block elements in the periodic table.

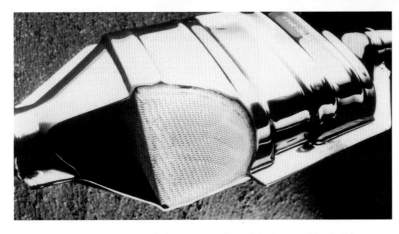

Electron configurations

As we go across the first transition series from scandium to zinc, the elements gain one proton in their nucleus and one extra electron. The electron configurations of the first transition series are given in Figure 8.6. Each additional electron is entering the 3d sub-shell. This extra electron helps to shield the 4s electrons from the increased nuclear charge. As a result, the effective nuclear charge remains fairly constant as the transition series is crossed from left to right. This means that the sizes of the atoms remain fairly similar, as do the first ionisation energies (Table 8.1).

Figure 8.6
Ground state electron configurations for the first transition series from Sc to Zn.
$[Ar] = 1s^2 2s^2 2p^6 3s^2 3p^6$

Element	Atomic number	Electron configuration		3d					4s
Sc	21	$[Ar]3d^1 4s^2$	[Ar]	↑					↑↓
Ti	22	$[Ar]3d^2 4s^2$	[Ar]	↑	↑				↑↓
V	23	$[Ar]3d^3 4s^2$	[Ar]	↑	↑	↑			↑↓
Cr	24	$[Ar]3d^5 4s^1$	[Ar]	↑	↑	↑	↑	↑	↑
Mn	25	$[Ar]3d^5 4s^2$	[Ar]	↑	↑	↑	↑	↑	↑↓
Fe	26	$[Ar]3d^6 4s^2$	[Ar]	↑↓	↑	↑	↑	↑	↑↓
Co	27	$[Ar]3d^7 4s^2$	[Ar]	↑↓	↑↓	↑	↑	↑	↑↓
Ni	28	$[Ar]3d^8 4s^2$	[Ar]	↑↓	↑↓	↑↓	↑	↑	↑↓
Cu	29	$[Ar]3d^{10} 4s^1$	[Ar]	↑↓	↑↓	↑↓	↑↓	↑↓	↑
Zn	30	$[Ar]3d^{10} 4s^2$	[Ar]	↑↓	↑↓	↑↓	↑↓	↑↓	↑↓

Table 8.1
The metallic (atomic) radii, electronegativity and ionisation energies for the elements scandium to zinc.

	Sc	Ti	V	Cr	Mn	Fe	Co	Ni	Cu	Zn
Metallic radius (nm)	0.16	0.15	0.14	0.13	0.14	0.13	0.13	0.13	0.13	0.13
Electronegativity (Pauling)	1.3	1.5	1.6	1.6	1.5	1.8	1.8	1.8	1.9	1.6
First ionisation energy (kJ mol^{-1})	631	658	650	653	717	759	758	737	745	906
Second ionisation energy (kJ mol^{-1})	1235	1310	1414	1592	1509	1561	1646	1753	1958	1733
Third ionisation energy (kJ mol^{-1})	2389	2653	2828	2987	3249	2958	3232	3394	3554	3833
Density (g cm^{-3})	2.99	4.54	5.96	7.20	7.20	7.87	8.90	8.90	8.92	7.14

1 Write the electron configuration for the Cu^+ and Cu^{2+} ions.
2 Why do you think there is a slight decrease in the metallic radius across the first transition series?
3 Write the general equations that represent the first, second and third ionisation energies for the first transition series (use M to represent any of the transition elements).
4 Why do you think that there is a slight increase in the ionisation energies going across the first transition series?

You will notice in Figure 8.6 that neither chromium nor copper follow the pattern of filling of the 3d sub-shell. In the case of chromium, the element completes its occupation of the 3d sub-shell by unpaired electrons, as 3d^5, at the expense of one of the 4s electrons. Copper completes its full 3d sub-shell, as 3d^{10}, also at the expense of one of its 4s electrons. This is due to a certain measure of stability associated with both a half-filled 3d^5 configuration and a full 3d^{10} electron configuration.

What is a transition element?

Because these elements exist within the d block of the periodic table, they are often referred to as 'd block elements' or 'd block metals'. However, this particular definition causes problems when the structure, properties and chemical reactions of these elements are considered. This has lead to the development of a more useful definition:

A transition element is one that forms at least one ion with a partially filled sub-shell of d electrons.

This definition excludes scandium (which is $3d^0$ in all its compounds) and zinc (which is $3d^{10}$ in its compounds). However, even though these elements are not true transition elements, they will be included in a study of these transition elements. This is because there is a similarity to the other transition elements in the chemistry of their compounds.

Based on this definition, the transition elements have many properties in common.

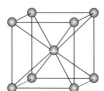

body-centred cubic structure
(e.g. iron, density = 7.87 g cm^{-3})

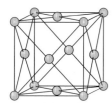

face-centred cubic structure
(e.g. copper, density
= 8.92 g cm^{-3})

Figure 8.7 (above)
Relating different close-packed structures to the density of iron and copper.

Figure 8.8 (below)
The position of the atoms in an alloy before and after a force has been applied and the layers have slipped over one another.

- They are silvery metals apart from copper. However, the slightly increasing electronegativity from scandium to copper means that the elements become slightly less metallic (see Table 8.1).
- They are harder and stronger than the metals of Groups 1 and 2.
- They have much higher melting and boiling points than the metals of Groups 1 and 2. This is because the metal atoms are strongly bonded together by the metallic bond. The strength of this bond is due to the fact that there is a strong electrostatic attraction between the *many* delocalised electrons available to it and the close-packed positive ion cores. This increased strength of metallic bonding and the rather more directional interatomic forces means that they have high enthalpies of fusion and vaporisation.
- The strong metallic bonds also, in part, create the high tensile strength which makes the transition elements of such great value in engineering (Figure 8.2a).
- They have much higher densities than the metals of Groups 1 and 2. This is because the atoms are relatively small and pack closely together (Figure 8.7).
- Most of these elements form alloys, that is solutions of one metal in another. They can be incorporated in each others lattices quite readily because they have similar sizes (Table 8.1). For example, brass is an alloy of copper and zinc. If the metals were of too different a size they would cause the metal to be considerably less malleable and ductile. In the case of transition element alloys this rarely happens. The alloying atoms fit quite well into the crystal lattice, so when a force is applied to the layers of atoms then they will still slide or **'slip'** over one another, but often a greater force is needed. The alloy is stronger than the pure metal (Figure 8.8).

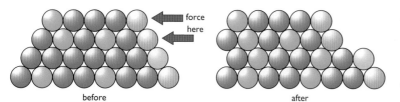

before

after

force here

- They are good conductors of heat.
- They are good conductors of electricity. This is because the 3d and 4s electrons are delocalised and free to move about the structure.
- They are generally less reactive metals than the metals found in the main groups of the periodic table.
- They form a range of brightly coloured compounds and ions (Figure 8.9).
- They form **complex ions**. In this type of ion, molecules or negative ions called **ligands** form dative covalent bonds with the transition metal ion.

Figure 8.9
Solutions of some transition metal compounds.

Chemical properties

Reactivity

Table 8.2 gives the $E^{\ominus}$ values for the $M^{2+}(aq)/M(s)$ systems of the majority of the first transition series.

Table 8.2
Standard electrode potential data for the first transition series.
$M^{2+}(aq) + 2e^- \rightarrow M(s)$

	Sc	Ti	V	Cr	Mn	Fe	Co	Ni	Cu	Zn
$E^{\ominus}$ (V)	−2.03*	−1.2	−1.2	−0.91	−1.18	−0.44	−0.28	−0.26	+0.34	−0.76

$*M^{3+}(aq) + 3e^- \rightarrow M(s)$

All of the $E^{\ominus}$ values for the transition elements are negative, except for copper. From this it can be concluded that under standard conditions these elements, except for copper, will react with dilute sulphuric acid or dilute hydrochloric acid, releasing hydrogen gas. For example:

$$Mn(s) + 2HCl(aq) \rightarrow MnCl_2(aq) + H_2(g)$$

manganese(II)
chloride

Figure 8.10
Stainless steel contains chromium which makes it resistant to corrosion.

5 The values for $E^{\ominus}$ for Na^+/Na and K^+/K are −2.71 V and −2.92 V, respectively. Use this information to explain why the relative reactivity of the transition elements is much less than that of sodium and potassium.

However, very few of these metals react in this way under ordinary conditions. This is due to the formation of an oxide layer which protects the metal surface from further reaction. The half-filled 3d and 4s sub-shells of chromium make it particularly resistant to corrosion. Chromium is alloyed with iron to make a stainless steel for many uses where there is a high risk of corrosion taking place (Figure 8.10).

The transition elements are much less reactive than the s block elements.

Variable oxidation states

When elements of the first transition series lose electrons to form positive ions, it is the 4s electrons that are lost first, followed by the 3d electrons. It should be noted that electrons are never lost from the 3p sub-shells.

The common oxidation states of these elements are shown in Table 8.3 overleaf, with the most common oxidation states given in bold type.

Notice the following points from Table 8.3.

- The oxidation states of scandium, titanium, vanadium, chromium and manganese increase to a maximum value (+7). In all these elements the maximum oxidation state corresponds to the removal of all the 3d and 4s electrons. To remove further electrons to increase the oxidation state beyond this value would involve the removal of 3p electrons, which are much more tightly bound.
- +1, +2 and +3 are among the most common oxidation states for each of the elements Sc to Cu. +3 is the most common oxidation state from Sc to Cr, then +2 is the most common from Mn onwards.
- The transition elements usually show their highest oxidation states when they are combined with oxygen or fluorine, as these are the most electronegative elements. For example, in the case of vanadium the 5+ oxidation state is found in the oxo anion, VO_3^-.
- When transition elements show a very high oxidation state of above +4, they form large ions such as chromate(VI), CrO_4^{2-}, or manganate(VII), MnO_4^-, which are covalent.

- In their lower oxidation states these elements form ionic compounds, whilst in their higher oxidation states they form covalent compounds involving, for example, the transition element covalently bonded to oxygen, as in the case of the MnO_4^- ion.
- Where the completely empty sub-shell ($3d^0$) or completely full sub-shell ($3d^{10}$) arrangements exist the solution of the transition element ion is colourless.
- Where the 3d sub-shells are partially filled the solution of the transition element ion is coloured. For example, V^{2+} ($3d^3$) in aqueous solution is lavender in colour, whilst Cr^{3+} ($3d^3$) is green (Figure 8.11).

Table 8.3
The common oxidation states in compounds of the first transition series elements.

Element	Electron configuration of element	Oxidation state	Colour in aqueous solution	Example
Scandium	$[Ar]3d^14s^2$	+3	Colourless	Sc_2O_3
Titanium	$[Ar]3d^24s^2$	+2		TiO
		+3	Violet	Ti_2O_3
		+4		TiO_2
Vanadium	$[Ar]3d^34s^2$	+2	Lavender	VO
		+3	Blue-green	V_2O_3
		+4	Blue	VO_2
		+5	Yellow	V_2O_5
Chromium	$[Ar]3d^54s^1$	+2		CrO
		+3	Green	Cr_2O_3
		+6	Yellow or orange	$K_2Cr_2O_7$, CrO_3
Manganese	$[Ar]3d^54s^2$	+2	Very pale pink	$MnCl_2$
		+3	Violet	Mn_2O_3
		+4		MnO_2
		+6*	Green	K_2MnO_4
		+7	Purple	$KMnO_4$
Iron	$[Ar]3d^64s^2$	+2	Pale green	FeO
		+3	Red-brown	$FeCl_3$
Cobalt	$[Ar]3d^74s^2$	+2	Pink	CoO
		+3		Co_2O_3
Nickel	$[Ar]3d^84s^2$	+2	Green	NiO
		+3		Ni_2O_3
		+4		NiO_2
Copper	$[Ar]3d^{10}4s^1$	+1*	Colourless	CuCl
		+2	Blue	CuO
Zinc	$[Ar]3d^{10}4s^2$	+2	Colourless	$ZnCl_2$

*Copper(I) and manganese(VI) ions disproportionate in aqueous solution. For example:

$$2Cu^+(aq) \rightarrow Cu(s) + Cu^{2+}(aq)$$

Manganese(VI) disproportionates much more slowly. (For explanation of disproportionation see Chapter 3, page 42.)

Figure 8.11
These transition element ions in solution have 3d sub-shells which are partially filled. They are lavender (V^{2+}) and green (Cr^{3+}), respectively.

It is possible to change the oxidation states of transition elements by using suitable oxidising agents or reducing agents under suitable conditions.

- If an acidified solution containing the ion VO_3^- (aq) (vanadium as vanadium(V)), which is yellow, is warmed with zinc powder (a reducing agent), the solution will change colour. First it will become green (this is a mixture of the yellow VO_3^- and blue of the ion containing V(IV) or V^{4+}), before it becomes blue due to V(IV) only as the ion VO^{2+}. It then changes to green due to V^{3+} ions before becoming lavender/mauve due to the formation of V^{2+} ions (Figure 8.12).

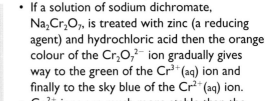

Figure 8.12
The colour changes that can be seen here are due to reduction of the acidified solution of VO_3^- (yellow), through V(IV) (blue), V(III) (green) to V(II) (lavender) by warming with zinc powder.

- If a solution of sodium dichromate, $Na_2Cr_2O_7$, is treated with zinc (a reducing agent) and hydrochloric acid then the orange colour of the $Cr_2O_7^{2-}$ ion gradually gives way to the green of the Cr^{3+}(aq) ion and finally to the sky blue of the Cr^{2+}(aq) ion.
- Co^{2+} ions are much more stable than the Co^{3+} ions in solution. However, a powerful oxidising agent such as hydrogen peroxide will oxidise the alkaline solution to Co^{3+}. Also hydrogen peroxide will oxidise the green Cr^{3+} ions in alkaline solution to Cr^{6+} in the yellow CrO_4^{2-} ion.

Complex ions

A further characteristic of transition elements is their tendency to form **complex ions**. Complex ions are composed of a central metal ion surrounded by anions or molecules called **ligands** (from the Latin 'to bind') (Table 8.4).

Table 8.4
Some common ligands.

Type of ligand		Formula
Anions	– halides	F^-, Cl^-, Br^-, I^-
	– sulphide	S^{2-}
	– nitrite	NO_2^-
	– hydroxide	OH^-
	– cyanide	CN^-
Molecules	– water	H_2O
	– ammonia	NH_3

Usually when metal atoms form compounds they do so by forming ionic bonds. However, complex ions are an exception to this since in these ions the metals form **dative** or **coordinate** bonds; complex ions are often called **coordination compounds**. Dative or coordinate bonds are **covalent bonds** in which both the bonding electrons come from the same atom or ion (see *Introduction to Advanced Chemistry*, Chapter 5, page 58). In complex ions the electron pairs from the ligands are donated into vacant orbitals of the transition element ions. The ligands are behaving as **Lewis bases**. A Lewis base is defined as an electron pair donor. The transition metal ions are behaving as electron pair acceptors or **Lewis acids**. A Lewis acid is defined as an electron pair acceptor. The number of dative or coordinate bonds from ligands to the central metal ion is known as the **coordination number**.

The most common coordination number for transition elements is six, as in these **cationic complexes** (positively-charged complexes): $[Fe(H_2O)_6]^{3+}$, $[Cr(H_2O)_6]^{2+}$, $[Ni(NH_3)_6]^{2+}$.

Table 8.5
Some complex ions with coordination numbers of 4 and 2.

Complex ion	Coordination number
$[CuCl_4]^{2-}$	4
$[Zn(NH_3)_4]^{2+}$	4
$[Ag(CN)_2]^-$	2
$[Ag(NH_3)_2]^+$	2

Examples of complex ions, both cationic and **anionic** (negatively-charged complexes), with coordination numbers of 4 and 2 are not uncommon (Table 8.5).

In the formulae of complex ions, the metal ion is written first, followed by the ligands. The total charge on the complex ion is the sum of the charge on the central metal ion and the charges on the ligands. When the ligands are neutral molecules, such as ammonia and water, the charge on the complex ion is that of the central metal ion. However, if the ligands are anionic then the charge will be different from that of the central metal ion, as in the complex ion $[CuCl_4]^{2-}$ (see Table 8.5).

6 What is the charge on the transition metal ions in the complexes shown in Table 8.5?

Ligands such as those shown in Table 8.4 form only one bond with the central metal ion and are said to be **unidentate**. Some ligands, however, are able to form two or more coordinate bonds with the central metal ion and are said to be **polydentate** (Figure 8.13). The resulting complexes are called **chelates** or **chelate complexes**. One of the most common polydentate ligands is the compound ethylenediaminetetraacetic acid, **edta**, which forms the edta^{4-} anion (Figure 8.13c). It has six available donor pairs of electrons and is a hexadentate ligand. This allows it to wrap itself around transition metal ions, as well as other metal ions, for example Ca^{2+} and Mg^{2+}, and form very stable complexes, such as that shown in Figure 8.14.

Figure 8.13
a The ethanedioate ion acts as a bidentate ligand.
b The 1,2-diaminoethane molecule acts as a bidentate ligand. This particular ligand forms donor bonds via the lone pairs on the N atoms in the NH$_2$ groups.
c The edta^{4-} ion acts as a hexadentate ligand, using lone pairs on both N atoms and O$^-$ groups.

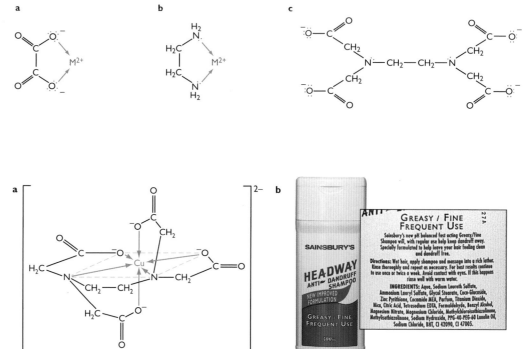

Figure 8.14
a The [Cu(edta)]$^{2-}$ complex.
b Shampoos contain the edta^{4-} anion. It helps soften the water by complexing with the Ca^{2+} ions found in hard water.

If the edta complex is to be used in a titration to quantitatively determine the concentration of a metal ion then there is a need to buffer the solutions used. For example, in the determination of the total hardness of water, that is, the determination of the total amount of calcium and magnesium present in water, the edta titration is carried out at a pH of 10 using an NH$_3$/NH$_4$Cl buffer (see Chapter 5, page 73).

Biological molecules that contain transition metal ions involved in complexing with polydentate ligands include vitamin B$_{12}$ and haemoglobin (Figure 8.15).

Figure 8.15
a This donor is donating blood, which contains haemoglobin.
b Part of the oxyhaemoglobin molecule. It consists of an Fe^{2+} ion at the centre of a ring structure. The ring is a tetradentate haem group. Also associated with the central ion is the protein, globin, and an oxygen molecule.

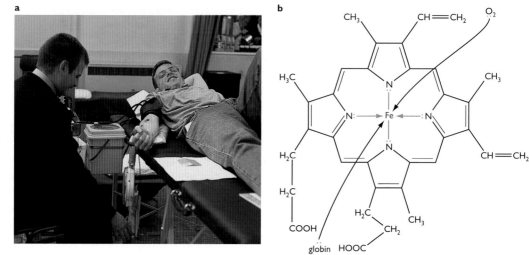

Naming of complexes

As you would expect there is a systematic way of naming complexes, which is recognised internationally. The naming process uses the following four rules.

- Identify the coordination number of the particular ligands surrounding the central metal ion. The following prefixes may be used: mono- (1), di- (2), tri- (3), tetra- (4), penta- (5), hexa- (6).
- Identify the ligands alphabetically. Use the ending **-o** for anions; for example, fluoro (F^-), chloro (Cl^-), cyano (CN^-), hydroxo (OH^-). Neutral ligands include ammonia, water and carbon monoxide; the name **ammine** is used for ammonia, **aqua** for water and **carbonyl** for carbon monoxide. Other neutral ligands keep their usual names.
- Identify the central metal ion and name it. Where the complex ion is neutral overall or positively charged, use the English name for the metal. If the overall charge on the complex ion is negative then the name ends in 'ate'. Note that some metals are given their Latin names; for example, ferrate (iron), plumbate (lead), cuprate (copper).
- Identify the oxidation state of the central metal ion and show this in roman numerals in brackets.

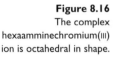
7 Name the following complex ions and state the coordination number of the complexes:
a $[Co(NH_3)_6]^{2+}$
b $[Cu(H_2O)_6]^{2+}$
c $[Fe(OH)(H_2O)_5]^{2+}$
d $[CrCl_2(H_2O)_4]^+$

Examples of some complexes with their names:

$[Co(H_2O)_6]^{2+}$	hexaaquacobalt(II) ion
$[CuCl_4]^{2-}$	tetrachlorocuprate(II) ion
$[CoCl_4]^{2-}$	tetrachlorocobaltate(II) ion
$[FeCN_6]^{4-}$	hexacyanoferrate(II) ion
$[CrCl_2(NH_3)_4]^+$	dichlorotetraamminechromium(III) ion

Shapes of complexes

The coordination number of the central metal atom or ion determines the shape of a complex. The most common coordination number for complex ions is 6. Complexes that have this coordination number usually have an **octahedral** arrangement of these relatively small ligands such as ammonia and water around the central metal atom or ion. The six electron pairs around the central metal atom or ion are repelled as far apart as possible from each other, and so give the shape shown in Figure 8.16.

Figure 8.16
The complex hexaamminechromium(III) ion is octahedral in shape.

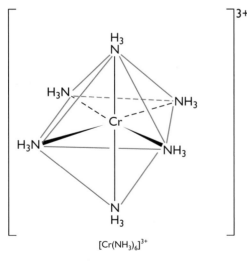

$[Cr(NH_3)_6]^{3+}$

If the six ligands are not identical then ***cis–trans* geometrical isomers** can occur. For example, in the tetraamminedichlorocobalt(III) ions shown in Figure 8.17 overleaf, the *cis* form has two chloride ligands adjacent to one another, whereas in the *trans* form the chloride ligands are opposite one another. (This should not be confused with the *cis–trans* isomerism observed in alkenes where the isomers are formed around the $C=C$ bond.)

The presence of the two forms of the complex ion in the compound tetraamminedichlorocobalt(III) chloride ($[CoCl_2(NH_3)_4]^+Cl^-$) gives rise to two different coloured crystals. The *cis* form is blue-violet whilst the *trans* form is green.

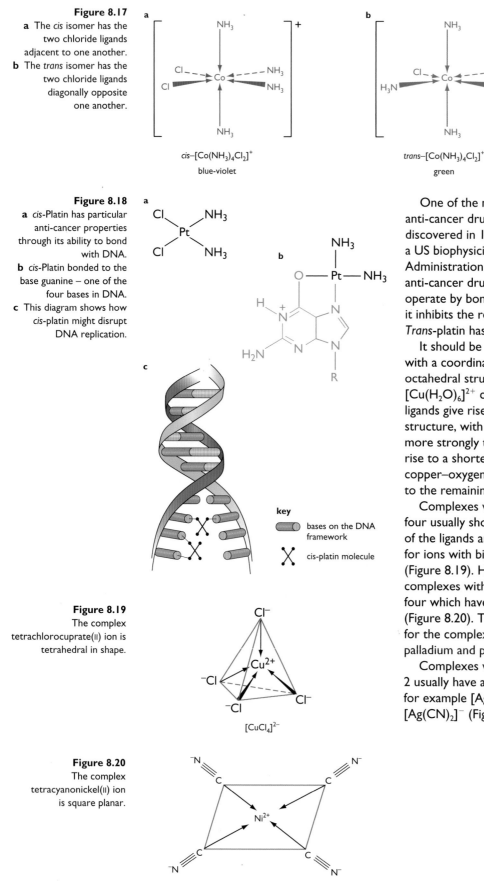

Figure 8.17
a The *cis* isomer has the two chloride ligands adjacent to one another.
b The *trans* isomer has the two chloride ligands diagonally opposite one another.

cis–[Co(NH₃)₄Cl₂]⁺
blue-violet

trans–[Co(NH₃)₄Cl₂]⁺
green

Figure 8.18
a *cis*-Platin has particular anti-cancer properties through its ability to bond with DNA.
b *cis*-Platin bonded to the base guanine – one of the four bases in DNA.
c This diagram shows how *cis*-platin might disrupt DNA replication.

key
bases on the DNA framework
cis-platin molecule

Figure 8.19
The complex tetrachlorocuprate(II) ion is tetrahedral in shape.

[CuCl₄]²⁻

Figure 8.20
The complex tetracyanonickel(II) ion is square planar.

One of the most effective and successful anti-cancer drugs is *cis*-platin (Figure 8.18), discovered in 1961 by Barnett Rosenberg, a US biophysicist. The US Food and Drug Administration approved it for use as an anti-cancer drug in 1978. It is believed to operate by bonding to DNA in such a way that it inhibits the replication of cancerous cells. *Trans*-platin has no anti-cancer properties.

It should be noted that not all complexes with a coordination number of 6 give a regular octahedral structure. For example, in the $[Cu(H_2O)_6]^{2+}$ complex ion, the arrangement of ligands give rise to a distorted octahedral structure, with four of the water ligands held more strongly than the other two. This gives rise to a shortening of four of the copper–oxygen dative bonds compared to the remaining two.

Complexes with a coordination number of four usually show a tetrahedral arrangement of the ligands around the central metal atom for ions with bigger ligands, such as Cl^- (Figure 8.19). However, there are some complexes with a coordination number of four which have a square planar structure (Figure 8.20). These are particularly common for the complexes of nickel, as well as palladium and platinum.

Complexes with a coordination number of 2 usually have a linear arrangement of ligands; for example $[Ag(NH_3)_2]^+$, $[Ag(S_2O_3)_2]^{3-}$ and $[Ag(CN)_2]^-$ (Figure 8.21).

Figure 8.21
a The diamminesilver(i) complex ion is linear. It is used in Tollen's reagent to test for the presence of aldehydes (see Chapter 9, page 141).
b The [Ag(CN)$_2$]$^-$ complex ion is present in the electrolyte solution used in silver plating of frames like that shown.

$$[\text{H}_3\text{N} \longrightarrow \text{Ag} \longleftarrow \text{NH}_3]^+$$

[Ag(NH$_3$)$_2$]$^+$

8 Draw diagrams to represent the structure of the following complexes and state the bond angles present:
a [Co(H$_2$O)$_6$]$^{2+}$
b [CoCl$_4$]$^{2-}$

Colours of transition element complex ions

Generally if light in the visible region of the spectrum hits a solid substance, part is absorbed, part will be transmitted (if the substance is transparent) and part may be reflected from the surface.

- If all the incident light is absorbed then the colour of the substance appears 'black'.
- If only light of certain wavelengths is absorbed the compound will appear coloured.
- If none of the white light is absorbed then the substance will appear 'colourless'.
- If all the incident light is reflected then the colour of the substance appears 'white'.

The vast majority of transition element compounds are coloured, both in solution and in the solid state (see Table 8.3 and Figure 8.22). The colour of these compounds can often be related to incompletely filled 3d sub-shells in the transition element ion present.

Figure 8.22
The transition element ions Co^{2+} and Ni^{2+} in the solid state and in solution have 3d sub-shells that are partially filled. They are pink and green in colour, respectively.

Figure 8.23
The shapes of the 3d atomic orbitals that make up the 3d sub-shell. They are all degenerate but have different orientations in space relative to one another.

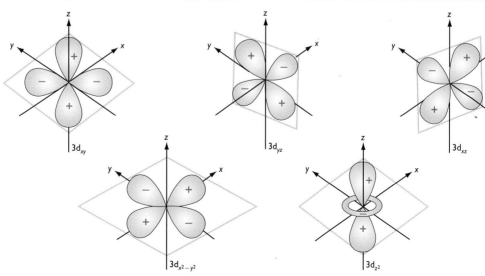

In an isolated transition element atom, the five 3d sub-shells are all oriented differently in space relative to one another (Figure 8.23) whilst all being of the same energy. That is, they are **degenerate**. In a complex ion, since the five 3d sub-shells are orientated differently in space relative to one another, an electron (or electrons) close to an incoming ligand will be repelled and hence the energy of such sub-shells is raised relative to the others. They are **non-degenerate**. The degeneracy of the 3d sub-shells is therefore lost and this gives rise to the situation shown in Figure 8.24 for the octahedral $[Cu(H_2O)_6]^{2+}$ complex ion.

Figure 8.24
The splitting of the 3d sub-shells in the Cu^{2+} ion in the octahedral $[Cu(H_2O)_6]^{2+}$ complex.

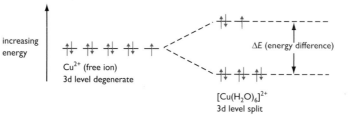

It can be seen from Figure 8.24 that two of the 3d sub-shells are raised in energy relative to the other three. The energy difference between the two sets of sub-shells is ΔE. ΔE is related to the frequency of the radiation necessary to cause an electron to be excited into the higher of the energy levels by the Planck equation:

$$\Delta E = h\upsilon$$

In the case of transition element ions, the frequency of the radiation absorbed is part of the visible region of the spectrum. The colour of the hexaaquacopper(II) ions is blue. This is because red light of the appropriate frequency is absorbed and the colour of the particular ion is complementary to that (Figure 8.25).

Different ligands affect the splitting of the 3d sub-shell. This explains, therefore, why the $[Cu(H_2O)_6]^{2+}(aq)$ ion is blue whereas the $[Cu(NH_3)_4(H_2O)_2]^{2+}(aq)$ ion is a very intense deep blue (Figure 8.26).

Figure 8.25
In the colour wheel the complementary colours are opposite one another.

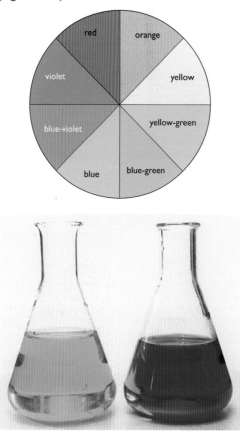

Figure 8.26
Different ligands affect the 3d splitting. The left-hand flask contains $[Cu(H_2O)_6]^{2+}(aq)$, whereas the right-hand flask contains $[Cu(NH_3)_4(H_2O)_2]^{2+}(aq)$.

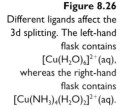

The ammonia ligands in the ammonia complex cause a bigger 3d sub-shell splitting than the water ligands do. This means that light of higher energy and frequency, and hence lower wavelength, is absorbed (Figure 8.27).

You will notice that the absorption spectra in Figure 8.27 show broad bands. This is in contrast to the absorption spectrum of a metal vapour, such as sodium, which is a series of dark lines that have sharply defined frequencies. The reasons for this difference are to do with the very simple environment experienced by gas phase sodium ions in a vapour compared to the situation experienced by complex ions in solution. The complex ions possess additional energies due to the vibration and rotation of the whole complex. The result of this is that there is a broadening of the bands in the absorption spectra of complex ions in solution.

Figure 8.27
The spectra show that $[Cu(NH_3)_4(H_2O)_2]^{2+}(aq)$ absorbs at lower wavelengths than $[Cu(H_2O)_6]^{2+}(aq)$.

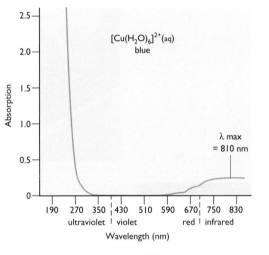

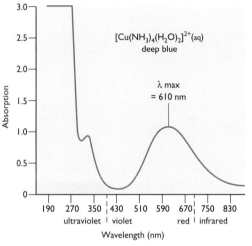

Figure 8.28
The student using this colorimeter is measuring the concentration of iron in solution.

It can be seen from Figure 8.27 that it is possible to decide the colour of a transition element complex by examining its absorption spectrum.

If we wish to measure the concentration of a transition element ion in solution then providing that the complex is coloured, the intensity of the colour can be measured using a colorimeter (Figure 8.28). The depth of colour of a solution is proportional to the concentration of the coloured substance present, at least for dilute solutions. If we pass light of a specific wavelength through a sample of a coloured substance then what we measure using a colorimeter is the amount of light transmitted through the sample.

For example, in the determination of the concentration of iron as iron(III), say in river water, potassium thiocyanate is added and a red colour is produced (see page 123). In order to convert the reading given by the colorimeter into the concentration of iron in the river water it is necessary to use a 'calibration curve'. The calibration curve in this case can be plotted by obtaining colorimeter readings for solutions of known concentrations of iron(III) ions.

Use of transition elements in pigments

Many transition element compounds are used as pigments because they are brightly coloured. For example, one of the best blue pigments ever made is copper phthalocyanine, better known as Monastral blue. This blue compound has the planar structure shown in Figure 8.29. You will notice that four of the nitrogen atoms act as ligands. Because of the extensive delocalised system it has a lower energy and so this compound is very stable. It is very resistant to heat, subliming at 580 °C, and also has excellent fastness to light. Other phthalocyanines have been made and cover the colour range from bright blue to green.

Figure 8.29
a The planar structure of copper phthalocyanine.
b The blue paint used on the Mallard steam train contains Monastral blue. It is also widely used to colour plastics, in printing inks and in enamels.

Figure 8.30
These yellow road markings contain a further transition element-based pigment – chrome yellow. Chrome yellow is lead chromate(VI). It is also used in printing inks. Chrome yellow is one of a range of metal chromates which are used to produce differing shades of yellow.

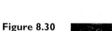

Properties of complex ions

The properties of complex ions as a whole are different from the simple ions that they are made of. For example, it is not generally possible to obtain positive chemical tests for the components of a complex ion due to the fact that the central metal ion is complexed with ligands by dative *covalent* bonds. However there are, in some specific cases, chemical tests to show the presence of complexed ions. For example, if copper(II) sulphate solution is added to a solution containing the hexacyanoferrate(II) complex ion, a brown precipitate of copper hexacyanoferrate(II) is formed:

$$2Cu^{2+}(aq) + [Fe(CN)_6]^{4-}(aq) \rightarrow Cu_2Fe(CN)_6(s)$$

Ligand exchange

A ligand may substitute for another ligand in a complex ion. This is known as a **ligand exchange** or **ligand substitution** reaction and will often produce a change in colour. If the ligands are uncharged and of a similar size there will be no change in the coordination number. For example:

$$[Co(H_2O)_6]^{2+}(aq) + 6NH_3(aq) \rightleftharpoons [Co(NH_3)_6]^{2+}(aq) + 6H_2O(l)$$
pink   pale yellow

$$[Cu(H_2O)_6]^{2+}(aq) + 4NH_3(aq) \rightleftharpoons [Cu(NH_3)_4(H_2O)_2]^{2+}(aq) + 4H_2O(l)$$
pale blue   deep blue

Figure 8.31
Sometimes ligand substitution reactions create a change in coordination number as well as a change in colour of the solution.
a The pink solution of $[Co(H_2O)_6]^{2+}(aq)$ is replaced by the blue solution of $[CoCl_4]^{2-}(aq)$.
b The blue solution of $[Cu(H_2O)_6]^{2+}(aq)$ is replaced by the green solution of $[CuCl_4]^{2-}(aq)$.

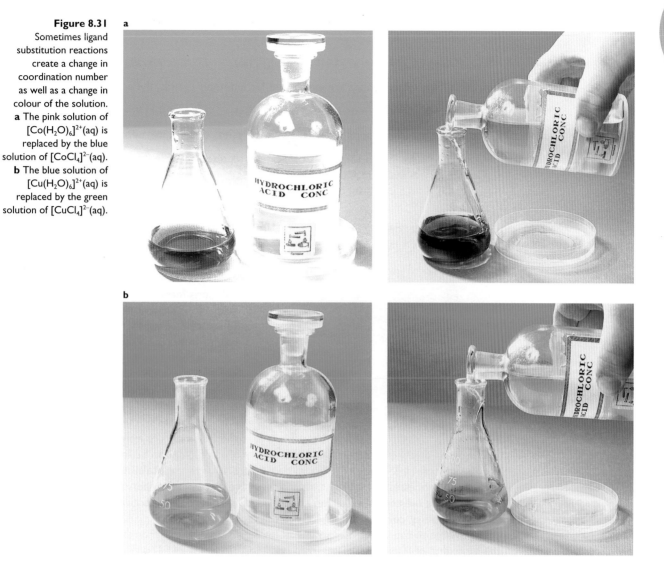

Ligand exchange reactions can sometimes cause a change in the coordination number of the complex as well as a change in colour (Figure 8.31). For example, using concentrated HCl:

$$[Co(H_2O)_6]^{2+}(aq) + 4Cl^-(aq) \rightleftharpoons [CoCl_4]^{2-}(aq) + 6H_2O(l)$$
$$\text{pink} \qquad\qquad\qquad\qquad \text{blue}$$

In this reaction the coordination number has changed from 6 to 4. This is because the Cl^- ligand is larger than the uncharged water ligands. This has also created a change in the shape of the complex from octahedral to tetrahedral. This sort of process can also be seen in the case of the chloride substitution reaction with $[Cu(H_2O)_6]^{2+}(aq)$ (again using concentrated HCl):

$$[Cu(H_2O)_6]^{2+}(aq) + 4Cl^-(aq) \rightleftharpoons [CuCl_4]^{2-}(aq) + 6H_2O(l)$$
$$\text{blue} \qquad\qquad\qquad\qquad \text{green}$$

Figure 8.32 (below)
This is a chemical test used to show the presence of iron(III) ions in solution.

A useful ligand exchange reaction is that used as the chemical test for the presence of iron(III) ions in solution. A small amount of potassium thiocyanate is added to a solution that may contain iron(III) ions. If a red coloration is produced then it is certain that iron(III) ions are present (Figure 8.32):

$$[Fe(H_2O)_6]^{3+}(aq) + SCN^-(aq) \rightleftharpoons [Fe(SCN)(H_2O)_5]^{2+}(aq) + H_2O(l)$$
$$\text{yellow/brown} \qquad\qquad\qquad \text{red}$$

The reduction potential of a transition element ion depends on the ligands surrounding it. For example:

$$[Co(H_2O)_6]^{3+}(aq) + e^- \rightleftharpoons [Co(H_2O)_6]^{2+}(aq) \qquad E^\ominus = +1.81\,V$$

However, if the water ligands are exchanged for ammonia ligands, the potential changes:

$$[Co(NH_3)_6]^{3+}(aq) + e^- \rightleftharpoons [Co(NH_3)_6]^{2+}(aq) \qquad E^\ominus = +0.11\,V$$

Complexing the central metal ion with different ligands changes its redox properties:

- $[Co(NH_3)_6]^{2+}(aq)$, which contains Co^{2+}, can be oxidised to Co^{3+} by air oxidation
- $[Co(NH_3)_6]^{3+}$ is stable in aqueous solution
- $[Co(H_2O)_6]^{3+}$ is a powerful oxidising agent and will oxidise iodide ions to iodine in an acidic solution.

Figure 8.33
Tinned fruit has quite a short shelf-life due to corrosion of the inside of the tin.

This change in $E^\ominus$ can have quite a marked effect on the corrosion that takes place in food cans made from steel coated in tin (Figure 8.33). In cans coated with tin and containing fruit, the Sn^{2+} ions form complexes with the carboxylic acid anions present in the fruit. This happens to such an extent that there is a marked change in the Sn^{2+}/Sn $E^\ominus$ value. The Sn^{2+}/Sn $E^\ominus$ value becomes much more negative than that of Fe^{2+}/Fe, the tin is oxidised in preference to the iron and so protects the iron from corrosion. Once all the tin has complexed with the anions of the carboxylic acids from the fruit then the iron will corrode.

For ligand displacement reactions, the stability of a complex can be expressed in terms of the equilibrium constant. These equilibrium constants are known as **stability constants**. However, since these processes are stepwise, that is they take place in a number of steps as the individual ligands around the central metal ion are replaced, it is more usual to give what is called the **overall stability constants** (K_{stab}). Since the range of values for K_{stab} is so large, ranging from the very large to the very small, a value of log K_{stab} is usually used. For example, the K_{stab} value for the formation of $[Co(NH_3)_6]^{2+}(aq)$ from $[Co(H_2O)_6]^{2+}(aq)$ is $24\,547\,dm^{18}\,mol^{-6}$, whilst the log K_{stab} is 4.39.

$$[Co(H_2O)_6]^{2+}(aq) + 6NH_3(aq) \rightleftharpoons [Co(NH_3)_6]^{2+}(aq) + 6H_2O(l)$$

$$K_{stab} = \frac{[[Co(NH_3)_6]^{2+}(aq)]}{[[Co(H_2O)_6]^{2+}(aq)]\,[NH_3(aq)]^6}$$

$$= 24\,547\,dm^{18}\,mol^{-6}$$

$$\log K_{stab} = 4.39$$

Values of log K_{stab} can be used to compare the stability of any two complexes formed from different ligands. Table 8.6 shows values of log K_{stab} for some complexes, relative to the aqua complex.

Table 8.6
Some log K_{stab} values.

Complex ion	log K_{stab}
$[Co(NH_3)_6]^{2+}$	4.39
$[CuCl_4]^{2-}$	5.62
$[Ni(NH_3)_6]^{2+}$	8.01
$[Ni(en)_3]^{2+}$	18.30
$[Ni(edta)]^{2-}$	18.60
$[Fe(edta)]^-$	25.10

Generally, greater values of log K_{stab} indicate that the equilibrium process involved is further over to the right, and the complex on the right-hand side of the equation is more stable than the aqua complex on the left-hand side of the equation. It can be seen from Table 8.6 that the polydentate ligands form more stable complexes than the monodentate ligands. The reason for this is that their formation there is an increase in the entropy caused by a single polydentate ligand replacing several water molecules. The ligand substitution reaction therefore increases the number of separate particles present and therefore increases the entropy of the system. For the significance of entropy see Chapter 4 (pages 54–7).

9 Use the information shown in Table 8.6 to state:
   a which of the complexes is the most stable and which is the least stable
   b which of the chelating ligands forms the most stable complexes.
   Explain your answers.

Redox reactions involving transition element ions

Figure 8.34
Acidified potassium manganate(VII) solution can be used in redox titrations to analyse chemical samples.

We have already seen earlier in this chapter that transition element ions show variable oxidation states (see Table 8.3). Some of these changes in oxidation states can be quite useful.

Redox titrations are used throughout the chemical industry (Figure 8.34). For example, the amount of iron in a steel sample can be determined using acidified potassium manganate(VII) solution.

The half-equation for the reduction of potassium manganate(VII) in acidic solution is:

$$MnO_4^-(aq) + 8H^+(aq) + 5e^- \rightarrow Mn^{2+}(aq) + 4H_2O(l) \qquad \text{(equation 1)}$$

The half-equation for the oxidation of iron(II) is:

$$Fe^{2+}(aq) \rightarrow Fe^{3+}(aq) + e^- \qquad \text{(equation 2)}$$

To produce an overall ionic equation for the oxidation of iron(II) by potassium manganate(VII) equation 2 has to be multiplied by 5 to give:

$$5Fe^{2+}(aq) \rightarrow 5Fe^{3+}(aq) + 5e^- \qquad \text{(equation 3)}$$

The overall ionic equation for this oxidation process is obtained by adding equations 1 and 3 to give:

$$MnO_4^-(aq) + 5Fe^{2+}(aq) + 8H^+(aq) \rightarrow Mn^{2+}(aq) + 5Fe^{3+}(aq) + 4H_2O(l) \qquad \text{(equation 4)}$$

Figure 8.35
The colour change in the $Fe^{2+}/KMnO_4$ titration is colourless to pink.

To carry out this analysis, the steel sample must be dissolved in dilute sulphuric acid. This renders the Fe^{2+} ions stable in aqueous solution so that they can be titrated with potassium manganate(VII). Because the potassium manganate(VII) is purple and the product of the titration is colourless, there is no need for an indicator because the potassium manganate(VII) acts as its own indicator (Figure 8.35).

10 Use the following standard electrode potential data to calculate the overall $E_{cell}^{\ominus}$ for the oxidation of $Fe^{2+}(aq)$ to $Fe^{3+}(aq)$ using an acidic solution of $KMnO_4$.

$MnO_4^-(aq) + 8H^+(aq) + 5e^- \rightarrow Mn^{2+}(aq) + 4H_2O(aq)$ $E^{\ominus} = +1.51\,V$
$Fe^{2+}(aq) + e^- \rightarrow Fe^{3+}(aq)$ $E^{\ominus} = +0.77\,V$

Example A sample of steel with a mass of 1.70 g was dissolved in dilute sulphuric acid in a volumetric flask and the solution made up to 250 cm³ with pure water. Several 25.00 cm³ samples of this solution containing Fe^{2+} ions were titrated with 0.02 mol dm⁻³ potassium manganate(VII) solution. 26.05 cm³ of potassium manganate(VII) solution was required for complete reaction. Calculate the percentage of iron in the steel sample. (A_r: Fe = 56)

$$\text{moles of } MnO_4^- = \frac{26.05 \text{ cm}^3}{1000} \times 0.02 \text{ mol dm}^{-3}$$

$$= 5.21 \times 10^{-4} \text{ mol}$$

From equation 4 on page 125, 5 mol of Fe^{2+} react with 1 mol of MnO_4^-. Therefore, the amount of Fe^{2+} in 25.00 cm³ of sample solution can be calculated.

$$\text{amount of } Fe^{2+} = 5.21 \times 10^{-4} \text{ mol} \times 5$$
$$= 2.61 \times 10^{-3} \text{ mol}$$

Hence the total amount of moles of Fe^{2+} present in 250 cm³ of the sample solution can be found.

$$\text{total amount of } Fe^{2+} = 2.61 \times 10^{-3} \text{ mol} \times 10$$
$$= 2.61 \times 10^{-2} \text{ mol}$$
$$\text{the mass of iron} = 2.61 \times 10^{-2} \text{ mol} \times 56 \text{ g mol}^{-1}$$
$$= 1.46 \text{ g}$$

$$\text{\% of iron in the sample} = \frac{1.46 \text{ g}}{1.70 \text{ g}} \times 100$$

$$= 85.8\%$$

11 1.00 g of a powdered sample of iron tablets (hydrated iron(II) sulphate) is dissolved in a little pure water. This solution is made up to 100 cm³ in a volumetric flask with dilute sulphuric acid. 20.20 cm³ of a 0.0100 mol dm⁻³ potassium manganate(VII) solution is needed to give the first permanent pink colour to the solution.

Calculate the percentage of iron in the iron tablets assuming that all the iron is present as Fe^{2+}.

12 Use your knowledge of oxidation states, developed in your earlier studies (see *Introduction to Advanced Chemistry*, Chapter 10, page 114), to work out the oxidation numbers for the chromium metal ions shown in the production of sodium chromate.

Another redox titration can be used to estimate the amount of iron in solution. This involves the use of acidified potassium dichromate(VI):

$$Cr_2O_7^{2-}(aq) + 6Fe^{2+}(aq) + 14H^+(aq) \rightarrow 2Cr^{3+}(aq) + 6Fe^{3+}(aq) + 7H_2O(l)$$

However, in this case an indicator is required. The indicator normally used is barium diphenylamine sulphonate. Since the indicator is sensitive to the presence of iron(III) ions the solution is treated before the titration with some phosphoric(V) acid, with which the iron(III) ions form a fairly stable complex.

Substances such as sodium chromate(VI) (Na_2CrO_4) can be made in the laboratory by the oxidation of a chromium(III) salt such as $CrCl_3$ by an alkaline solution of hydrogen peroxide (an oxidising agent):

$$2CrCl_3(aq) + 4NaOH(aq) + 3H_2O_2(aq) \rightarrow 2Na_2CrO_4(aq) + 2H_2O(l) + 6HCl(aq)$$
$$\text{green} \qquad\qquad\qquad\qquad\qquad\qquad \text{yellow}$$

Sodium chromate can be used as an indicator in the titration used to estimate the amount of chloride ions in solution. The colour change at the end-point of the titration is yellow to red. The red colour is generated by the formation of a precipitate of Ag_2CrO_4, which is formed as a result of the reaction:

$$2Ag^+(aq) + CrO_4^{2-}(aq) \rightarrow Ag_2CrO_4(s)$$

Other metal chromates are used extensively as pigments and in printing inks (see page 122).

Note that the addition of an acid to an aqueous solution of a chromate results in the formation of the dichromate anion, $Cr_2O_7^{2-}$, a condensed ion, and the elimination of water. The colour of the solution changes from the yellow of the chromate ion to orange of the dichromate ion. The reaction is easily reversed by the addition of the OH^- ion.

$$2CrO_4^{2-}(aq) + 2H^+(aq) \rightarrow Cr_2O_7^{2-}(aq) + H_2O(l)$$
$$\text{yellow} \qquad\qquad\qquad\qquad \text{orange}$$

Acidity of transition metal ions in solution

Transition metal ions behave as weak acids in aqueous solution. This is due to the relatively small metal cation polarising the water ligands and so causing the loss of a proton. For example, a very weakly acidic solution is created with the $[Cu(H_2O)_6]^{2+}$ ion in aqueous solution:

$$[Cu(H_2O)_6]^{2+}(aq) + H_2O(l) \rightleftharpoons [Cu(OH)(H_2O)_5]^+(aq) + H_3O^+(aq)$$

The acidity of the solution is more marked with a 3+ cation which has a greater charge to size ratio compared to a 2+ ion and so causes an increased polarisation of the water ligands. For example, a weakly acidic solution is created in the aqueous solution of $[Fe(H_2O)_6]^{3+}$ (Figure 8.36):

$$[Fe(H_2O)_6]^{3+}(aq) + H_2O(l) \rightleftharpoons [Fe(OH)(H_2O)_5]^{2+}(aq) + H_3O^+(aq)$$

Figure 8.36
The 3+ charge on the iron ion increases the polarisation of the water ligands and so the solution is weakly acidic in aqueous solution.

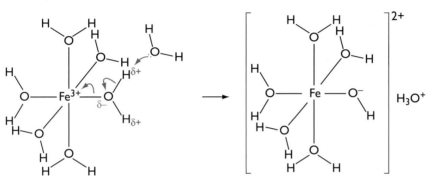

Precipitation reactions

The hydroxides of transition element ions can be precipitated from their aqueous solutions by the addition of a base, such as hydroxide ions. These precipitates are usually gelatinous due to the fact that they are hydrated, but some are granular and they are often coloured. The colour of the precipitate can be used to identify the metal present (Table 8.7 and Figure 8.37).

Table 8.7
Transition element hydroxides.

Cation present	Precipitate	Colour of precipitate
Chromium(III)	$Cr(OH)_3$	Green
Manganese(II)	$Mn(OH)_2$	White (rapidly turns brown)
Iron(II)	$Fe(OH)_2$	Pale green
Iron(III)	$Fe(OH)_3$	Rusty brown
Cobalt(II)	$Co(OH)_2$	Pink
Nickel(II)	$Ni(OH)_2$	Green
Copper(II)	$Cu(OH)_2$	Blue
Zinc	$Zn(OH)_2$	White

Figure 8.37
The colours of these precipitates can be used to identify the transition element present in solution. Can you work out which transition elements are shown here?

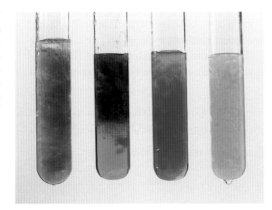

13 Write balanced equations for the formation of the
following from the transition element ion in solution:
  a copper(II) hydroxide
  b cobalt(II) hydroxide
  c iron(II) hydroxide.

How are these precipitates formed? When hydroxide ions are added to a hexaaqua transition element complex then more water ligands are deprotonated (lose hydrogen ions) and the hydroxide precipitate is produced. For example:

$$[Co(H_2O)_6]^{2+}(aq) + 2OH^-(aq) \rightleftharpoons [Co(OH)_2(H_2O)_4](s) + 2H_2O(l)$$

In the case of zinc, the precipitate disappears when an excess of sodium hydroxide is added. This is because zinc hydroxide is amphoteric (see Chapter 6, page 89) and the tetrahydroxozincate ion, $[Zn(OH)_4]^{2-}$(aq), is formed in the presence of excess hydroxide ions:

$$Zn(OH)_2(s) + 2OH^-(aq) \rightarrow [Zn(OH)_4]^{2-}(aq)$$

If OH^-(aq) is added to either the $[Fe(H_2O)_6]^{3+}$ ion or the $[Cr(H_2O)_6]^{3+}$ ion, then a similar process of deprotonation and subsequent precipitation takes place.

$$[Fe(H_2O)_6]^{3+}(aq) + 3OH^-(aq) \rightleftharpoons [Fe(OH)_3(H_2O)_3](s) + 3H_2O(l)$$
$$[Cr(H_2O)_6]^{3+}(aq) + 3OH^-(aq) \rightleftharpoons [Cr(OH)_3(H_2O)_3](s) + 3H_2O(l)$$

However, in the case of chromium a further reaction takes place, resulting in the dissolving of the green precipitate of $[Cr(OH)_3(H_2O)_3]$(s):

$$[Cr(OH)_3(H_2O)_3](s) + 3OH^-(aq) \rightleftharpoons [Cr(OH)_6]^{3-}(aq) + 3H_2O(l)$$
hexahydroxochromate(III) ion

This shows the amphoteric nature of chromium(III) hydroxide. (This is similar to the behaviour of aluminium hydroxide.)

If aqueous ammonia is used as the base the precipitate may also disappear. Initially, OH^-(aq) from the equilibrium:

$$NH_3(aq) + H_2O(l) \rightleftharpoons NH_4^+(aq) + OH^-(aq)$$

causes deprotonation of the hydrated ion, as mentioned above. A ligand exchange process then takes place and an ammine complex is formed with the central metal ion (Figure 8.38). For example:

$$[Cu(OH)_2(H_2O)_4](s) + 4NH_3(aq) \rightarrow [Cu(NH_3)_4(H_2O)_2]^{2+}(aq) + 2OH^-(aq) + 2H_2O(l)$$
pale blue hydrated precipitate   deep-blue solution

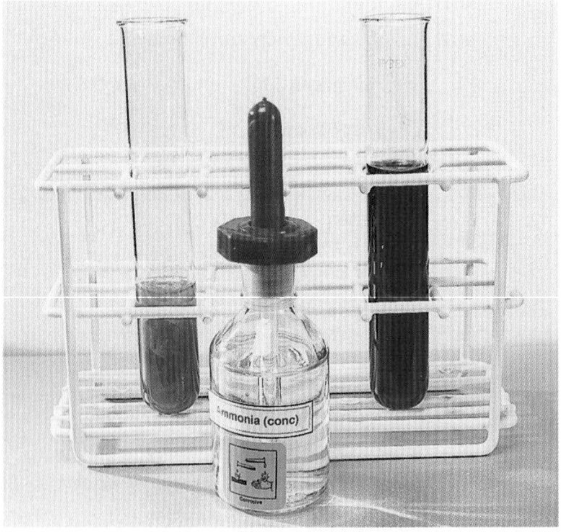

Figure 8.38
The blue precipitate of $Cu(OH)_2$ disappears and is replaced by a deep-blue solution containing $[Cu(NH_3)_4(H_2O)_2]^{2+}$(aq). This particular reaction is used as a chemical test for the presence of copper ions in solution.

This process also takes place with Co^{2+}(aq) and Ni^{2+}(aq).

However, in the case of Co^{2+} when $[Co(NH_3)_6]^{2+}$(aq) is formed as the yellow complex ion it rapidly turns brown due to oxidation and the formation of $[Co(NH_3)_6]^{3+}$(aq). Generally, if ammonia solution is added to an $[M(H_2O)_6]^{3+}$ complex ion in solution then a ligand displacement reaction usually takes place and $[M(NH_3)_6]^{3+}$ is produced.

If the base is the CO_3^{2-}(aq) ion from aqueous sodium carbonate solution, then the carbonate reacts with the aqueous transition element ions to produce a precipitate of the insoluble carbonate. Generally:

$$[M(H_2O)_6]^{2+}(aq) + CO_3^{2-}(aq) \rightarrow MCO_3(s) + 6H_2O(l)$$

The colours of these carbonate precipitates are similar to those of the transition element hydroxides.

Due to the acidic nature of the $[M(H_2O)_6]^{3+}$ ion, carbon dioxide gas is produced on the addition of sodium carbonate.

For example:

$$2[Cr(H_2O)_6]^{3+}(aq) + 3CO_3^{2-}(aq) \rightleftharpoons 2[Cr(OH)_3(H_2O)_3](s) + 3CO_2(g) + 3H_2O(l)$$
$$2[Fe(H_2O)_6]^{3+}(aq) + 3CO_3^{2-}(aq) \rightleftharpoons 2[Fe(OH)_3(H_2O)_3](s) + 3CO_2(g) + 3H_2O(l)$$

Note that $[Al(H_2O)_6]^{3+}$(aq) also behaves in a similar manner.

Catalytic properties of transition elements and their compounds

Transition elements and their compounds are important catalysts. Some industrial processes that are catalysed by transition elements and transition element compounds are shown in Table 8.8.

Table 8.8
Some important transition element catalysts.

Process	Product	Catalyst
Haber	Ammonia (NH_3)	Iron
Contact	Sulphuric acid (H_2SO_4)	Platinum–rhodium alloy or V_2O_5
Oxidation of ammonia	Nitric acid (HNO_3)	Platinum–rhodium alloy
Hydrogenation of unsaturated oils to harden them	Saturated oils	Nickel

Figure 8.39
Finely divided nickel is used as the catalyst for the hydrogenation of unsaturated oils to manufacture solid margarine with a lower amount of unsaturation.

The examples shown in Table 8.8 all involve **heterogeneous catalysis** – the catalyst is in a different physical state to the reactants. For example, in the hydrogenation of unsaturated fats a nickel catalyst is used (Figure 8.39). It is important for this type of catalyst to have a large surface area for contact with the reactants. It is for this reason that solid catalysts are used in a powdered form, as a wire mesh or supported on a porous material such as a ceramic support. For example, in a catalytic converter the transition elements used (rhodium, palladium and platinum) are powdered and spread over a ceramic support material full of tiny holes. Using a support medium helps to cut down the cost of the catalyst, which aids the economics of the process it is used in. During this type of catalysis the reactant particles are adsorbed onto the surface at the active site following collision with the catalyst. Molecular rearrangement takes place and the products are then desorbed from the surface. The strength of the adsorption process at the active site on the surface of the catalyst helps to determine the activity of the catalyst. For example, tungsten has too strong an adsorption whilst silver has too weak an adsorption for these two transition elements to be considered as catalysts. However, nickel and platinum have an optimum adsorption for many reactions that need to be catalysed, so they are used quite widely.

Catalysts can be poisoned so that they no longer function at full efficiency. This is the reason why in many industrial processes the reactants are purified before they enter a reaction chamber that contains a catalyst. In heterogeneous catalysis the particles which poison the catalyst are adsorbed more strongly onto the catalyst surface than the actual reactant particles. This is the reason why cars fitted with catalytic converters cannot use leaded petrol. The lead atoms are adsorbed onto the surface of the catalytic converter, poisoning it.

Chemists think that these processes take place because of the availability of 3d electrons and 4s electrons on the surface of the catalyst. This allows catalysts to form weak bonds with the reactants. After the reaction has occurred at the surface these weak bonds break and the product molecules are released.

Transition elements are also involved in **homogeneous catalysis** – the catalyst is in the same physical state as the reactants. This type of catalyst works by forming an intermediate compound with the reactants, which then breaks down to give the products. You will have already met many reactions of this type. They occur in the human body where the reactions are catalysed by enzymes.

In the case of transition elements, homogeneous catalysts are usually in the form of the aqueous ion and so will catalyse some reactions in the aqueous phase. In these reactions their variable oxidation state allows them to take part in the sequence of reaction stages and emerge unchanged at the end of the reaction. For example, the oxidation of iodide ions by peroxodisulphate ions is catalysed by iron(II) ions. The reaction:

$$I^-(aq) + \tfrac{1}{2}S_2O_8^{2-}(aq) \rightarrow \tfrac{1}{2}I_2(aq) + SO_4^{2-}(aq)$$

or

$$2I^-(aq) + S_2O_8^{2-}(aq) \rightarrow I_2(aq) + 2SO_4^{2-}(aq)$$

proceeds via the following alternative lower activation energy reaction pathway involving iron(II) ions:

$$2Fe^{2+}(aq) + S_2O_8^{2-}(aq) \rightarrow 2Fe^{3+}(aq) + 2SO_4^{2-}(aq)$$
$$2Fe^{3+}(aq) + 2I^-(aq) \rightarrow 2Fe^{2+}(aq) + I_2(aq)$$

As you can see, the iron(II) is oxidised to iron(III) in the first stage and the iron(III) is reduced to iron(II) in the second stage (Figure 8.40). The iron(II) is therefore regenerated within the reaction sequence.

Figure 8.40
The iron(II) ions provide a reaction pathway of lower activation energy.

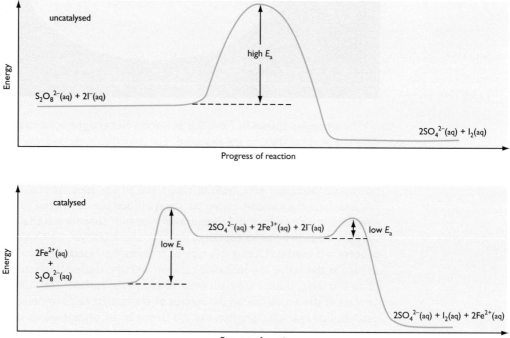

A further reaction in which transition metal ions act as a catalyst is the reaction between potassium manganate(VII) and ethanedioate ions. This is a standard method of estimating ethanedioates in solution:

$$2MnO_4^-(aq) + 5C_2O_4^{2-}(aq) + 16H^+(aq) \rightarrow 2Mn^{2+}(aq) + 10CO_2(g) + 8H_2O(l)$$

This is an example of **autocatalysis**, as one of the products of the reaction (Mn^{2+}) is a catalyst for the reaction.

Transition elements are important trace elements

Approximately 99% of the human body is made up of four elements: carbon, hydrogen, nitrogen, oxygen. The remaining 1% contains other elements or trace elements. Certain transition elements are vital trace elements, which help to make up this 1%.

- Iron is found in haemoglobin in your blood (see Figure 8.15). It is responsible for the transport of oxygen by the blood.
- Copper is essential in certain enzymes that are involved in oxidation processes.
- Zinc is also needed for the activity of certain enzymes.

Crystal hydrates

Many transition element compounds produce **hydrates** when they crystallise from aqueous solution. A hydrate is a substance which incorporates water into its crystal structure. This water is referred to as the **water of crystallisation**. The shape and colour of the crystal hydrate is very much dependent on the presence of this water of crystallisation. Some examples of crystal hydrates are given in Table 8.9 and Figure 8.41.

Table 8.9
Examples of crystal hydrates containing transition elements.

Crystal hydrate	Formula
Cobalt(II) chloride hexahydrate	$CoCl_2.6H_2O$
Copper(II) sulphate pentahydrate	$CuSO_4.5H_2O$
Iron(II) sulphate heptahydrate	$FeSO_4.7H_2O$
Nickel(II) sulphate heptahydrate	$NiSO_4.7H_2O$

Figure 8.41 (left)
Hydrate crystals (clockwise from top left): cobalt(II) chloride, calcium nitrate, nickel(II) sulphate, chromium potassium sulphate, copper(II) sulphate and manganese sulphate.

Figure 8.42 (right)
Adding water to anhydrous copper(II) sulphate, to produce the pentahydrate with the distinctive blue colour.

14 With reference to the discussion of the formation of colour in transition element complexes on page 119, describe how you think the blue colour of the $CuSO_4.5H_2O$ hydrate arises.

When many of these hydrates are heated, the water of crystallisation is driven off. For example, if crystals of copper(II) sulphate pentahydrate are heated strongly they lose their water of crystallisation. Anhydrous copper(II) sulphate remains as a white powder.

$$CuSO_4.5H_2O(s) \rightarrow CuSO_4(s) + 5H_2O(g)$$
$$\text{blue} \qquad\qquad \text{white}$$

When water is added carefully to anhydrous copper(II) sulphate it turns blue and the pentahydrate is produced (Figure 8.42). This is an extremely exothermic process:

$$CuSO_4(s) + 5H_2O(l) \rightarrow CuSO_4.5H_2O(s)$$

- Use a CD-ROM or database to retrieve and compare chemical and physical data related to the first and second row transition elements.
- Use a spreadsheet and graphical display to investigate trends and variations in the properties of the first row transition elements.

Number

- Select relevant numerical data related to the first row transition elements from a database, ensuring correct units.

● **Skills task** Use the Internet and presentation software to produce a presentation describing how complexes of the element platinum are being used as treatments for cancer.

CHECKLIST After studying Chapter 8 you should know and understand the following terms.

- **Transition element:** An element that forms at least one ion with a partially filled sub-shell of d electrons.
- **Ligands:** Neutral molecules that possess lone pairs of electrons, for example water, or negative ions, for example chloride ion, that donate electrons to metal ions via a dative covalent bond.
- **Complex ions:** Formed when ligands surround a metal ion, forming dative covalent bonds to it.
- **Dative covalent bonds (coordinate bonds):** The sharing of a pair of electrons by two atoms, but with both electrons being donated by one of the atoms.
- **Coordination number:** The number of dative covalent or coordinate bonds which form between the ligands and the central metal ion of a complex ion.
- **Cationic complexes:** Positively charged complexes, for example $[Ni(H_2O)_6]^{2+}$.
- **Anionic complexes:** Negatively charged complexes, for example $[CuCl_4]^{2-}$.
- **Unidentate ligands:** Ligands that form only a single dative covalent bond with the central metal ion, for example Cl^-.
- **Polydentate ligands:** Ligands that form two or more dative covalent bonds with the central metal ion, for example ethanedioate ion.
- **Chelate complexes:** Complexes formed between a central metal ion and polydentate ligands.
- **Geometrical isomers:** *cis* and *trans* isomers can be formed if the ligands in four or six coordinate complexes are not identical.
- **Degenerate sub-shells:** Sub-shells that are of the same energy, for example, the $5 \times 3d$ sub-shells.
- **Calibration curve:** A graph of absorption against concentration, produced by the absorption of electromagnetic radiation by a series of standard solutions of known concentration. The graph can then be used to determine the concentration of unknown solutions.
- **Ligand exchange reaction:** A reaction that involves one ligand substituting for another in a complex ion.
- **Overall stability constant:** The value obtained from multiplying together the equilibrium constants for the individual ligand exchange reactions. Symbol: K_{stab}.
- **Heterogeneous catalysis:** A reaction in which the catalyst is in a different physical state to the reactants.
- **Homogeneous catalysis:** A reaction in which the catalyst is in the same physical state as the reactants.

Examination questions

1 Chemical reactions can be affected by homogeneous or by heterogeneous catalysts.

a Explain what is meant by the term **homogeneous** and suggest the most important feature in the mechanism of this type of catalysis when carried out by a transition-metal compound. (2)

b In aqueous solution, $S_2O_8^{2-}$ ions can be reduced to SO_4^{2-} ions by I$^-$ ions.
  i Write an equation for this reaction.
  ii Suggest why the reaction has a high activation energy, making it slow in the absence of a catalyst.
  iii Iron salts can catalyse this reaction. Write two equations to show the role of the catalyst in this reaction. (4)

c Below is a sketch showing typical catalytic efficiencies of transition metals from Period 5 (Rb to Xe) and Period 6 (Cs to Rn) when used in heterogeneous catalysis.

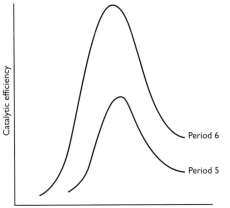

Successive transition metals

  i Identify two metals which lie at opposite ends of these curves and explain why they show rather low catalytic efficiency.
  ii Suggest why these curves pass through a maximum. (5)

d In catalytic converters which clean up petrol engine exhaust gases, a catalyst promotes the reduction of nitrogen oxides using another polluting gas as reductant.
  State a suitable catalyst for this task, identify the reductant, and write an equation for the reaction that results. (3)

AQA, A level, Specimen Paper 6421, 2001/2

2 In 1869 Mendeleev put chromium and lead in the same group of the periodic table; chromium in Group 4A, and lead in Group 4B. Group 4 now usually refers to the elements carbon to lead, and chromium is classified as a transition element. Nevertheless chromium and lead do have some properties in common. Both have amphoteric oxides; both have more than one oxidation state, and for both their highest oxidation state is strongly oxidising.

The following electrode potential data is required in this question:

$$
\begin{array}{lr}
 & E^{\ominus}\ (V) \\
Cr^{3+} + 3e^- \rightleftharpoons Cr & -0.74 \\
O_2 + 2H^+ + 2e^- \rightleftharpoons H_2O_2 & +0.68 \\
Cr_2O_7^{2-} + 14H^+ + 6e^- \rightleftharpoons 2Cr^{3} + 7H_2O & +1.33
\end{array}
$$

a Chromium(III) may be converted to chromium(VI) by heating a strongly alkaline solution of chromium(III) ions with hydrogen peroxide.

$$Cr_2(SO_4)_3(aq) \xrightarrow[\text{step I}]{\substack{\text{KOH} \\ \text{in excess}}} \substack{\text{green} \\ \text{solution}} \xrightarrow[\text{step 2}]{H_2O_2,\ \text{heat}} K_2CrO_4(aq)$$

The yellow solution of potassium chromate(VI) is boiled until the excess of hydrogen peroxide has been destroyed, the solution then being made acidic with ethanoic acid to give a solution of potassium dichromate(VI).

$$K_2CrO_4(aq) \xrightarrow[\text{step 3}]{H^+} K_2Cr_2O_7(aq)$$

Show by means of equations the processes which are occurring in steps 1, 2 and 3 of the reaction scheme, and suggest in terms of data given above why the excess of hydrogen peroxide must be destroyed before the solution is acidified in step 3. (8)

b Ammonium dichromate(VI), $(NH_4)_2Cr_2O_7$, decomposes on heating to chromium(III) oxide, nitrogen and water in a reaction which gives out enough heat to be self-sustaining. Write an equation representing the reaction, and by a consideration of the oxidation states of the atoms concerned show that the reaction is an oxidation of the cation by the anion. (4)

c Define the term **disproportionation**, and use the data given to show whether chromium(III) ions will disproportionate in acidic solution. (4)

d Explain why lead(IV) is an oxidising agent and give an example of a reaction in which a lead(IV) compound acts as an oxidising agent. (4)

e Illustrate by means of equations the amphoteric nature of chromium(III) oxide and lead(II) oxide. (5)

London, A level, Synoptic Paper (CH6), June 1999

3 Stainless steel contains chromium which reduces corrosion, mainly by forming a very thin layer of chromium(III) oxide on the surface. Stainless steel is made by adding chromium in the last stage of steelmaking.

a Steel is usually made from blast furnace iron by the BOS process. Oxygen is blown through the molten iron and basic oxides (for example, calcium oxide) are added. Carbon escapes as carbon monoxide. Other elements (for example, silicon) form oxides which react with the basic oxides present. Thus they can be removed as slag.
  i Write equations that show how silicon is removed by the BOS process. (2)
  ii Suggest why the chromium is added **after** the oxygen blow rather than before it. (1)

b One way in which the steel may be analysed to find its chromium content is to dissolve out the chromium as the complex $[Cr(NH_3)_6]^{3+}$ ion which is green.
  i What is the oxidation state of chromium in the complex ion? (1)
  ii Draw a diagram to show the shape of the $[Cr(NH_3)_6]^{3+}$ ion, labelling the metal ion and a ligand molecule. Show clearly which part of the ligand molecule forms a bond with the metal ion. (2)

c Describe **in outline** how you would use a colorimeter to find the concentration of the green complex ion in solution, given a sample of the pure green solid. (5)

d In a separate experiment to find the iron content of a sample of steel, 1.40 g of the metal were dissolved by boiling with acid. The iron in the solution was then all converted into the +2 oxidation state and the solution was made up to 100.0 cm³ using dilute sulphuric acid. 10.0 cm³ of this solution were titrated against a 0.0200 mol dm⁻³ solution of potassium manganate(VII). 24.2 cm³ of the potassium manganate(VII) solution were required to reach the end-point of the titration. The reaction occurring in the titration is:

$$5Fe^{2+}(aq) + MnO_4^-(aq) + 8H^+(aq) \rightarrow$$
$$5Fe^{3+}(aq) + Mn^{2+}(aq) + 4H_2O(l)$$

  i Calculate the mass of iron in the sample of steel. (A_r: Fe, 56.0) (3)
  ii Calculate the percentage by mass of iron in the steel. (1)

OCR, A level, Specimen Paper A7887, Sept 2000

4 1,2-Diaminoethane, $NH_2CH_2CH_2NH_2$, is a **bidentate ligand**.
  a Explain the term **bidentate ligand**. (2)
  b There are three isomeric complexes with the formula $[Cr(NH_2CH_2CH_2NH_2)_2Cl_2]^+$, all having the same basic shape.
    i State the shape of these complexes.
    ii Draw structures of these three complexes, Complex I, Complex II and Complex III, to show the differences between them.

iii Which of the complexes you have drawn will have a dipole? (5)

OCR, A level, Specimen Paper A7882, Sept 2000

5 a Explain what is meant by ligand exchange. (1)
  b Describe all the colour changes and observations that take place when an aqueous solution of ammonia is gradually added to a solution of $Cu^{2+}(aq)$, until the ammonia is in excess. Write equations for these transformations. (4)
  c Blood gets its colour from oxygen-carrying molecules with organic groups surrounding a transition metal ion. In humans this transition metal is iron, and the blood is red. In horseshoe crabs, the metal is copper and the blood is blue, and in sea squirts the metal is vanadium and the blood is green.
    The sketch below shows the major absorption peak for human blood. Copy this sketch showing and labelling the corresponding absorption peaks for the blood of horseshoe crabs and sea squirts. (2)

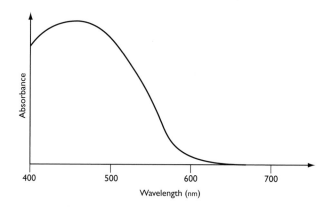

  d A 0.0100 mol sample of an oxochloride of vanadium, $VOCl_x$ required 20.0 cm³ of 0.100 mol dm⁻³ acidified potassium manganate(VII) for oxidation of the vanadium to its +5 oxidation state.
    i Calculate how many moles of potassium manganate(VII) were reacted.
    ii How many moles of electrons were removed by the MnO_4^- ions?
    iii Determine the change in oxidation state of the vanadium.
    iv Deduce the value of x in the formula $VOCl_x$. (4)

OCR, A level, Specimen Paper A7882, Sept 2000

6 Cobalt(II) forms the following coloured complexes with water molecules and chloride ions: $[Co(H_2O)_6]^{2+}$ and $[CoCl_4]^{2-}$.
  Describe how different ligands, H_2O and Cl^- affect the stereochemistry and colour of these complexes. (In this question, 1 mark is available for the quality of written communication.) (9)

OCR, A level, Specimen Paper A7882, Sept 2000

III
ORGANIC CHEMISTRY

Carbon is found naturally in three allotropic forms: amorphous, graphite and diamond. More recently a new allotrope of carbon was formed named buckminsterfullerene.

9 Aldehydes and ketones

STARTING POINTS
- Aldehydes and ketones are carbonyl compounds.
- All aldehydes contain the —CHO functional group.
- Aldehydes are formed by the oxidation of primary alcohols.
- All ketones contain a carbonyl group attached to two alkyl groups.
- Ketones are formed by the oxidation of secondary alcohols.
- Aldehydes and ketones are found in nature and they have uses as solvents and in the manufacture of polymers.

Figure 9.1
a Ethanal, an aldehyde, is used in the manufacture of poly(ethenyl ethanoate), which is used as an essential constituent in some plastic emulsion paints.
b Propanone, a ketone, is an important solvent both in industry and in cosmetics. This nail varnish remover contains propanone.
c Benzaldehyde, an aromatic aldehyde, is used to make the 'almond essence' used in Bakewell tarts.

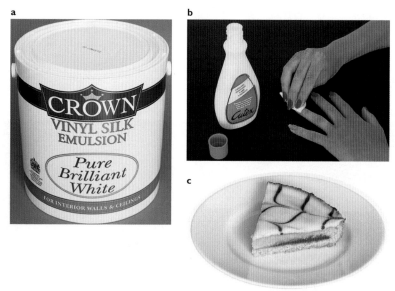

Aldehydes and ketones are homologous series of compounds that both contain the —C=O group or **carbonyl group**. They are often called **carbonyl compounds** and they are structurally quite similar.

Aldehydes are a homologous series that you have already encountered in your earlier study of Organic Chemistry. They are formed in the first stage of the oxidation of primary alcohols (see page 139) and have the general formula

$$\begin{array}{c} O \\ \parallel \\ R-C-H \end{array}$$

This formula is usually written RCHO: aldehydes have the carbonyl group at the end of a chain. In this general formula, R represents an alkyl group, an aryl group (for example the phenyl group, C_6H_5—) or hydrogen.

Ketones are a homologous series of compounds that you have also met earlier in your study of Organic Chemistry. Ketones have the carbonyl group in a non-terminal position in the chain. They are formed as the only product of the oxidation of secondary alcohols (see page 139). Ketones have the general formula

$$\begin{array}{c} O \\ \parallel \\ R-C-R' \end{array}$$

This formula is usually written RCOR', where R and R' are alkyl or aryl groups. It can be seen that the carbonyl group is joined to two other carbon atoms.

Naming aldehydes and ketones

Table 9.1 shows the names and formulae of the first few aldehydes and ketones.

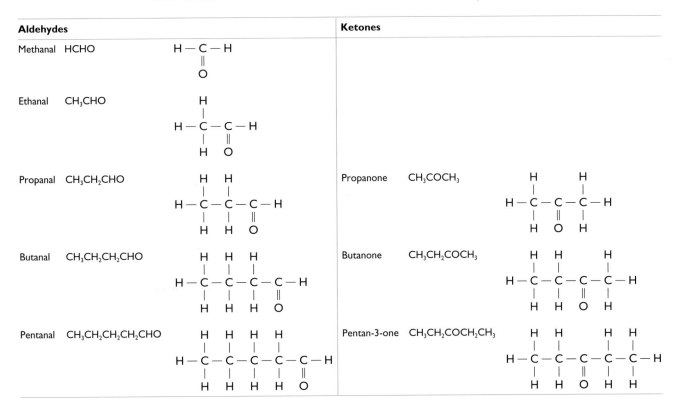

Aldehydes			Ketones		
Methanal	HCHO				
Ethanal	CH₃CHO				
Propanal	CH₃CH₂CHO		Propanone	CH₃COCH₃	
Butanal	CH₃CH₂CH₂CHO		Butanone	CH₃CH₂COCH₃	
Pentanal	CH₃CH₂CH₂CH₂CHO		Pentan-3-one	CH₃CH₂COCH₂CH₃	

Table 9.1 (above)
The first few aldehydes and ketones.

Aldehydes are named with the ending '**-al**'. This ending comes after the alkane chain stem that indicates the number of carbon atoms in the chain; this includes the carbon atom present in the carbonyl group. CH_3CHO is called ethanal, CH_3CH_2CHO is called propanal, and so on.

Ketones are named with the ending '**-one**'. Again this ending comes after the alkane chain stem that indicates the number of carbon atoms in that chain. The name also has a number, if necessary, to indicate the position of the carbonyl group. $CH_3COCH_2CH_3$ is called butanone and $CH_3CH_2COCH_2CH_3$ is called pentan-3-one. You will have noticed that there is the possibility of an isomer of this ketone related to the position of the carbonyl group, that is, pentan-2-one – $CH_3COCH_2CH_2CH_3$.

1 Name the following compounds:
   a $CH_3CH_2CH_2CH_2CH_2CH_2CHO$
   b $CH_3CH_2CH_2CH_2CH_2COCH_2CH_3$
2 Give the formulae of the following compounds:
   a heptan-3-one
   b hexanal.

Aldehydes and ketones are found in nature. Simple sugars such as glucose and fructose exist in solution as an equilibrium between two forms, a straight-chain form and a ring form (Figures 9.2 and 9.3). The straight-chain form of glucose contains an aldehyde group; glucose is an aldose sugar. The straight-chain form of fructose contains a ketone group; fructose is a ketose sugar.

Figure 9.2 (left)
Straight-chain and ring forms of the sugar glucose exist as an equilibrium in solution.

Figure 9.3 (right)
Straight-chain and ring forms of the sugar fructose exist as an equilibrium in solution.

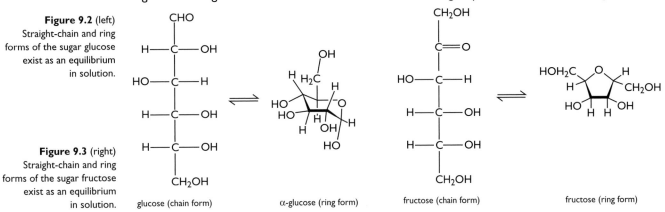

glucose (chain form) α-glucose (ring form) fructose (chain form) fructose (ring form)

Aldehydes containing aryl groups are found in nature; for example, benzaldehyde contributes to the flavouring of fruits such as apricots and plums (Figure 9.4).

Figure 9.4
Benzaldehyde, the simplest aromatic aldehyde, contributes to the flavouring of several fruits, including apricots and plums. Benzaldehyde is also used to make cinnamic acid, used extensively in the manufacture of perfumes, cosmetics and food flavourings.

benzaldehyde

Physical properties of aldehydes and ketones

Figure 9.5 (above)
The carbonyl group is polar.

The carbonyl group contains a polar bond (Figure 9.5). This has a significant effect on the physical properties of both aldehydes and ketones. Table 9.2 shows the boiling points and molecular masses for some aldehydes, ketones and alkanes. You can see that the smaller aldehydes and ketones have considerably higher boiling points than alkanes of a similar relative molecular mass. For example, propane has the same relative molecular mass as ethanal; however, ethanal is a volatile liquid at room temperature whereas propane is a gas.

Table 9.2
Comparing aldehydes, ketones and alkanes.

	Name	Shortened structural formula	M_r	Boiling point (°C)
Aldehydes	Methanal	HCHO	30	−21
	Ethanal	CH_3CHO	44	20.9
	Propanal	CH_3CH_2CHO	58	48.9
	Butanal	$CH_3CH_2CH_2CHO$	72	75.7
Ketones	Propanone	CH_3COCH_3	58	56.2
	Butanone	$CH_3CH_2COCH_3$	72	79.7
Alkanes	Methane	CH_4	16	−164
	Ethane	CH_3CH_3	30	−87
	Propane	$CH_3CH_2CH_3$	44	−42
	Butane	$CH_3CH_2CH_2CH_3$	58	0

Figure 9.6
Water forms hydrogen bonds with the carbonyl group.

The polarity of the carbonyl group means that the smaller aldehydes and ketones are completely miscible with water – water forms hydrogen bonds with the carbonyl group (Figure 9.6). The larger aldehydes and ketones, however, are much less soluble. This is due to the decrease in the polar nature of the molecules relative to their increasing hydrocarbon nature – hydrogen bonding is less likely to take place between the molecules and water.

Making aldehydes and ketones

Oxidation

Figure 9.7
Orange potassium dichromate(VI) slowly turns green.

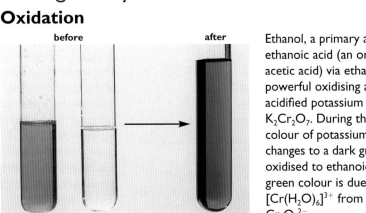

before   after

acidified potassium   ethanol
dichromate(VI)

ethanoic acid + Cr^{3+}

Ethanol, a primary alcohol, can be oxidised to ethanoic acid (an organic acid also known as acetic acid) via ethanal (an aldehyde) by powerful oxidising agents such as warm acidified potassium dichromate(VI) solution, $K_2Cr_2O_7$. During the reaction the orange colour of potassium dichromate(VI) gradually changes to a dark green as the ethanol is oxidised to ethanoic acid (Figure 9.7). The green colour is due to the formation of $[Cr(H_2O)_6]^{3+}$ from the reduction of Cr(VI) in $Cr_2O_7^{2-}$.

$$CH_3CH_2OH(l) + \underset{\substack{\text{from acidified} \\ \text{potassium dichromate(VI)}}}{[O]} \xrightarrow{\text{heat}} \underset{\text{ethanal}}{CH_3CHO(aq)} + H_2O(l)$$
$$\underset{\text{ethanol}}{}$$

[O] indicates oxygen from an oxidising agent. If the aldehyde is required then it can be distilled away during the reaction (Figure 9.8).

Figure 9.8
Distillation apparatus used to prepare an aldehyde by oxidation of an alcohol.

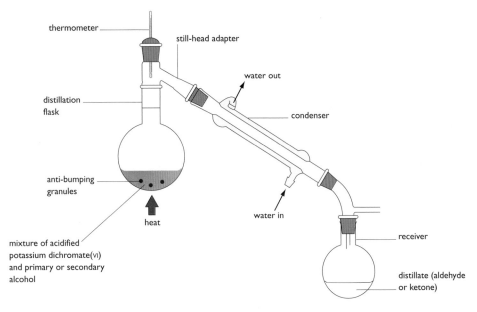

thermometer

still-head adapter

water out

distillation flask

condenser

anti-bumping granules

heat

water in

mixture of acidified potassium dichromate(VI) and primary or secondary alcohol

receiver

distillate (aldehyde or ketone)

A similar, but slower, oxidation process takes place if wine or beer is left open to the air. Ethanol will eventually turn to vinegar (the sharp taste of ethanoic acid) due to oxidation of the ethanol in the alcoholic drink.

Secondary alcohols such as propan-2-ol are oxidised to ketones. For example, propan-2-ol will undergo oxidation with acidified potassium dichromate(VI), $K_2Cr_2O_7$:

3 a Write the structural formula for butan-2-one.
b Write a balanced chemical equation for the production of butan-2-one from butan-2-ol.

$$\underset{\text{propan-2-ol}}{CH_3CH(OH)CH_3(l)} + \underset{\substack{\text{from acidified} \\ \text{potassium dichromate(VI)}}}{[O]} \xrightarrow{\text{heat}} \underset{\text{propanone}}{CH_3COCH_3(aq)} + H_2O(l)$$

Tertiary alcohols are resistant to oxidation, so 2-methylpropan-2-ol is very difficult to oxidise. This is because tertiary alcohols do not have a hydrogen atom attached to the carbon atom that has the —OH group attached.

Some reactions of aldehydes and ketones

Figure 9.9
Nucleophilic addition reactions to aldehydes and ketones.

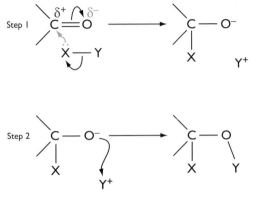

Addition

The carbonyl group, C=O, is polarised due to the electronegative nature of the oxygen atom. This polarisation makes the $\delta+$ carbon atom of the carbonyl group susceptible to attack by nucleophiles, which attack and bond to the carbonyl carbon atom causing the π bond between the C and O atoms of the group to break. With the general nucleophile, X—Y, a **nucleophilic addition reaction** (sometimes called heterolytic nucleophilic addition) occurs (Figure 9.9).

4 a Write the structural formula for butan-1-ol.
b Write a balanced chemical equation for the production of butan-1-ol from butanal.
5 a Write the structural formula for hexan-3-one.
b Write a balanced chemical equation for the production of hexan-3-one from hexan-3-ol.
6 a Write a balanced chemical equation for the addition reaction between butanal and hydrogen cyanide.
b What is the name of the product of this reaction? Write the structural formula for this substance.
7 a Write a balanced chemical equation for the addition reaction between pentan-3-one and hydrogen cyanide.
b What is the name of the product of this reaction? Write the structural formula for this substance.

Reduction

The oxidation process shown earlier can be reversed by warming with a reducing agent such as lithium tetrahydridoaluminate, $LiAlH_4$, with dry ethoxyethane (ether) as solvent. Other reducing agents that can be used include sodium tetraborohydride, $NaBH_4$ (with water as solvent), or hydrogen gas under pressure with a nickel catalyst.

$$CH_3CHO(l) + \underset{\substack{\text{from lithium} \\ \text{tetrahydridoaluminate}}}{2[H]} \rightarrow \underset{\text{ethanol}}{CH_3CH_2OH(l)}$$

$$\underset{\text{ethanal}}{}$$

$$CH_3COCH_3(l) + \underset{\substack{\text{from lithium} \\ \text{tetrahydridoaluminate}}}{2[H]} \overset{\text{heat}}{\longrightarrow} \underset{\text{propan-2-ol}}{CH_3CH(OH)CH_3(aq)}$$

$$\underset{\text{propanone}}{}$$

These reduction reactions can be regarded as addition reactions of hydrogen to the C=O group. You will have met similar addition reactions in your earlier study of alkenes where the reaction is electrophilic addition (see *Introduction to Advanced Chemistry*, Chapter 16, page 179).

A further addition reaction

Both aldehydes and ketones will undergo addition reactions with hydrogen cyanide, producing 2-hydroxynitriles (also known as cyanohydrins). For example, ethanal will form 2-hydroxypropanenitrile.

$$\underset{\text{ethanal}}{CH_3CHO} + HCN \rightarrow \underset{\text{2-hydroxypropanenitrile}}{CH_3CH(OH)CN}$$

You will have noticed that the stem name of the product has changed. This is because an extra carbon atom has been introduced into the product. This reaction can be used to increase the length of the carbon chain in organic molecules. This is very useful in organic synthesis.

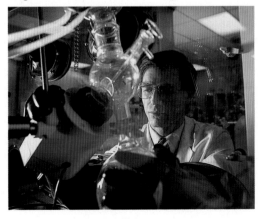

Figure 9.10
2-Hydroxy-2-methyl-propanenitrile is an intermediate in the manufacture of the monomer used to make Perspex. The use of Perspex in this glove box allows the researchers to see the experiments they are manipulating.

In the case of propanone, 2-hydroxy-2-methylpropanenitrile is produced, which is an intermediate in the manufacture of the monomer for Perspex (Figure 9.10).

$$\underset{\text{propanone}}{CH_3COCH_3} + HCN \rightarrow \underset{\substack{\text{2-hydroxy-2-} \\ \text{methylpropanenitrile}}}{(CH_3)_2C(OH)CN}$$

Because of the polarity of the carbonyl group, the mechanism for this type of reaction involves **nucleophilic addition**. This is unlike the electrophilic addition that takes place in the case of alkenes.

Testing for aldehydes and ketones

Figure 9.11
Ethanal reacts with 2,4-dinitrophenylhydrazine to form an orange 2,4-dinitrophenylhydrazone precipitate.

before

after

Aldehydes and ketones react with compounds that contain the $-NH_2$ group. The reaction involves addition across the carbonyl double bond, followed by the elimination of water. The water molecule is eliminated as a double bond forms between the carbonyl carbon and the nitrogen of the $-NH_2$ group. This is an example of a **condensation reaction** or an **addition–elimination reaction**. When a solution of 2,4-dinitrophenylhydrazine (Brady's reagent) is added to an aldehyde or a ketone, an orange/yellow precipitate is formed, which is a 2,4-dinitrophenylhydrazone (Figure 9.11).

The products of this type of condensation reaction, the 2,4-dinitrophenylhydrazones, are all easily recrystallised and all have well-defined melting points that are tabulated in chemical data books. They are useful, therefore, in identifying individual aldehydes and ketones. It should be noted that other compounds with C=O groups, such as carboxylic acids and esters, do not undergo this reaction.

8 Draw the structure of the product produced when butanone reacts with 2,4-dinitrophenylhydrazine.

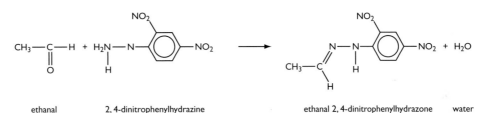

ethanal   2, 4-dinitrophenylhydrazine   ethanal 2, 4-dinitrophenylhydrazone   water

Distinguishing between aldehydes and ketones

The fact that the $-CHO$ group of aldehydes will readily undergo oxidation, as compared to ketones, enables us to distinguish between the two types of compound:

$$-CHO + [O] \rightarrow -COOH$$

before   after   before   after

Figure 9.12
A red precipitate of copper(I) oxide (Cu_2O) is produced as the aldehyde is oxidised by the Fehling's solution.

Figure 9.13
A silver mirror is produced as the aldehyde is oxidised by Tollen's reagent.

Two reactions are used for this purpose.

- **Fehling's solution** – made by mixing a solution of copper(II) sulphate with an alkaline solution of a salt of 2,3-dihydroxybutanoic acid. This results in a deep-blue complex of copper(II). This complex ion is reduced by aldehydes to copper(I) oxide, which is a red precipitate (Figure 9.12). This test can be used to show the presence of a reducing sugar, such as glucose.

  It should be noted that Benedict's reagent is often preferred because sodium carbonate is used to produce the alkaline conditions, rather than sodium hydroxide, and it is therefore safer.

- **Tollen's reagent** – a solution of silver nitrate made up in an excess of aqueous ammonia. This solution contains the linear $[Ag(NH_3)_2]^+$ complex ion. This is reduced by an aldehyde to form silver, which deposits on the wall of the test-tube and is seen as a 'silver mirror' (Figure 9.13).

Figure 9.14
Ethanal will undergo the triiodomethane reaction.

The triiodomethane (iodoform) reaction

Aldehydes and ketones that possess at least one hydrogen atom on the carbon atom adjacent to the carbonyl group react with iodine in the presence of sodium hydroxide. The reaction results in the formation of yellow crystals of triiodomethane (also known as iodoform) (Figure 9.14).

$$CH_3CHO(l) + 3I_2(aq) + 4NaOH(aq) \rightarrow CHI_3(s) + HCOONa(aq) + 3NaI(aq) + 3H_2O(l)$$
ethanal   triiodomethane

When using this reaction for identification purposes, it should be noted that alcohols which contain the group $-CH(OH)CH_3$, for example, ethanol and propan-2-ol, will also give a positive triiodomethane reaction. This is because during the reaction this grouping of atoms in the alcohol are oxidised by the iodine to give the group $-COCH_3$.

● **Key skills**   ICT
- Carry out an investigation looking at the concentration of alcohol in white wine using acidified potassium dichromate(VI) as an oxidising agent. Use a data-logger to obtain the values for absorption for various concentrations of alcohol with the oxidising agent to plot a calibration curve. Use this calibration curve to obtain the concentration of alcohol in the white wine.

● **Skills task**   Use the Internet to obtain the structures of two aldehydes and two ketones, other than those in this chapter, which are found in nature.

CHECKLIST   After studying Chapter 9 you should know and understand the following terms.

- **Carbonyl group:** $C=O$ group, found in aldehydes and ketones (as well as carboxylic acids and their derivatives, see Chapter 10).
- **Nucleophilic addition reaction:** The attack of a nucleophile on a carbon atom attached to a more electronegative element, in this case oxygen, leading to an addition reaction.
- **Condensation reaction:** Often called addition–elimination reactions because they involve molecules joining together (addition) followed by the elimination of a small molecule, often water.
- **Fehling's solution:** Made by mixing together a solution of copper(II) sulphate with an alkaline solution of a salt of 2,3-dihydroxybutanoic acid. The resulting solution is used to identify aldehydes by the production of a red precipitate of copper(I) oxide (Cu_2O).
- **Tollen's reagent:** A solution of silver nitrate made up in an excess of aqueous ammonia. This solution leaves a silver deposit, a 'silver mirror', in the presence of an aldehyde.

● Examination questions

1 The two functional groups in compound **A**,

⬡—CHO , behave independently.

a State what would be observed if a few drops of compound **A** were added to Fehling's solution and heated. Give the structure of the organic reaction product. (2)

b Using RCHO to represent compound **A**, write an equation for the reaction between RCHO and hydrogen cyanide. State the type of reaction taking place and outline a mechanism. (5)

AQA, A level, Specimen Paper 6421, 2001/2

2 Citral is a colourless natural product, which gives lemons their characteristic flavour and smell. Its structural formula is:

$$CH_3-C=CH-CH_2-CH_2-C=CH-CHO$$
(with CH_3 groups on each $=C$ carbon)

a i How would you show that citral has a carbonyl group, C=O?
Reagent
Observation (2)
ii How would you show that citral is an aldehyde?
Reagent
Observation (2)

b Citral has geometric isomers. Draw them and explain why they are not easily interconvertable. (3)

London, AS/A level, Module 2, Jan 2000

3 a Compound **W** can be converted into three different organic compounds as shown by the reaction sequence below. Give the structures of the new compounds **X**, **Y** and **Z**. (3)

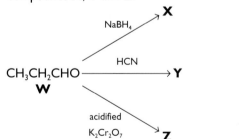

b Outline a mechanism for the formation of **Y**. (4)

NEAB, A/AS level, CH03, Mar 1998

4 β-ionone occurs in oil of violets and is used as a fragrance. Its structure is:

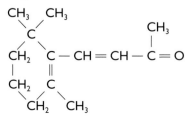

a Would this molecule exhibit optical activity? Give a reason for your answer. (2)

b Could this molecule exhibit geometric isomerism? Give a reason for your answer. (1)

c Give the structure of the product formed when β-ionone reacts with:
i hydrogen bromide (2)
ii 2,4-dinitrophenylhydrazine (2)
iii sodium tetrahydridoborate(III), $NaBH_4$, in ethanol solution at room temperature. (1)

d Would β-ionone have any reaction with an ammoniacal solution of silver nitrate? Give a reason for your answer. (2)

e i Describe how you would carry out the triiodomethane (iodoform) test on an unknown compound in the laboratory. (3)
ii What observation would indicate a positive result of this test? (1)
iii Would β-ionone give this positive result? Give a reason for your answer. (1)

London, A level, CH6, June 1998

5 a Give the structure of the organic product of the reaction between $CH_3COCH_2CH_3$ and aqueous $NaBH_4$. (1)

b

X

Compound **X** reacts separately with HBr, with HCN and with an excess of hydrogen in the presence of a Ni catalyst.
i Name the type of reaction and give the structure of the product formed in the reaction of compound **X** with HBr.
ii Name the type of reaction and give the structure of the product formed in the reaction of compound **X** with HCN.
iii Suggest the structure of the fully saturated product formed when compound **X** reacts with an excess of hydrogen in the presence of a Ni catalyst. (5)

NEAB, A/AS level, CH03, June 1998

6 a Outline the reaction of propanone with the following reagents. Give the equation for the reaction, the conditions, and the name of the organic product.
i hydrogen cyanide (3)
ii sodium tetrahydridoborate(III) (sodium borohydride) (3)

b i Give the mechanism for the reaction in part **a i**. (3)
ii What type of mechanism is this? (1)

iii What feature of the carbonyl group makes this type of mechanism possible? Explain how this feature arises. (2)

iv Explain briefly, by reference to its structure, why ethene would not react with HCN in a similar way. (1)

London, A level, Module Test 4, June 1998

7 This question concerns the compounds in the following reaction scheme:

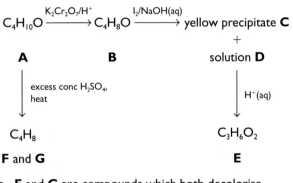

a **F** and **G** are compounds which both decolorise bromine water. **F** has two stereoisomers.

i What functional group is present in both **F** and **G**? (1)

ii Give the structural formulae of both stereoisomers of **F**. (2)

iii Explain how these two isomers arise. (2)

iv Write the structural formula of **G**. (1)

b **B** cannot be oxidised by acidified potassium dichromate(VI) solution.

i Write the structural formula of **B**. (1)

ii Draw the general structural features of molecules which can be detected by the reaction with iodine and alkali. (2)

iii Give the structure of the substance in solution **D**, and of the product **E**. (2)

c The mass spectrum of **A** gives peaks at *m/e* 29 and 45, amongst others, That at 45 is the largest. This spectrum shows no molecular ion peak, which would be expected at *m/e* 74. **A** is chiral.

i Give the structural formula for **A**. (1)

ii Identify the ions responsible for the peaks at *m/e* 45 and 29, and hence suggest why the molecule shows no molecular ion peak. (3)

d The infrared spectrum of **A** is shown below. The very broad peak at 3500 cm^{-1} is due to the presence of an —OH group. This peak becomes much narrower when diluted with benzene and moves to 3600 cm^{-1}.

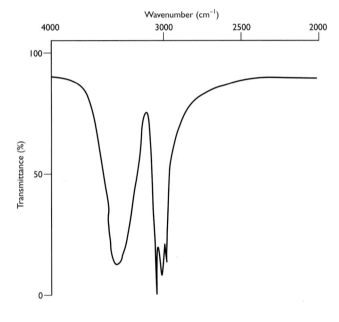

Suggest why the —OH absorption peak changes as **A** is diluted with benzene. (2)

London, A level, Module Test 4, Jan 1998

10 Carboxylic acids and their derivatives

STARTING POINTS ● Carboxylic acids are formed by the oxidation of primary alcohols or aldehydes.
● Carboxylic acids are weak acids.
● Esters are sweet-smelling substances formed by the reaction of an alcohol with a carboxylic acid.

● Carboxylic acids

Figure 10.1
The carboxyl group.

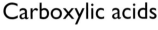
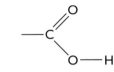

Carboxylic acids contain the **carboxyl group**, **—COOH**. This group contains both the carbonyl group —C=O and the alcohol group —OH attached to the same carbon atom (Figure 10.1).

Many carboxylic acids occur naturally (Figure 10.2). The simplest carboxylic acid is methanoic acid, HCOOH, which is produced by ants. Lactic acid, a more complex carboxylic acid, is formed both when milk ferments (the production of yoghurt) and in muscles when exercising.

Figure 10.2
a The sting from red ants contains methanoic acid.
b Citrus fruits such as lemons and oranges contain citric acid.
c Athletes will have lactic acid building up in their muscles as they race.

a
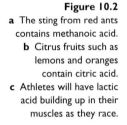

b

methanoic acid

citric acid

c

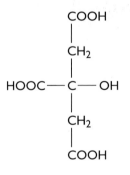

lactic acid

Naming carboxylic acids

As with all homologous series, the name of unbranched or branched carboxylic acids is taken from the name of the alkane which has the same number of carbon atoms in the longest chain of the acid. The carbon atom of the carboxyl group is usually carbon number one in terms of the naming of the molecules. Table 10.1 gives the names, structures and properties of the first four aliphatic carboxylic acids. Carboxylic acids have relatively high melting and boiling points compared to their parent alkanes, due to the formation of hydrogen bonds between pairs of the carboxylic acid molecules (**dimers**) in organic solvents (Figure 10.3). Pure ethanoic acid, for example, has a high melting point of 16.7 °C, due to the hydrogen bonded dimers it contains.

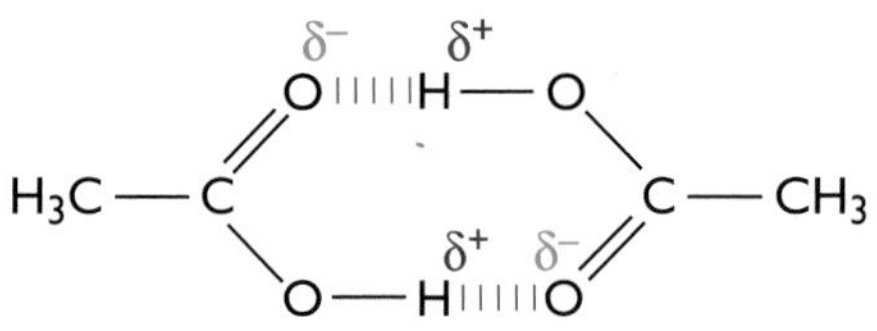

Figure 10.3
Hydrogen-bonded dimers are formed in organic solvents, which leads to the relatively high melting and boiling points of carboxylic acids.

If the temperature drops below this value the ethanoic acid freezes and becomes a solid. Due to this tendency to change to a solid, pure ethanoic acid is known as glacial ethanoic acid (Figure 10.4).

Table 10.1

Name	Formula	Structure	Melting point (°C)	Boiling point (°C)
Methanoic acid	HCOOH	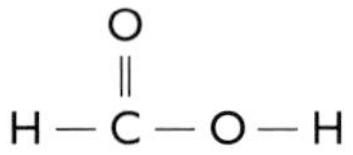	8.5	110.7
Ethanoic acid	CH₃COOH	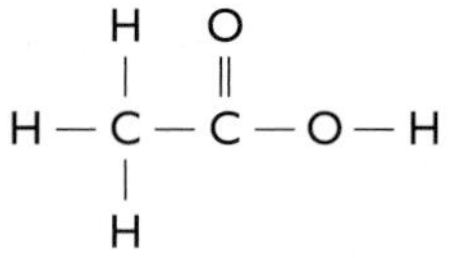	16.7	118.0
Propanoic acid	CH₃CH₂COOH	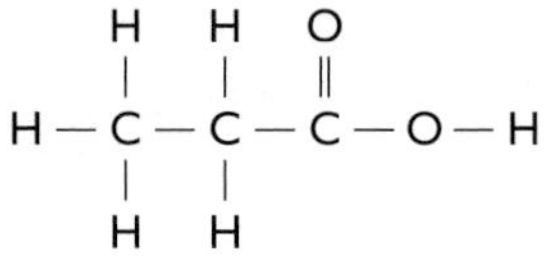	−21.7	141.0
Butanoic acid	CH₃CH₂CH₂COOH	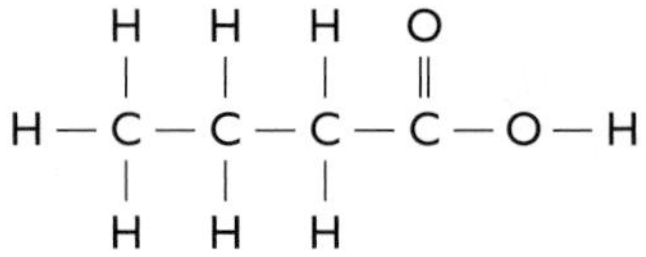	−5.0	163.0

Figure 10.4 (above) Strong hydrogen bonding gives pure ethanoic acid a high melting point. In cold weather it can be seen to freeze.

Figure 10.5
Hydrogen bonds form between ethanoic acid molecules and water molecules, leading to a high solubility.

It should be noted that the trend in the melting points of the first four carboxylic acids is a very atypical trend for any homologous series.

The ability to form hydrogen bonds also explains the high solubility of the smaller carboxylic acids in water (Figure 10.5). Larger carboxylic acids, those with a long hydrocarbon chain attached to the carboxylic acid group, are not as soluble as smaller carboxylic acid molecules.

Dicarboxylic acid molecules, with two —COOH groups, exist. Ethanedioic acid, for example, is a toxic acid found in rhubarb leaves (Figure 10.6).

Aromatic carboxylic acids, such as benzoic acid (also known as benzenecarboxylic acid) also exist.

Figure 10.6
Rhubarb leaves contain ethanedioic acid (also known as oxalic acid), a dicarboxylic acid. It is this acid that makes rhubarb leaves poisonous.

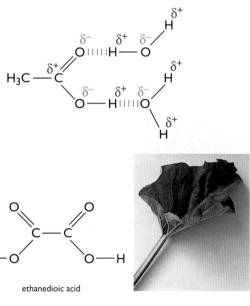

ethanedioic acid

All carboxylic acids have characteristic and usually unpleasant odours. A common example is ethanoic acid, which is responsible for the pungent smell of vinegar. The smell of body odour is caused by the formation of a mixture of various long-chain carboxylic acids. Each individual produces their own mixture of acids: these can be used by specially trained tracker dogs who can identify the trace from a piece of a person's clothing and follow it to the person themselves (Figure 10.7).

Figure 10.7
Tracker dogs have a very acute sensitivity to smells produced by the carboxylic acids in a person's body odour.

1 Write down the structures of the following carboxylic acids:
a hexanoic acid
b propanedioic acid
c 1,4-benzene dicarboxylic acid.

Chemical properties of carboxylic acids

Carboxylic acids are weak acids, meaning that they do not dissociate completely in solution:

$$CH_3COOH + H_2O \rightleftharpoons CH_3COO^- + H_3O^+$$

Methanoic acid is the strongest acid of the simple carboxylic acids. Acidity decreases with the length of the carbon chain (Table 10.2). Carboxylic acids with long chains are insoluble in water and, therefore, do not form acidic solutions. They will, however, react with bases to form salts and water.

Table 10.2
pK_a values of some carboxylic acids.

Carboxylic acid	Formula	pK_a
Methanoic acid	HCOOH	3.75
Ethanoic acid	CH_3COOH	4.76
Chloroethanoic acid	$ClCH_2COOH$	2.85
Dichloroethanoic acid	$Cl_2CHCOOH$	1.25
Trichloroethanoic acid	Cl_3CCOOH	0.66

When discussing the relative acidities of the carboxylic acids both the strength of the O—H bond in the carboxylic group and the stability of the negative ion produced when the acid dissociates need to be considered.

Methanoic acid ($pK_a = 3.75$) is a stronger acid than ethanoic acid ($pK_a = 4.76$) because of the differences between the structures of the two molecules. The methyl group in the ethanoic acid molecule has a tendency to push electrons away from itself towards the carboxyl group. This is known as a **positive inductive effect**. This effect increases the strength of the O—H bond in the molecule, decreasing the extent of dissociation. In the ethanoate ion, formed by dissociation, the positive inductive effect increases further the size of negative charge on the carboxylate oxygen.

This reduces the stability of the ion and increases the chance that the molecule will re-form, decreasing the concentration of H^+(aq) in the solution and leading to a less acidic solution (Figure 10.8). The methyl group is not present in the methanoic acid molecule and so no further negative charge is pushed onto the carboxyl group. This leads to the methanoate ion being more stable than the ethanoate ion, and consequently methanoic acid is the stronger acid.

Figure 10.8
The stability of the ethanoate ion formed on dissociation of ethanoic acid molecules is reduced due to the positive inductive effect of the methyl group. The effect increases the size of the negative charge on the oxygen atom destabilising it. In addition, the inductive effect increases the strength of the O—H bond in the ethanoic acid molecule, making dissociation less likely.

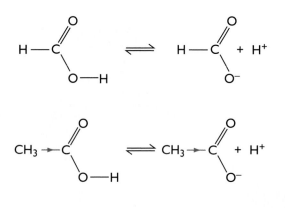

The acidity of carboxylic acids can be modified by substituting hydrogen atoms in the hydrocarbon chain with other atoms. For example, the substitution of one of the hydrogen atoms in ethanoic acid with the electronegative atom chlorine produces chloroethanoic acid, which is a stronger acid than ethanoic acid. Further substitution produces the even stronger acids dichloroethanoic acid and trichloroethanoic acid (see Table 10.2).

Figure 10.9
The presence of the electronegative chlorine atoms enhances the acidity of the halogenated ethanoic acid.

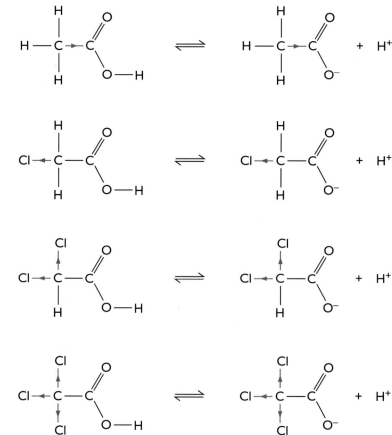

2 Discuss, with the aid of diagrams, which will be the stronger acid – bromoethanoic acid or chloroethanoic acid.

The introduction of electronegative chlorine atoms into the ethanoic acid molecule increases acidity (Figure 10.9). Chlorine atoms show a **negative inductive effect**, withdrawing negative charge from other parts of the acid molecules, and so reducing the strength of the O—H bond, and from the carboxylate group in the anions formed, stabilising them. The more chlorine atoms are present, the stronger the acid becomes.

Preparation of carboxylic acids

There are various routes for the preparation of a carboxylic acid:

- oxidation of a primary alcohol
- oxidation of aldehydes
- acid hydrolysis of nitriles
- acid hydrolysis of esters
- acid hydrolysis of amides.

Oxidation of a primary alcohol

A primary alcohol such as ethanol can be oxidised, initially to the aldehyde ethanal, then on to ethanoic acid. Oxidation can be carried under reflux using either acidified potassium manganate(VII) solution or acidified potassium dichromate(VI) solution:

$$CH_3CH_2OH + 2[O] \rightarrow CH_3COOH + H_2O$$

(see *Introduction to Advanced Chemistry*, Chapter 19, page 207).

Oxidation of aldehydes

Instead of starting from primary alcohols, carboxylic acids can be prepared by oxidising aldehydes, such as propanal, using acidified potassium manganate(VII) or acidified potassium dichromate(VI):

$$CH_3CH_2CHO + [O] \rightarrow CH_3CH_2COOH$$

(see page 139).

Acid hydrolysis of nitriles

A nitrile such as propanenitrile can be hydrolysed under reflux and acid conditions to give propanoic acid:

$$CH_3CH_2CN(l) + 2H_2O(l) + HCl(aq) \rightarrow CH_3CH_2COOH(aq) + NH_4Cl(aq)$$

Acid hydrolysis of esters

Acid hydrolysis of an ester, such as ethyl ethanoate, gives a carboxylic acid, ethanoic acid, and an alcohol, ethanol, as products:

3 Write equations for the following reactions:
a the oxidation of propan-1-ol
b the acid hydrolysis of propanamide
c the hydrolysis of the ester HCOOCH_2CH_3.

$$CH_3COOCH_2CH_3(aq) + H_2O(l) \rightleftharpoons CH_3COOH(aq) + CH_3CH_2OH(aq)$$

Acid hydrolysis of amides

Acid hydrolysis of an amide, such as ethanamide, results in the breaking of the C—N bond in the amide, producing, in this instance, ethanoic acid:

$$CH_3CONH_2(aq) + H_2O(l) + HCl(aq) \rightarrow CH_3COOH(aq) + NH_4Cl(aq)$$

Reactions of carboxylic acids

Reactions as acids

Carboxylic acids are proton donors and undergo all the reactions of mineral acids, such as hydrochloric acid and sulphuric acid:

- with alkalis:

$$CH_3COOH(aq) + NaOH(aq) \rightarrow CH_3COO^-Na^+(aq) + H_2O(l)$$

- with bases:

$$2CH_3COOH(aq) + MgO(s) \rightarrow (CH_3COO^-)_2Mg^{2+}(aq) + H_2O(l)$$

- with metals

$$CH_3COOH(aq) + Ca(s) \rightarrow (CH_3COO^-)_2Ca^{2+}(aq) + H_2(g)$$

- with metal carbonates

$$2CH_3COOH(aq) + Na_2CO_3(aq) \rightarrow 2CH_3COO^-Na^+(aq) + CO_2(g) + H_2O(l)$$

- with metal hydrogencarbonates

$$CH_3COOH(aq) + NaHCO_3(aq) \rightarrow CH_3COO^-Na^+(aq) + CO_2(g) + H_2O(l)$$

The reactions above occur much more slowly with carboxylic acids than with strong acids of the same concentration.

4 Write equations for the reactions of:
a sodium metal with ethanoic acid
b propanoic acid with potassium hydroxide
c ethanoic acid with copper(II) oxide.

Figure 10.10
This salt of a long-chain carboxylic acid has both hydrophobic and hydrophilic sections.

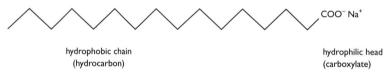

hydrophobic chain
(hydrocarbon)

hydrophilic head
(carboxylate)

The salts of long-chain carboxylic acids are used in the production of detergents (Figure 10.10). The hydrophobic, hydrocarbon, section of a carboxylate ion is insoluble in water but it is more soluble in organic substances. The ionic end is soluble in water but not in organic substances. When added to water that has grease in it, the hydrophobic, hydrocarbon, section is attracted to the grease and becomes embedded in it. On the other hand, the hydrophilic end is not attracted to the grease but is strongly attracted to the water molecules. When the water is stirred, the grease becomes completely surrounded by the detergent. The grease is therefore 'solubilised' and removed from the greasy dish or dirty clothes. The grease remains in the water and is unable to reattach itself (Figure 10.11).

Figure 10.11
The removal of grease.

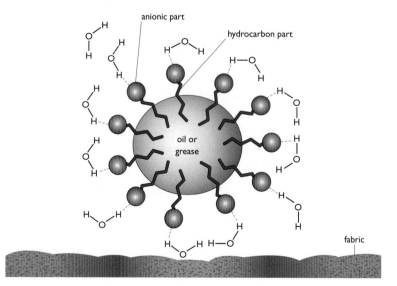

Reduction reactions

Carboxylic acids can be prepared by the oxidation of primary alcohols. Carboxylic acids can, therefore, be reduced to primary alcohols. The reduction can be carried out using lithium tetrahydridoaluminate, $LiAlH_4$, in dry ether solution or sodium borohydride in aqueous solution. The process would be carried out by heating under reflux:

$$RCOOH(l) + 4[H] \rightarrow RCH_2OH(l) + H_2O(l)$$

[H] represents the hydrogen obtained from the reducing agent which enables the reduction process to occur.

Reaction with phosphorus(v) chloride

In the reaction of a carboxylic acid with phosphorus(v) chloride, PCl_5, the —OH group of the acid is replaced with a chlorine atom, forming another acid derivative, an acid chloride. Acid chlorides are much more reactive than the corresponding carboxylic acid and are used extensively in industry as they undergo faster reactions (see page 155).

Esterification

An **ester** is formed by replacing the —OH group in a carboxylic acid (or the Cl atom in an acid chloride, see page 155) with an oxygen atom bonded to a carbon chain. The general structure of an ester is shown below.

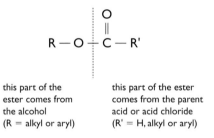

| this part of the ester comes from the alcohol (R = alkyl or aryl) | this part of the ester comes from the parent acid or acid chloride (R' = H, alkyl or aryl) |

Esterification takes place by heating a carboxylic acid and an alcohol under reflux in the presence of an acid catalyst such as concentrated sulphuric acid. The reaction is reversible, so the ester produced needs to be separated from the reaction mixture.

The ester methyl propanoate can be produced starting with propanoic acid and methanol:

$$CH_3OH(l) \quad + \quad CH_3CH_2COOH(l) \quad \rightleftharpoons \quad CH_3O\overset{\overset{\displaystyle O}{\|}}{-}CCH_2CH_3(l) \quad + \quad H_2O(l)$$

methanol   propanoic acid   methyl propanoate

Esters

Naming esters

Unlike the other homologous series considered so far, which contain a single carbon chain, esters contain two carbon chains separated by an oxygen atom. The carbon chain that contains the carbonyl group ($>C=O$) has a name taken from the carboxylic acid from which it was formed. The other carbon chain, separated from the carbonyl group by an oxygen atom, takes its name from the alcohol from which it was made.

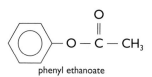

$$CH_3CH_2 - O - \overset{\overset{\displaystyle O}{\|}}{C} - CH_2CH_3$$

ethyl propanoate

$$CH_3CH_2 - O - \overset{\overset{\displaystyle O}{\|}}{C} - H$$

ethyl methanoate

$$CH_3CH_2CH_2 - O - \overset{\overset{\displaystyle O}{\|}}{C} - CH_3$$

propyl ethanoate

Aromatic esters also exist, for example:

phenyl ethanoate

Uses of esters

Because of their strong and pleasant smells, esters are used in perfumes. Esters have important uses in the food industry, where they are used in flavouring foods and in providing the pleasant smells associated with some of them. Almost all esters employed in the food industry are produced by esterification reactions involving carboxylic acids (or acid chlorides) with alcohols.

Local anaesthetics such as benzocaine and procaine are esters; both are made by the esterification of 4-aminobenzoic acid. Benzocaine is prepared by the esterification reaction of 4-aminobenzoic acid with ethanol (Figure 10.12).

Another major use of esters is in the production of polyesters; this is discussed in Chapter 13, page 196.

Figure 10.12
Benzocaine is just one of several esters used as local anaesthetics.

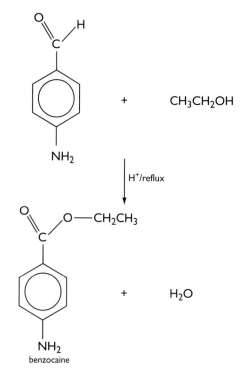

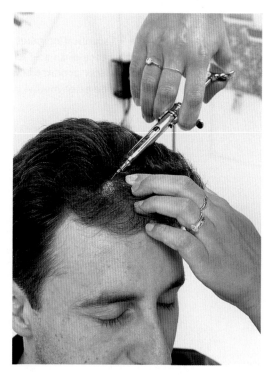

Reactions of esters

Hydrolysis

The hydrolysis reaction of esters is the reverse of the esterification reaction, in that the products of ester hydrolysis are the reactants of the esterification reaction. In the hydrolysis reaction the water molecule acts as a nucleophile, attacking the carbonyl carbon atom of the ester. The reaction is slow due to the poor nucleophilic properties of the water molecule, but can be speeded up using an acid catalyst.

$$RCOOR' + H_2O \overset{H^+(aq)}{\rightleftharpoons} RCOOH + R'OH$$

Base-catalysed hydrolysis of esters

5 Write chemical equations for the reactions of:
a methanoic acid with propan-1-ol
b ethanoic acid with ethane-1,2-diol
c benzoic acid with methanol
d methyl ethanoate with potassium hydroxide solution
e ethyl methanoate with hydrochloric acid solution.

When a base such as sodium hydroxide is used to hydrolyse an ester a sodium salt of the acid is produced along with an alcohol. For example:

$$CH_3COOCH_2CH_3 + NaOH \rightarrow CH_3COO^-Na^+ + CH_3CH_2OH$$

Base-catalysed hydrolysis produces much higher yields of the products than acid-catalysed hydrolysis. This is due to the hydroxide ions being more powerful nucleophiles than water molecules. In addition, the process is not an equilibrium process: the sodium hydroxide reacts with the acid formed to give the sodium salt of the acid, preventing the reverse reaction.

Base-catalysed hydrolysis of an ester is known as saponification, because such a reaction is used in the manufacture of soap. The difference is that in the manufacture of soap much larger, naturally occurring, esters are used. Soaps have been made for thousands of years and it is known that the Romans used base-catalysed hydrolysis of natural esters, **fats** and **oils**, to produce soap.

Fats, oils and soap manufacture

Naturally occurring fats and oils are esters of the alcohol propane-1,2,3-triol (also known as glycerol) (Figure 10.13).

Figure 10.13
Propane-1,2,3-triol.

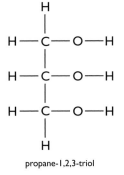

propane-1,2,3-triol

Fats and oils are formed by the reaction of propane-1,2,3-triol with long-chain carboxylic acids, such as those shown in Table 10.3. There is no structural difference between a fat and an oil. The two terms merely distinguish between the physical state of the substances at room temperature: a fat is a solid and an oil is a liquid. These carboxylic acids in nature are usually referred to as **fatty acids**. The fatty acids have been given trivial names after their sources.

Table 10.3

Structure	Traditional name	Source
$CH_3(CH_2)_{14}COOH$	Palmitic acid	Palm trees
$CH_3(CH_2)_{16}COOH$	Stearic acid	Suet
$CH_3(CH_2)_7CH{=}CH(CH_2)_7COOH$	Oleic acid	Olive trees

Figure 10.14 (left)
The triester formed from palmitic acid and propane-1,2,3-triol.

Figure 10.15 (right)
A general mixed triester.

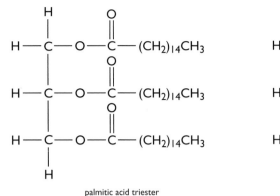

palmitic acid triester

Since propane-1,2,3-triol has three —OH groups, carboxylic acids can form ester linkages with all three to form a **triester**. For example, the triester formed from propan-1,2,3-triol and palmitic acid has the structure shown in Figure 10.14.

Mixed triesters can be formed in which the three acid groups are not the same: this often happens in nature (Figure 10.15).

Vegetable oils tend to have a greater proportion of unsaturated fatty acids, those with many C=C bonds per molecule, than animal fats. Saturated fats, those with no C=C bonds, are linked with the thickening of arteries. It is now recognised that unsaturated fats and, in particular, polyunsaturated fats are much better for health.

Figure 10.16 (below)
It is now recognised that spreads which have a relatively high content of polyunsaturated fats are much better for your health.

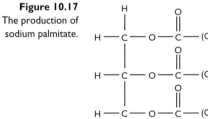

The consumption of low-fat spreads and margarines has greatly increased (Figure 10.16). These products contain vegetable oils that have been **hydrogenated** to remove some of the double bonds. This is done using hydrogen gas under pressure and a nickel catalyst at 180 °C. Not all the double bonds in the polyunsaturated fatty acid are hydrogenated – just those needed to produce the correct physical properties required of a margarine or low-fat spread.

Any fat or oil can be identified by carrying out a hydrolysis reaction, which breaks it down into propane-1,2,3-triol and its fatty acids. The hydrolysis is usually carried out in concentrated sodium hydroxide solution, to give propane-1,2,3-triol and the sodium salts of the fatty acids (Figure 10.17). This is how soap is manufactured. These sodium salts are one of the main ingredients of soap. Sodium palmitate, for example, is one of the main components of Palmolive soap, sodium oleate is the other (Figure 10.18).

Figure 10.17
The production of sodium palmitate.

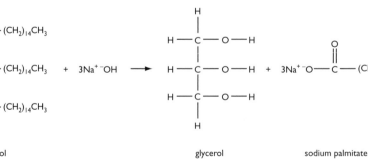

triester formed from glycerol and palmitic acid

glycerol

sodium palmitate

Figure 10.18
Palmolive soap contains sodium palmitate, a product of the base-catalysed hydrolysis of a triester.

6 Give the structure of the triester made by reacting stearic acid with glycerol.

Acid chlorides

The acid chlorides (also known as acyl chlorides) are another homologous series of compounds that are derivatives of carboxylic acids. They are formed from carboxylic acids in reactions which cause the replacement of the —OH group in the acid with a chlorine atom.

Acid chlorides are much more reactive than the carboxylic acids from which they are derived. This is due to the presence of the electronegative chlorine atom which increases the polarity of the carbonyl group, making the carbonyl carbon more susceptible to attack by nucleophiles (Figure 10.19).

Figure 10.19
a An acid chloride.
b The increased reactivity of the acid chloride compared to the parent carboxylic acid is due to the presence of the electronegative chlorine atom.

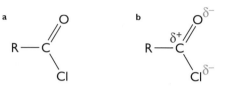

The acid chlorides are named after the acid from which they are derived, by adding —**oyl**. For example:

CH_3COCl ethanoyl chloride
CH_3CH_2COCl propanoyl chloride

Preparation of acid chlorides

Carboxylic acids can be converted into acid chlorides by reaction with thionyl chloride, $SOCl_2$, or with phosphorus(V) chloride, PCl_5. The reaction is hazardous, due to the corrosive and toxic nature of both the reactants and products, and is carried out by heating under reflux. The acid chloride produced can be separated by distillation.

$$CH_3COOH + SOCl_2 \rightarrow CH_3COCl + SO_2 + HCl$$

$$CH_3COOH + PCl_5 \rightarrow CH_3COCl + POCl_3 + HCl$$

Reactions of acid chlorides

Acid chlorides undergo nucleophilic addition–elimination reactions, also known as condensation reactions. The mechanism of a general reaction of this type is shown in Figure 10.20. The carbonyl carbon in an acid chloride molecule is very electron-deficient, due to the electronegative oxygen atom to which it is bonded, and this makes it vulnerable to attack by a nucleophile. When the nucleophile attacks and forms a bond to this carbon atom the carbonyl double bond is broken, leaving the oxygen atom with a negative charge. Chloride is then lost from the intermediate ion and the carbonyl double bond is re-formed.

Figure 10.20
The mechanism of a nucleophilic addition–elimination reaction.

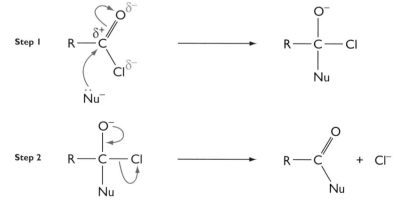

Hydrolysis of acid chlorides

The reaction of an acid chloride molecule with water produces a carboxylic acid along with hydrogen chloride gas (which can be seen as white fumes). The HCl will dissolve in the water to produce hydrochloric acid.

$$CH_3COCl + H_2O \rightarrow CH_3COOH + HCl$$

If the hydrolysis is carried out in alkaline conditions the salt of the carboxylic acid is formed:

$$RCOCl + 2KOH \rightarrow RCOO^-K^+ + H_2O + KCl$$

Reaction with alcohols

Acid chlorides react with alcohols to form esters. The advantage of preparing esters by this method is that the reaction goes to completion, unlike the reaction between carboxylic acids and alcohols which results in an equilibrium. As a result this is the main route by which esters are formed in industry.

$$CH_3COCl + CH_3CH_2OH \rightarrow CH_3COOCH_2CH_3 + HCl$$

Reaction with ammonia

7 Write equations for the following reactions:
a propanoic acid with thionyl chloride
b methanoic acid with phosphorus(V) chloride
c propanoyl chloride with ammonia
d propanoyl chloride with sodium hydroxide solution.

Acid chlorides react with ammonia to produce another acid derivative, a primary amide (see Chapter 11, page 164). The nitrogen atom in the ammonia molecule has a lone pair of electrons, which enables the ammonia molecule to act as a nucleophile.

$$CH_3COCl + NH_3 \rightarrow CH_3CONH_2 + HCl$$

When this reaction is carried out it is usual to ensure that an excess of ammonia is used, as it can neutralise the hydrogen chloride formed in the reaction.

Amines can also react with acid chlorides to form *N*-substituted amides. The amide produced in this reaction is a secondary amide:

$$CH_3COCl + CH_3NH_2 \rightarrow CH_3CONHCH_3 + HCl$$

Acid anhydrides

Acid anhydrides are another homologous series of acid derivatives. They are formed by the reaction between an acid chloride and the sodium salt of a carboxylic acid. For example, ethanoic anhydride can be prepared as shown by the equation below:

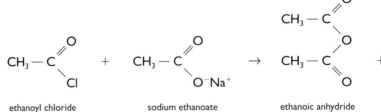

ethanoyl chloride   sodium ethanoate   ethanoic anhydride

Acid anhydrides are important chemicals in the chemical industry as they are used as **acylating agents**. Acylating agents are compounds used to introduce acyl groups

into molecules when they react.

Reactions of acid anhydrides

Acid anhydrides react with water, ammonia, amines and alcohols in a similar way to acid chlorides, producing the same organic products. Instead of hydrogen chloride being produced, a carboxylic acid is the other product:

- with water

$$(CH_3CO)_2O + H_2O \rightarrow 2CH_3COOH$$

- with ammonia

$$(CH_3CO)_2O + NH_3 \rightarrow CH_3CONH_2 + CH_3COOH$$

- with amines

$$(CH_3CO)_2O + CH_3NH_2 \rightarrow CH_3CONHCH_3 + CH_3COOH$$

- with alcohols

$$(CH_3CO)_2O + C_2H_5OH \rightarrow CH_3COOCH_2CH_3 + CH_3COOH$$

Acid anhydrides are not as reactive as acid chlorides but are safer to use, hence their industrial importance.

Aspirin manufacture

One of the major uses of ethanoic anhydride is in the preparation of the analgesic, aspirin. Aspirin is synthesised from salicylic acid (2-hydroxybenzoic acid) (Figure 10.21). The pain-killing properties of salicylic acid have been known of for hundreds of years. It can be obtained from willow bark. Salicylic acid, however, although an effective painkiller has the side-effect of causing irritation of the stomach due to its acidity. Aspirin is an ester of salicylic acid and it has less effect on the stomach whilst retaining the pain-killing properties (Figure 10.22).

Ethanoyl chloride could be used to manufacture aspirin but the anhydride is preferred as it is cheaper than the acid chlorides and not as reactive: this allows the reaction to be controlled more precisely.

Figure 10.21 (below) The manufacture of aspirin using ethanoic anhydride.

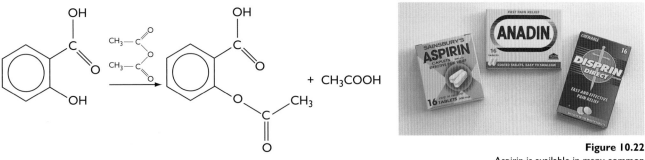

Figure 10.22
Aspirin is available in many common over-the-counter painkillers.

Key skills   ICT

- Use a data-logger to produce and compare titration curves for the reactions of ethanoic and propanoic acids with sodium hydroxide. From the curves identify a suitable indicator which could be used.

Number

- Calculations involving the determination of the percentage of ethanoic acid in different vinegars.

Skills task

Using the information in this chapter and other sources, including the Internet, produce an A4 sheet to highlight the advantages and disadvantages of aspirin compared to other painkillers such as paracetamol and ibuprofen.

> **CHECKLIST** After studying Chapter 10 you should know and understand the following terms.
>
> - **Carboxyl group:** The —COOH functional group is found in all carboxylic acid molecules.
> - **Carboxylic acids:** Compounds with the general formula R—COOH, where R represents an alkyl or aryl group or a hydrogen atom.
> - **Esterification:** The reversible reaction between a carboxylic acid and an alcohol in the presence of an acid catalyst resulting in the formation of an ester.
> - **Esters:** Compounds with the general formula RCOOR', where R and R' are alkyl or aryl groups.
> - **Fats and oils:** Esters of propane-1,2,3-triol (glycerol). A fat is a solid and an oil is a liquid at room temperature.
> - **Acid chlorides:** Compounds with the general formula R—COCl, where R represents an alkyl or aryl group.
> - **Acid anhydrides:** Compounds with the general formula $(RCO)_2O$, where R represents an alkyl or aryl group.

Examination questions

1 **a** Write an equation for the formation of ethyl ethanoate from ethanoyl chloride and ethanol. Name and outline the mechanism for the reaction taking place. (6)

 b Explain why dilute sodium hydroxide will cause holes to appear in clothing made from polymers such as Terylene but a poly(phenylethene) container can be used to store sodium hydroxide. (2)

AQA, A level, Specimen Paper 6421, 2001/2

2 A student prepared benzoic acid, C_6H_5COOH, by hydrolysing methyl benzoate, $C_6H_5COOCH_3$, using the following method.

- Dissolve 4.0 g of sodium hydroxide in water to make 50 cm³ of an alkaline solution.
- Add the aqueous sodium hydroxide to 2.70 g of methyl benzoate in a 100 cm³ flask and set up the apparatus for reflux.
- Reflux this mixture for 30 minutes.
- Distil the mixture and collect the first 2 cm³ of distillate.
- Pour the residue from the flask into a beaker and add dilute sulphuric acid until the solution is acidic.
- Filter the crystals obtained and recrystallise from hot water to obtain the benzoic acid.

The overall equation for this hydrolysis is:

$$C_6H_5COOCH_3 + H_2O \rightarrow C_6H_5COOH + CH_3OH$$

The student obtained 1.50 g of benzoic acid, C_6H_5COOH.

 a Name the functional group that reacts during this hydrolysis. (1)

 b i Calculate how many moles of methyl benzoate were used.

  ii What was the concentration, in mol dm⁻³, of the aqueous sodium hydroxide used?

 iii Calculate the percentage yield of the C_6H_5COOH obtained by the student.

 iv Suggest why the percentage yield was substantially below 100%. (9)

 c i Why was the residue from the flask acidified before recrystallising?

  ii Why were the crystals recrystallised? (2)

 d Infrared spectroscopy can be used to monitor the progress of a chemical reaction.

  i Predict the key identifying features of the infrared spectra of methyl benzoate and its hydrolysis products, benzoic acid and methanol.

  ii How could you use infrared spectroscopy to show that the ethanol did **not** contain any benzoic acid. (6)

OCR, A level, Specimen Paper A7882, Sept 2000

3 **a** Acyl halides such as ethanoyl chloride, CH_3COCl, are useful synthetic intermediates because they are quite reactive.

  i Suggest synthetic steps, giving reagents and conditions, by means of which ethanol could be converted to ethanoyl chloride. (5)

  ii Ethanoyl chloride, CH_3COCl, reacts violently with water at room temperature, whereas chloroethane, CH_3CH_2Cl, reacts extremely slowly with water under the same conditions. Suggest a reason why CH_3COCl is **more** susceptible to nucleophilic attack than CH_3CH_2Cl. (2)

 b Ethyl ethanoate $CH_3COOCH_2CH_3$ can be made either from ethanol and ethanoic acid or from ethanol and ethanoyl chloride.

  i Give the equations for the two reactions. (2)

  ii Suggest why the reaction with ethanoyl chloride gives a better yield of the ester. (1)

London, A level, Module 4, June 1999

4 a Write an equation for the reaction between ethanoyl chloride and dimethylamine. Name and outline the mechanism of this reaction. (6)

b Aspirin is manufactured by the reaction of 2-hydroxybenzenecarboxylic acid with ethanoic anhydride. Write an equation for this reaction and give two reasons why ethanoic anhydride, rather than ethanoyl chloride, is used. (4)

NEAB, A level, CH06, Mar 1998

5 Consider the following reaction scheme:

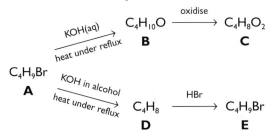

The compound **A** is not chiral, but **E** is. **C** will liberate carbon dioxide from sodium hydrogencarbonate solution. **B** and **C** will react with one another under suitable conditions.

a Draw the structural formulae for **A**, **B**, **C** and **E**. (4)

b Give a qualitative test for the functional group in **D**. (2)

c Write the equation for the reaction of **C** with sodium hydrogencarbonate. (2)

d Give the structure of the organic compound formed by reaction between **B** and **C**. (1)

e Suggest suitable reagents and conditions for the conversion of **B** to **C**. (2)

f Substance **C** has K_a value of $1.51 \times 10^{-5}\,mol\,dm^{-3}$.
i Find the pH of a $0.100\,mol\,dm^{-3}$ solution of **C**. (3)
ii What property is shown by a solution which contains a mixture of **C** and its sodium salt? (1)
iii Calculate the pH of a solution formed by adding $5.5\,g$ of the sodium salt of **C** to $500\,cm^3$ of a solution of **C** of concentration $0.100\,mol\,dm^{-3}$. (4)

London, A/AS level, Module Test 2, Jan 1998

6 a Write an equation for the formation of ethyl ethanoate from ethanoic acid and ethanol in the presence of a strong acid catalyst.
State the type of reaction taking place. (2)

b Write an equation for the formation of ethyl ethanoate from ethanoyl chloride and ethanol. Name and outline the mechanism of this reaction. (6)

c Suggest two reasons why the reaction in part **b** is a more effective way of preparing ethyl ethanoate than that in part **a**. (2)

d The proton NMR spectrum of ethyl ethanoate has peaks at δ 4.1, 2.0 and 1.2. Given that the peak at δ 2.0 is a singlet, deduce the splitting patterns of the other two peaks. You may find the proton chemical shift data provided in the table below helpful. (2)

Type of proton	δ (p.p.m.)
RCH$_3$	0.7–1.2
R$_2$CH$_2$	1.2–1.4
R$_3$CH	1.4–1.6
RCOCH$_3$	2.1–2.6
ROCH$_3$	3.1–3.9
RCOOCH$_3$	3.7–4.1

NEAB, A level, CH06, June 1998

7 A carboxylic acid **A** contains 40.0% carbon, 6.70% hydrogen and 53.3% oxygen by mass. When $10.0\,cm^3$ of an aqueous solution of **A**, containing $7.20\,g\,dm^{-3}$, was titrated against $0.050\,mol\,dm^{-3}$ sodium hydroxide, the following pH readings were obtained.

Volume NaOH (cm^3)	0.0	2.5	5.0	7.5	10.0	14.0	15.0	16.0	17.5	20.0	22.5
pH	2.5	3.2	3.5	3.8	4.1	4.7	5.2	9.1	11.5	11.8	12.0

a i Plot a graph of pH (on the *y* axis) against volume of NaOH (on the *x* axis).
Use the graph to determine the end-point of the titration. Hence calculate the relative molecular mass of **A**. (8)
ii Calculate the value of K_a for **A** and state its units. (4)

b Calculate the molecular formula of **A**. Given that **A** contains one asymmetric carbon atom, deduce its structure. Briefly indicate your reasoning. (4)

c The mass spectrum of **A** shows major peaks at *m/e* values of 15, 30, 45 and 75. Suggest the formula for the species responsible for each of these four peaks. (4)

d Describe a series of tests you would perform in order to confirm the structure obtained in part **b**, given that you already know that it is an acid. (5)

London, A level, CH6, June 1998

11 Organic nitrogen-containing molecules

STARTING POINTS ● Ammonia is a base.
● Amino acids are naturally occurring organic compounds that possess —NH$_2$ and —COOH groups.
● Proteins are biopolymers formed from amino acids by condensation reactions.

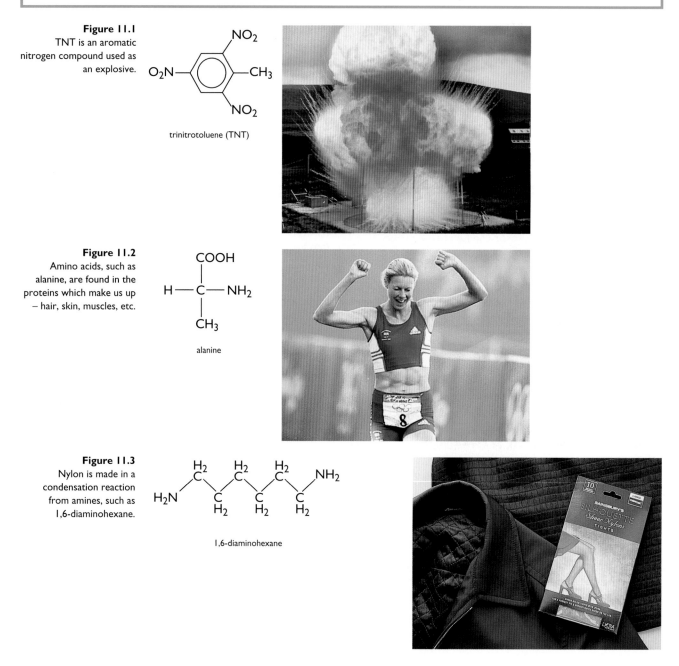

Figure 11.1
TNT is an aromatic nitrogen compound used as an explosive.

trinitrotoluene (TNT)

Figure 11.2
Amino acids, such as alanine, are found in the proteins which make us up – hair, skin, muscles, etc.

alanine

Figure 11.3
Nylon is made in a condensation reaction from amines, such as 1,6-diaminohexane.

1,6-diaminohexane

The presence of a nitrogen atom in an organic molecule gives rise to many important and diverse homologous series. In this chapter we look at the major homologous series of nitrogen-containing compounds.

Amines

Amines are derivatives of ammonia. These are compounds in which one or more of the hydrogens in ammonia have been replaced by an alkyl group. Amines have similar chemical properties, therefore, to ammonia.

Amines are of one of three types.

- If a single hydrogen has been replaced then a **primary amine** is formed. For example, methylamine (Figure 11.4a).
- A **secondary amine** is formed when two of the hydrogens from ammonia are replaced. For example, dimethylamine (Figure 11.4b).
- If alkyl groups replace all three hydrogens then a **tertiary amine** is formed. For example, trimethylamine (Figure 11.4c).

Figure 11.4
a Methylamine is a primary amine.
b Dimethylamine is a secondary amine.
c Trimethylamine is a tertiary amine.

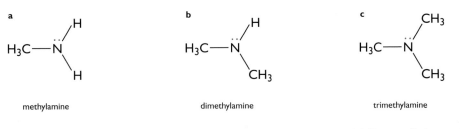

methylamine   dimethylamine   trimethylamine

The alkyl groups do not necessarily need to be of the same type. If different alkyl groups are present then the names of the alkyl groups are given in *alphabetical* order. For example:

$$CH_3 — N — CH_2CH_3$$
$$|$$
$$CH_2CH_2CH_3$$

is called ethylmethylpropylamine.

On occasions it is necessary to use the *amino* prefix to state the presence of an —NH_2 group in a compound. For example:

$$CH_3CH_2CHCH_3$$
$$|$$
$$NH_2$$

can be called 2-aminobutane.

1 Name the following amines:
  a $(CH_3)_2NH$
  b $(CH_3)_2NCH_2CH_3$
  c $CH_3CH_2CH_2CH_2 NHCH_2CH_3$
2 Draw the structural formulae of the following amines:
  a triethylamine
  b diethylmethylamine
  c ethyldimethylamine.

Aromatic amines are compounds in which an —NH_2 group is bonded to the benzene ring. The most common is phenylamine, $C_6H_5NH_2$:

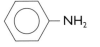

Properties of amines

Physical properties

Figure 11.5
Hydrogen bonds can form between amines and water molecules.

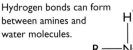

The smaller amines are gases or volatile liquids (low boiling points) with a powerful characteristic odour of rotting fish. They are soluble in water. The solubility arises from their ability to form hydrogen bonds with water molecules (Figure 11.5). This ability stems from the electronegative nature of the nitrogen atom in the amine molecule.

The larger the amine molecule the lower the solubility. This is because the larger the carbon chain the greater the disruption to the hydrogen bonding, which is not compensated for by bonding with the carbon chain of the amine.

Chemical properties

Basic properties

The nitrogen atom in amines has a lone pair of electrons. This enables it to form a dative covalent bond with, for example, a hydrogen ion (H^+). A base is an electron pair donor and when it donates a lone pair of electrons to a hydrogen ion it also acts as a base by accepting the proton:

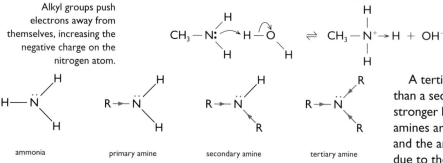

Figure 11.6 (below) Alkyl groups push electrons away from themselves, increasing the negative charge on the nitrogen atom.

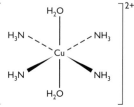

ammonia   primary amine   secondary amine   tertiary amine

A tertiary amine is generally a stronger base than a secondary amine, which in turn is a stronger base than a primary amine. Primary amines are stronger bases than both ammonia and the aromatic amine phenylamine. This is due to the fact that alkyl groups push electrons away from themselves, making the negative charge on the nitrogen atom greater (Figure 11.6). The greater the negative charge the more likely that it will act as an electron pair donor (that is, a base).

As bases, amines will react with an acid to give a salt. For example:

$$CH_3NH_2 + HCl \rightarrow CH_3NH_3{}^+Cl^-$$

The free amine can be obtained by the addition of sodium hydroxide to a solution of the salt.

As a ligand

Figure 11.7 (below) The lone pair of electrons on the nitrogen atom of ammonia enables it to act as a **ligand**, donating a pair of electrons to the central copper ion to form a dative covalent bond.

$$[Cu(NH_3)_4(H_2O)_2]^{2+}$$

The lone pair of electrons on the nitrogen atom in an amine can be donated to a transition metal ion to form a complex ion in very much the same way as ammonia reacts (Figure 11.7):

$$[Cu(H_2O)_6]^{2+} + 4NH_3 \rightarrow [Cu(NH_3)_4(H_2O)_2]^{2+} + 4H_2O$$

A similar complex can be formed with methylamine and aqueous copper ions:

$$[Cu(H_2O)_6]^{2+} + 4CH_3NH_2 \rightarrow [Cu(CH_3NH_2)_4(H_2O)_2]^{2+} + 4H_2O$$

The bond between the amine and the transition metal ion is a dative covalent bond.

Nucleophilic properties

The nitrogen atoms in amines have lone pairs of electrons, and so amines are able to act as nucleophiles: they are able to donate electron pairs to areas of positive charge. For example, halogenoalkanes undergo nucleophilic substitution reactions with amines (Figure 11.8). This method can be used to prepare secondary amines:

Figure 11.8 The nucleophilic substitution reaction between bromoethane and methylamine.

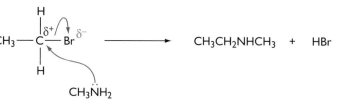

The products react together to produce the salt:

$$CH_3CH_2 - \overset{\overset{\displaystyle CH_3}{|}}{\underset{\underset{\displaystyle H}{|}}{N^+}} - H \quad Br^-$$

In this example a primary amine, CH_3NH_2, has reacted with a halogenoalkane to form a secondary amine. The secondary amine that has been formed also contains a nitrogen atom with a lone pair of electrons, so this will also react with the halogenoalkane.

This time a tertiary amine is formed, again as the salt:

$$CH_3CH_2Br + CH_3CH_2NHCH_3 \rightarrow CH_3CH_2NCH_3 + HBr$$
$$\overset{|}{\underset{CH_2CH_3}{}}$$

3 Write balanced chemical equations for the following reactions of amines:
a propylamine with ethanoyl chloride
b methylamine with hydrochloric acid
c ethylmethylamine with iodomethane.

Properties of aromatic amines

Aromatic amines, such as phenylamine, have different properties to those of the aliphatic amines discussed above. The reason for the different properties is that the lone pair of electrons on the nitrogen atom of the amine group becomes part of the delocalised electron structure of the benzene ring. This makes the lone pair of electrons unavailable. Aromatic amines are:

- weaker bases than aliphatic amines as the lone pair cannot be donated as easily
- poorer nucleophiles than aliphatic amines, leading to slower reactions
- less able to act as ligands.

Quaternary ammonium salts

Figure 11.9
The general structure of a quaternary ammonium salt.

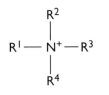

Quaternary ammonium salts are formed when all the hydrogen atoms in an ammonium ion are replaced by alkyl groups (Figure 11.9). They are white crystalline ionic solids which are water soluble.

A surfactant molecule is one with an ionic or polar group directly attached to a hydrocarbon chain. The molecule has, therefore, a hydrocarbon chain that is water-hating (hydrophobic) and an ionic or polar part that is water-loving (hydrophilic). Surfactants help to separate grease from dirty dishes and then keep the grease away from the dishes so that the grease molecules do not reattach themselves. Unsurprisingly, surfactants are one of the important substances present in washing-up liquids (Figure 11.10). Surfactants are also present in soaps, bath salts, shampoos and hair conditioners.

Figure 11.10
Surfactant molecules play an important part in washing-up liquids.

The keratin molecules in hair become anionic when washed in alkaline solutions, such as those present in many hair shampoos. Quaternary ammonium salts, which are cationic surfactants, are found in hair conditioners. The positive charge on the quaternary ammonium salt allows it to bond to the keratin. By using a suitable quaternary ammonium salt with a large side-chain, hair seems to spread out, giving it extra body (Figure 11.11).

Figure 11.11
Cetyltrimethylammonium bromide, a quaternary ammonium salt, is used in hair conditioners. The large side-chain on this ion gives hair extra body.

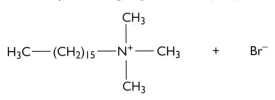

cetyl trimethylammonium bromide

Amides

Amides can be prepared by reacting ammonia or an amine with an acid chloride (or any other acid derivative). The lone pair of electrons on the nitrogen atom is attracted to the $\delta+$ carbonyl carbon. Reacting an acid chloride with ammonia gives a **primary amide**:

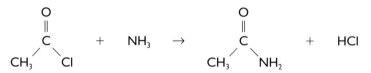

With a primary amine, such as methylamine, a **secondary amide** is formed:

$$\text{CH}_3\text{COCl} + \text{CH}_3\text{NH}_2 \rightarrow \text{CH}_3\text{CONHCH}_3 + \text{HCl}$$

Secondary amides contain the group:

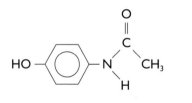

The highlighted group is known as the **peptide group**, or **peptide link**. This group is present both in proteins and in some condensation polymers, the polyamides (see Chapter 13, page 194).

Paracetamol – a useful amide

Paracetamol is a widely used and successful analgesic (painkiller) and has antipyretic (fever-reducing) properties. It has the structure:

paracetamol

It was first synthesised in 1878 by the reduction of 4-nitrophenol using tin in glacial ethanoic acid. The 4-aminophenol produced by the reducing action of tin, in acid conditions, undergoes a further reaction with the ethanoic acid to produce paracetamol. Other synthetic routes, many of which are guarded to prevent competition, have superseded this method.

Paracetamol was first made available in the United Kingdom as 500 mg tablets in 1956. Initially, it was only available by prescription, but it is now the most purchased over-the-counter analgesic in the UK. If paracetamol is taken as recommended it has few side-effects. However, if a large overdose is taken and left untreated, severe damage to the liver will result, and even death.

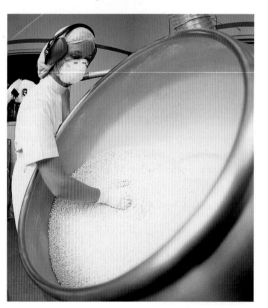

Figure 11.12
The commercial manufacture of paracetamol tablets – the most used painkiller in the UK.

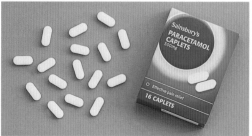

Reactions of amides

Reaction with water (hydrolysis)

When amides are reacted with water the C—N bond is broken. The reaction occurs in both acidic and alkaline conditions. The products formed depend on the conditions used.

$$CH_3COOH \quad + \quad NH_4^+$$

↑ in acidic conditions

$$CH_3-\overset{\overset{O}{\|}}{C}-NH_2 \; + \; H_2O \; \rightarrow \; \left[CH_3COOH \; + \; NH_3\right]$$

↓ in alkaline conditions

$$CH_3COO^- \quad + \quad NH_3$$

- With HCl, the products are CH_3COOH and $NH_4^+Cl^-$.
- With NaOH, the products are $CH_3COO^-Na^+$ and NH_3.

Hydrolysis of the peptide group occurs in the body when proteins are broken down to produce the individual amino acids from which they are composed (see page 170). In the body, enzymes catalyse the breakdown of the C—N bond.

Reaction with phosphorus(V) oxide

Reaction with phosphorus(V) oxide is a dehydration reaction, with the phosphorus(V) oxide acting as the dehydrating agent. In the reaction water is lost from an amide molecule, resulting in the formation of a compound known as a **nitrile**.

$$RCONH_2 \xrightarrow{\;P_2O_5\;} RCN + H_2O$$

Nitriles are important intermediates in many organic synthesis reactions.

- Nitriles can be *reduced* to give amines using reducing agents such as lithium tetrahydridoaluminate(III) or hydrogen gas:

$$RCN + 4[H] \xrightarrow{\;LiAlH_4 \text{ or } H_2\;} RCH_2NH_2$$

- Nitriles can be *hydrolysed* in either acid or alkaline conditions to form carboxylic acids or salts of carboxylic acids:

$$RCN + 2H_2O + HCl \rightarrow RCOOH + NH_4Cl$$

$$RCN + H_2O + NaOH \rightarrow RCOO^-Na^+ + NH_3$$

Reaction with bromine in alkaline solution

The reaction of amides with bromine, often called the **Hofmann degradation reaction** is important synthetically. The reaction of the amide with bromine and then with aqueous sodium hydroxide under reflux results in the formation of an amine with *one fewer carbon atom* than the starting amide. This type of reaction is important in the manufacture of new molecules. An equation for the general reaction is shown below:

$$RCONH_2 + Br_2 + 4KOH \rightarrow RNH_2 + 2KBr + K_2CO_3 + 2H_2O$$

4 a Hydrolysis of ethanamide, CH_3CONH_2, with a solution of hot sodium hydroxide gives two products, one of them a gas. Name the gas produced and give a chemical test to prove the presence of this gas.
   b Hydrolysis of ethanamide with a solution of hydrochloric acid does not give a gas. Name the two products formed from this reaction and indicate whether the resulting solution would be acidic or alkaline.

5 Identify the two functional groups in a paracetamol molecule.
6 Suggest a possible problem associated with the preparation of paracetamol starting from 4-aminophenol and ethanoyl chloride.
7 Draw the structures of the expected organic products from the reaction of paracetamol with:
   a chlorine in the presence of ultraviolet light
   b dilute hydrochloric acid.
8 Give a simple chemical test which would allow you to identify the presence of the $CH_3CO—$ group in the paracetamol molecule.
9 Paracetamol forms a weakly acidic solution which has a pK_a of 9.5 at 20 °C.
   a Calculate a value of K_a for paracetamol at 20 °C.
   b Account for the acidic character of paracetamol.

Amino acids

Proteins are a very important part of our bodies. Skin, nails, hair, tendons, ligaments and muscles contain protein molecules. Proteins are essential for the correct function of our immune system and for the transport of oxygen to our vital organs (Figure 11.13). Proteins known as **enzymes** catalyse the chemical reactions which go on inside our bodies and are essential for life. Many different protein molecules exist in our bodies.

The huge range of proteins are all made up from around 20 amino acid molecules, bonded together in different arrangements, as dictated by our DNA (Figure 11.14).

Figure 11.13 (above) Haemoglobin, a protein, transports oxygen in the blood.

Figure 11.14 (right) The 20 α-amino acids from which our protein molecules are composed.

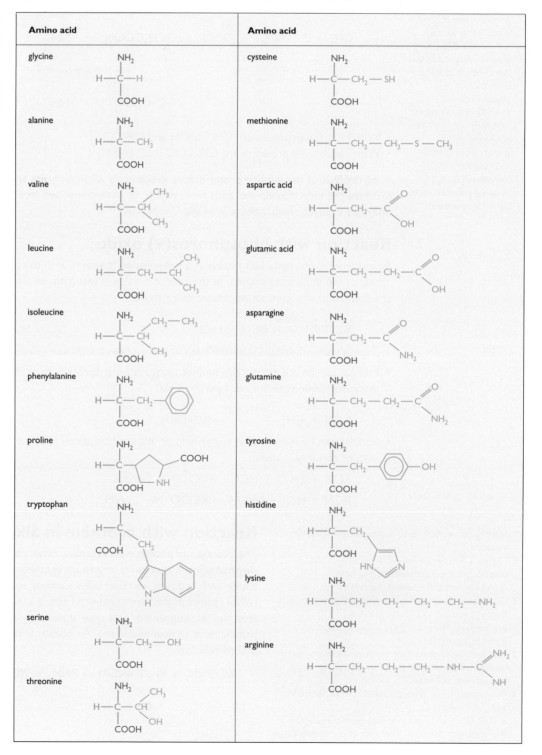

Of these 20 amino acids, 11 are called non-essential amino acids whilst 9 are essential. The essential amino acids, such as leucine, can only be obtained through diet. The non-essential amino acids can be synthesised in the body by modification of an essential amino acid, although they can also be taken in as food.

Amino acids contain the **amino group**, **—NH₂**, and the **carboxylic acid group**, **—COOH**. They are examples of **bifunctional compounds**, because they have two functional groups. This is limited to the naturally occurring amino acids which have the general formula:

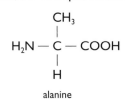

Amino acids and optical isomerism

Optical isomerism is a type of stereoisomerism. Optical isomerism arises because it is possible to arrange four different groups around a central carbon atom in two different ways. One of the isomers or **enantiomers** is a **non-superimposable** mirror image of the other isomer. This property is known as **chirality**, from the Greek word meaning 'handedness'. A carbon atom that has four different groups of atoms bonded to it is known as a **chiral carbon atom**. All amino acid molecules, apart from glycine, show chirality.

Take, for example, alanine:

$$H_2N - \underset{\underset{H}{|}}{\overset{\overset{CH_3}{|}}{C}} - COOH$$

alanine

Figure 11.15
Alanine exists as two non-superimposable mirror images.

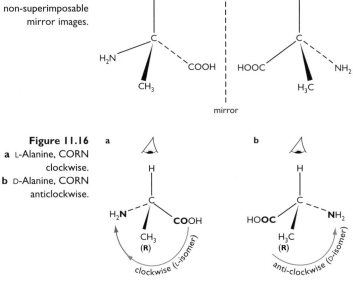

Figure 11.16
a L-Alanine, CORN clockwise.
b D-Alanine, CORN anticlockwise.

The central carbon atom is a chiral carbon as it has four different groups attached to it. It can exist, therefore, as two enantiomers (Figure 11.15).

In our bodies only one enantiomer of an amino acid is produced, known as the L-enantiomer. The non-superimposable enantiomer is called the D-enantiomer. The L- and D-enantiomers can be distinguished from one another using the CORN rule. The letters CO, R and N represent three of the four different groups attached to the chiral carbon atom:

COOH **R** **N**H₂

To use this method of identification you need to look down through the H—C bond of the amino acid molecule. If whilst looking down through the H—C bond the word CORN is spelled in a clockwise direction then the enantiomer is a L-enantiomer. A D-enantiomer would have the groups arranged in an anticlockwise manner (Figure 11.16).

The two enantiomers of the same molecule will undergo the same chemical reactions and most of their physical properties will be the same. They do differ, however, in the way they behave in the presence of other chiral molecules.

If an amino acid such as alanine was made in a laboratory, a 50:50 mixture of the D and L versions would be manufactured. Such a mixture is called a **racemic mixture**.

The two enantiomers of the same molecule will have different effects on plane-polarised light. This effect can be shown using an instrument called a polarimeter. Light that has been passed through a polarising filter becomes polarised. In polarised light the light waves vibrate in only one direction (Figure 11.17, overleaf). A second polarising filter can be aligned at the other end of the instrument to allow the polarised light to pass through. When enantiomers are placed between the two polarising filters it is found that the polarised light no longer emerges from the second filter. The enantiomers rotate the plane of plane-polarised light. In order to allow the light to emerge from the polarimeter the second filter needs to be rotated. Enantiomers that rotate the plane of polarised light clockwise (to the right) are D-enantiomers and those which rotate the plane anticlockwise (to the left) are L-enantiomers.

10 Draw 3D diagrams to represent:
a L-serine
b D-aspartic acid.

Figure 11.17
Plane-polarised light is unaffected if no enantiomer is present. If an enantiomer is placed between the filters, the plane-polarised light is rotated.

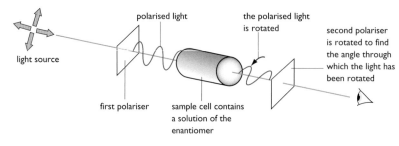

Properties of amino acids

The *basic* amino group and the *acidic* carboxylic acid group in amino acids interact. The carboxylic acid group donates a proton to the amino group, which accepts the proton, and a **zwitterion** is formed.

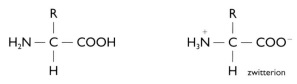

- Amino acids are solids at room temperature due to the ionic bonds that exist between the zwitterions.
- Solutions of amino acids are neutral provided that they contain only one —COOH group and one —NH₂ group.
- A zwitterion contains regions of positive and negative charge.
- Zwitterions are present in aqueous solutions of the amino acids. Very few molecules, as opposed to ions, are present.
- The presence of these ionic groups means that amino acids are soluble in water.

11 Draw the structure of the zwitterions of the following amino acids:
a glycine (R = H)
b aspartic acid (R = CH₂COOH)
c serine (R = CH₂OH)
Which of the above will be the most acidic?

Any acidity of amino acids in water is due to the $-NH_3^+$ group, whereas any basic properties are as a result of the $-COO^-$ group. The pH of an amino acid solution, therefore, depends on the 'strength' of these two groups within a particular zwitterion. For example, the simplest α-amino acid glycine, H_2NCH_2COOH, has a pH just below 7 as the zwitterion is slightly more acidic than basic. Some amino acids, however, contain more than one acid or amine group, which affects the pH of their solutions markedly.

Buffering action of amino acids

There is little change in the pH of an amino acid solution, due to the neutralising effect of the zwitterion, when small quantities of acid or alkali are added. The amino acid solution is effectively acting as a buffer solution (see Chapter 5, page 72).

- Addition of alkali:

$$H_3N^+-CHR-COO^- + OH^- \rightarrow H_2N-CHR-COO^- + H_2O$$

On the addition of alkali the zwitterion acts as an acid and donates a proton to the hydroxide ion, forming water. This removes the hydroxide ion from solution and maintains the pH.

- Addition of acid:

$$H_3N^+-CHR-COO^- + H^+ \rightarrow H_3N^+-CHR-COOH$$

When an acid is added, the hydrogen ions from the acid are attracted to the carboxylate group of the zwitterion, removing the hydrogen ions from solution.

12 Write chemical equations for the reactions between:
a glycine and hydrochloric acid
b aspartic acid and sodium hydroxide
c lysine and hydrochloric acid.

Proteins

Figure 11.18
The reaction between glycine and alanine to form a dipeptide.

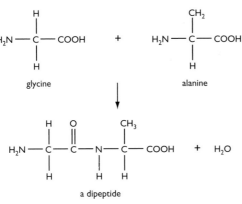

glycine   alanine

a dipeptide

13 Write structural formulae for the dipeptide formed from each of the reactions below:
a a molecule of valine and a molecule of serine
b a molecule of alanine and a molecule of leucine.

Protein molecules are natural polymers, polyamides, and are formed as a result of the reaction between amino acid molecules. The amino group of one amino acid carries out a nucleophilic attack on the carboxylic acid group of another amino acid, as a result of the lone pair of electrons on the nitrogen atom (Figure 11.18). A **dipeptide** is formed when two amino acids react with one another.

A **polypeptide** is a general classification used for small protein molecules composed of ten or fewer amino acids. Several polypeptides react together to form a protein molecule.

Protein molecules are 'put together' in the body following instructions that are chemically inscribed in our body's DNA. DNA is present in the nuclei of all our cells. **DNA** has a double helix structure and is composed of two polynucleotide chains with sugar–phosphate backbones (Figure 11.19).

Figure 11.19
Part of the DNA double helix.

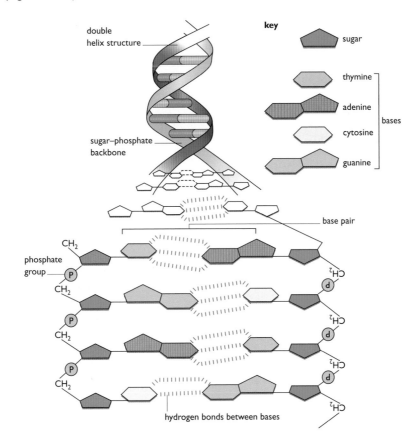

Each sugar unit on the DNA backbone has one of four bases bonded to it. The four bases are adenine (A), thymine (T), cytosine (C) and guanine (G). The two polynucleotide chains are hydrogen bonded to one another through the pairing of adenine with thymine and cytosine with guanine. This means that the order of the bases on one chain determines the order on the other chain. Other molecules, known as **RNA** molecules, use the coding on one of the DNA chains to produce protein molecules from amino acids which are present in the cells.

Structure of proteins

Figure 11.20
Prolonged refluxing in acidic conditions will hydrolyse a protein molecule back to the amino acids from which it was made.

The sequence in which amino acids are bonded together to form the protein molecules is known as the **primary structure**. Taking the protein molecule and breaking it down by hydrolysis of the peptide links in the molecule can determine this primary structure. This would normally be done by prolonged heating under reflux, under acidic conditions (Figure 11.20).

After hydrolysis, the amino acids present in the product mixture can be identified by chromatography. Many different forms of chromatography can be employed, but paper chromatography is perhaps the simplest (see Chapter 14, page 213). The chromatographic process is run twice, using two different solvents, with the solvents travelling at 90° to one another on the same piece of paper (Figure 11.21). This is 2D chromatography.

Figure 11.21
2D paper chromatography can be used to identify the amino acids present, from a hydrolysed protein molecule.

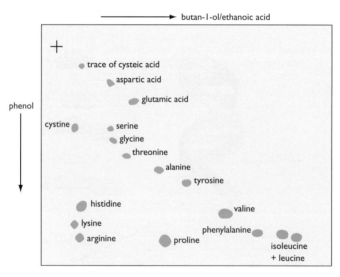

Figure 11.22
A method used to identify the sequence of amino acids in a protein.

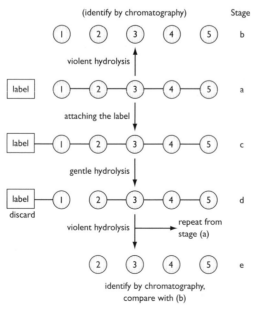

2D chromatography identifies which amino acids are present in the protein molecule but does not tell us the order in which they were bonded to one another. To do this a 'special molecule' is used which bonds itself to the end amino acid in the protein and makes it more susceptible to hydrolysis (Figure 11.22). The 'special molecule' and the end amino acid can be separated from the remainder of the protein chain, which can then be fully hydrolysed and analysed chromatographically to find out which amino acid is missing. The process is then repeated to determine the sequence of amino acids in the entire protein molecule.

It has been essential to determine the amino acid sequence of some molecules, such as insulin (Figure 11.23), in order to produce the whole protein molecule synthetically; in the case of insulin this is for the treatment of diabetes (Figure 11.24).

Figure 11.23
Knowing the sequence of amino acids in an insulin molecule has enabled the production of synthetic insulin for the treatment of diabetic patients.

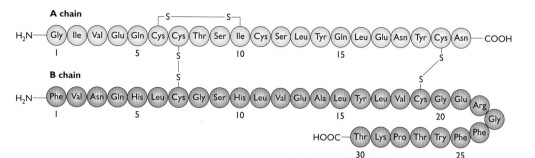

Figure 11.24
Regular injections of insulin are used in the treatment of diabetes.

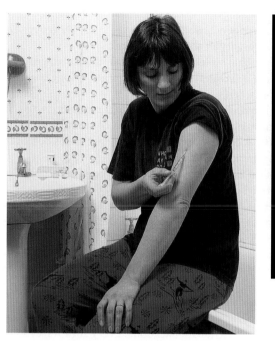

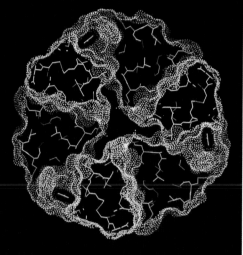

Figure 11.25
A computer model of the quaternary structure of an insulin hexamer.

The **secondary structure** of a protein molecule is the shape adopted by the chain of amino acids by virtue of free rotation around some bonds in the protein and the restrictions placed on the rotations by inter- and intra-molecular hydrogen bonds. Parts of the molecule may coil up or adopt helix or sheet structures as a result.

The **tertiary structure** of a protein describes how the helices and sheets of an individual protein molecule fold up further and are eventually stabilised by more inter- and intra-molecular bonding to produce the final 3D shape of the molecule.

The **quaternary structure** of a protein results when several protein molecules join together to form a larger structure (Figure 11.25).

● **Key skills** ICT

- Ions are formed during the hydrolysis of ethanamide. The rate of the hydrolysis process can be determined by using a data-logger to measure the conductivity of the solution. Use this information to determine whether acidic or alkaline conditions produce a faster rate of hydrolysis.
- Use modelling software to help visualise the primary, secondary, tertiary and quaternary structure of proteins.

● **Skills task** Use the Internet and other sources to obtain information about the function of insulin in the human body. Use the information to produce a Powerpoint presentation about insulin and diabetes.

CHECKLIST   After studying Chapter 11 you should know and understand the following terms.

- **Amines:** Compounds formed by the substitution of one or more hydrogen atoms in ammonia with alkyl or aryl groups. The number of alkyl or aryl groups present determines whether the amine is primary, secondary or tertiary.
- **Ligands:** Molecules or ions with lone pairs of electrons that form dative covalent bonds with a central metal ion to form a complex ion.
- **Quaternary ammonium salts:** Salts formed when all the hydrogen atoms in an ammonium ion, NH_4^+, are replaced by alkyl groups.
- **Amides:** Nitrogen compounds with the general formula, $R—CONH_2$ where R represents an alkyl or aryl group.
- **Nitriles:** Nitrogen compounds which contain the $—C\equiv N$ group. Their general formula is R—CN, where R represents an alkyl or aryl group.
- **Amino acids:** Bifunctional molecules that contain the amino group ($—NH_2$) and a carboxylic acid group ($—COOH$). They join together in long chains to make proteins by condensation reactions.
- **Enzymes:** Protein molecules that act as biological catalysts.
- **Chiral carbon atom:** Has four different groups of atoms bonded to it.
- **Optical isomerism:** A type of stereoisomerism that arises due to the presence of a chiral carbon atom within a molecule.
- **Enantiomers:** Optical isomers that are mirror image forms of the same molecule. They have opposite effects on plane-polarised light.
- **Zwitterions:** Ions that possess both a positive and a negative charge. They are formed in amino acids when the amino group accepts protons from the acid groups which donate them.
- **Peptides:** Compounds made up of chains of amino acids.
- **Dipeptide:** The simplest peptide, produced when two amino acids react with one another.
- **Polypeptide:** A general classification used to describe small protein molecules composed of ten or fewer amino acids.
- **DNA:** Deoxyribonucleic acid. It is a double helix made up of two polynucleotide chains. Both chains have a sugar–phosphate backbone. Every sugar unit has one of the four bases, adenine, cytosine, guanine and thymine, linked to it. DNA belongs to a group of biopolymers called the nucleic acids. DNA is involved in the polymerisation of amino acids in a specific order to form particular protein molecules required by a cell.
- **RNA:** Ribonucleic acid. RNA molecules are single-chained and are formed by transcription from DNA. They contain the four bases adenine, cytosine, guanine and uracil. These molecules are involved in the production of proteins in cells.
- **Proteins:** Long-chained molecules formed by condensation reactions between amino acids producing the peptide link, $—CONH—$, between the amino acid residues.
- **Primary structure:** The sequence in which the amino acid molecules are bonded together in a protein.
- **Secondary structure:** The way in which the long-chained protein molecules are arranged and held in place by hydrogen bonding, both within and between chains. This causes parts of the molecule to coil up or to adopt helical or sheet structures as a result.
- **Tertiary structure:** How the helices and sheets of an individual protein molecule fold up and are eventually stabilised by inter- and intra-molecular bonding to produce the final 3D shape.
- **Quaternary structure:** The structure formed by several protein molecules joining together to form a larger structure.

Examination questions

1 State the type of isomerism shown by 2-hydroxypropanoic (**lactic**) acid, $CH_3CH(OH)COOH$, and point out the structural feature of the molecule which causes the existence of two isomers. With the aid of diagrams, show the structural relationship between the two isomers and state how these isomers can be distinguished. (5)

AQA, A level, Specimen Paper 6421, 2001/2

2 **a** Explain why ethylamine is a Brønsted–Lowry base. (2)
  b Why is phenylamine a weaker base than ethylamine? (2)
  c Ethylamine can be prepared from the reaction between bromoethane and ammonia.
   i Name the type of reaction taking place.
   ii Give the structures of **three** other organic substitution products which can be obtained from the reaction between bromoethane and ammonia. (4)
  d Write an equation for the conversion of ethanenitrile into ethylamine and give one reason why this method of synthesis is superior to that in part **c**. (2)

AQA, A level, Specimen Paper 6421, 2001/2

3 **Aspartame** is an artificial sweetener which is about 200-times sweeter than sucrose. It is a methyl ester of a dipeptide formed from two α-amino acids: aspartic acid and phenylalanine.

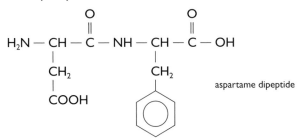

aspartame dipeptide

  a Explain the meaning of the term **dipeptide**. (2)
  b Draw a skeletal formula for the aspartame dipeptide. (2)
  c One way to show the composition of aspartame is to hydrolyse it and then separate and identify the substances formed.
   i Briefly describe the procedure which is used to hydrolyse a protein or a dipeptide. (2)
   ii Three compounds are formed when aspartame is hydrolysed. Name these **three** compounds and explain what further information you would need to establish the structure of aspartame. (4)
  d Both aspartic acid and phenylalanine exist in two optical isomer forms, labelled the L form and the D form. Only the L amino acids occur naturally in proteins. Many of the chemical reactions of L and D amino acids are identical, but in the body they can behave differently. For example, D-phenylalanine tastes sweet, but L-phenylalanine is bitter.

i Copy the diagram of the structure of phenylalanine which follows, and place an asterisk (*) next to the atom which indicates to you that there are two optical isomers. (1)

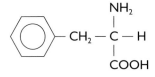

ii Copy the mirror line below, then draw diagrams to illustrate the two optical isomers of phenylalanine. (2)

mirror

  e A solution of phenylalanine in water will contain few molecules like the one shown in part **d i**. Draw a structure for the form of phenylalanine most likely to be present in an aqueous solution. (2)
  f A disadvantage of using aspartame as a sweetener is that it breaks down much faster than sucrose. In soft drinks stored in cans or bottles, about 10% of the aspartame is destroyed each month. Suggest a reason why aspartame breaks down in this way. (1)

OCR, A level, Specimen Paper A7887, Sept 2000

4 Three naturally-occurring amino acids are valine, glutamic acid and cystine.

$$CH(CH_3)_2$$
$$H_2N - C - COOH$$
$$H$$
valine

$$CH_2CH_2COOH$$
$$H_2N - C - COOH$$
$$H$$
glutamic acid

$$CH_2SH$$
$$H_2N - C - COOH$$
$$H$$
cystine

  a Draw a tripeptide composed from each of these three amino acids. (2)
  b Describe how a section of a protein containing these three amino acids can contribute to ordered secondary and tertiary structures of a protein with an α-helix.
   Your answer should include diagrams and should discuss the relevant bonds and forces that stabilise each structure. (In this question, 1 mark is available for the quality of written communication.) (11)

OCR, A level, Specimen Paper A7882, Sept 2000

5 3-Amino-4-methylbenzenesulphonic acid can be obtained from methylbenzene in three steps:

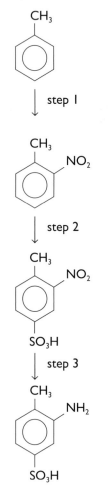

a For **each** step, name the type of reaction taking place and suggest a suitable reagent or combination of reagents. (6)

b Identify the reactive inorganic species present in step 2 and outline a mechanism for this reaction. (5)

NEAB, A level, CH06, June 1998

6 A weakly acidic compound **A** ($C_8H_9NO_2$) gives no reaction with a mixture of sodium nitrite and hydrochloric acid at low temperature. On heating **A** under reflux with excess dilute hydrochloric acid hydrolysis occurs giving two compounds **B** (C_6H_8NClO) and **C** ($C_2H_4O_2$). After complete separation from all other components of the mixture, **C** was found to react with sodium carbonate, liberating a colourless gas.

When **B** was cooled to about 3 °C and sodium nitrite solution added at the same temperature, a colourless solution **D** ($C_6H_5ClN_2O$) was formed. Addition of phenol to this solution gave a coloured precipitate **E**.

The mass spectra of **A**, **B** and **D** all show a peak at $m/e = 76$.

a Deduce structures for compounds **A** to **E** inclusive. (5)

b Show how you have used the data to reach your conclusions, writing equations where appropriate. (8)

London, A level, CH6, June 1998

Aromatic chemistry

- Aromatic hydrocarbons are a family of unsaturated organic molecules based on rings of carbon atoms.
- The most important aromatic hydrocarbon is benzene.
- A σ (sigma) bond is a single covalent bond formed when two atomic orbitals on adjacent molecules overlap in a linear manner.
- A π (pi) bond is a bond formed by the sideways overlap of two atomic p orbitals.
- Delocalised electrons are electrons which become 'detached' from their atom and spread throughout the structure of the molecule.
- An electrophile is a positively charged species which attacks electron-rich areas of molecules.
- A carbocation is any ion that contains a carbon atom with a positive charge on it.

Figure 12.1
Aromatic compounds have many uses. These include the manufacture of medicines, polymers and dyes.

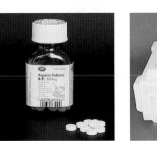

a

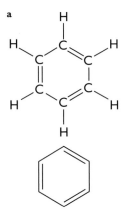

b

Figure 12.2
a The Kekulé structure for benzene, proposed in 1862.
b Friedrich Kekulé, 1825–1896.

The term 'aromatic' in 'aromatic chemistry' is perhaps misleading in that it implies that members of this group of organic chemicals have pleasant odours. Many have very unpleasant and, in some cases, dangerous vapours. Many of these substances are toxic and/or carcinogenic to humans. Aromatic compounds are those that are based on a similar structure, that of benzene. In recent times this group of compounds has been systematically named **arenes**.

In 1825 the young Michael Faraday identified a previously unknown hydrocarbon that subsequently became known as **benzene**. The molecular formula of the benzene molecule, the fundamental component of all arenes, was determined in 1834 as being C_6H_6. There then followed about 30 years of discussion as to the possible structure of this molecule, one which had so few hydrogen atoms relative to the number of carbon atoms. (The *alkane* with six carbon atoms, hexane, is C_6H_{14}.) In 1862 a German chemist, Friedrich August Kekulé, proposed that the benzene molecule was a six-membered ring with alternating single and double bonds between the carbon atoms (Figure 12.2).

For many years the Kekulé structure for benzene was accepted because it explained the chemical reactions of the benzene molecule. In the 1900s, however, as experimental techniques became more advanced and values such as the C—C bond length and bond energy became known, the Kekulé structure began to show flaws.

The problems with the Kekulé structure of benzene are the following.

- If the Kekulé structure for benzene had been correct and it contained both double and single covalent bonds between carbon atoms, then it should undergo *electrophilic addition reactions*. In practice, the benzene molecule very rarely undergoes addition reactions. Most of the reactions on the ring were found to be *substitution reactions*.

- It was shown by X-ray diffraction techniques that all the carbon–carbon bond lengths in the benzene molecule were 0.139 nm. This is a value which lies between that of a C=C bond (0.133 nm) and of a C—C bond (0.154 nm). This evidence showed that all the bonds in the benzene ring were the same and that they were neither single nor double bonds.

- If benzene contained three C=C bonds then it should undergo three hydrogenation reactions to eventually form cyclohexane. It was known that the enthalpy of hydrogenation, $\Delta H^{\ominus}_{hyd}$, of a single C=C bond was about $-120\,kJ\,mol^{-1}$. Kekulé's structure would contain three C=C bonds, so on hydrogenation the enthalpy change was expected to be around $-360\,kJ\,mol^{-1}$ (Figure 12.3). Experimentally, however, a value of $-208\,kJ\,mol^{-1}$ is obtained (Figure 12.4). This showed that benzene does not contain three C=C bonds and that the benzene molecule is approximately $152\,kJ\,mol^{-1}$ more stable than expected by the Kekulé structure.

Figure 12.3 (left)
The hydrogenation of benzene should give an enthalpy of hydrogenation of $-360\,kJ\,mol^{-1}$.

Figure 12.4 (right)
Enthalpy level diagrams showing the enthalpy of hydrogenation of Kekulé's benzene and actual benzene.

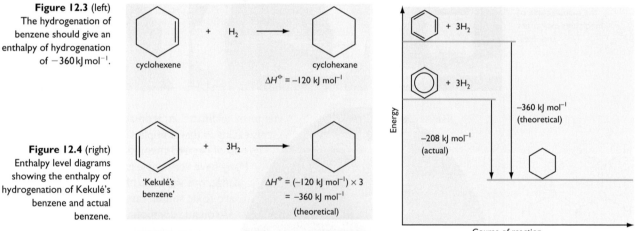

Figure 12.5
Linus Pauling (1901–1994) provided the final missing links to state the structure of benzene accurately.

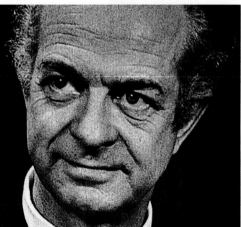

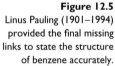
1 Draw a dot-cross diagram to show the bonding present in a benzene molecule.

In 1931 Linus Pauling (Figure 12.5) proposed that benzene was actually a hybrid of two structures, brought about by **resonance**. Pauling's idea was that the actual structure for the benzene molecule was midway between two Kekulé structures – a **resonance hybrid**. A resonance hybrid is a structure that shows characteristics of *both* structures, as shown in Figure 12.6. Note that the structure is switching continually between the two structures. The modern representation of benzene is abbreviated to the structure shown in Figure 12.7.

Figure 12.6
The actual structure of the benzene molecule is a blend of the two Kekulé structures.

Figure 12.7
The benzene molecule is represented by this symbol.

Bonding in the benzene molecule

The benzene ring has a planar structure. All the bonds between the carbon atoms in the ring have the same length and same bond enthalpy. The bonds between the carbon atoms in the benzene ring are a mixture of a σ **bond**, formed by the linear overlap of two atomic 2s orbitals, one from each carbon atom, and a π **bond**, formed by the sideways overlap of two 2p atomic orbitals (Figure 12.8). The sideways overlap of the 2p atomic orbitals forms a ring of negative charge above and below the plane of the molecule. The electrons that form these bonds are free to move around and are known as **delocalised electrons**. It is the presence of these delocalised electrons that gives the benzene molecule unexpected stability and they account for the fact that the molecule undergoes **electrophilic attack**.

Figure 12.8
a σ bonded skeleton and p orbitals.
b σ skeleton and π bonds.

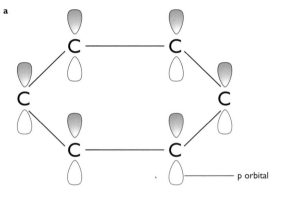

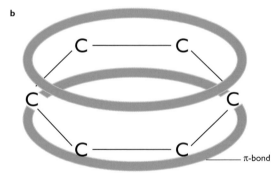

Naming aromatic compounds

When benzene reacts it does so by **electrophilic substitution reactions**. When these reactions occur, the hydrogen atoms on the ring are substituted with other atoms or groups of atoms.

When benzene undergoes mono-substitution the atom substituted onto the ring can bond to any of the carbon atoms, as they are all identical. Some examples of molecules produced by mono-substitution are shown in Figure 12.9.

Figure 12.9
Mono-substituted benzene derivatives.

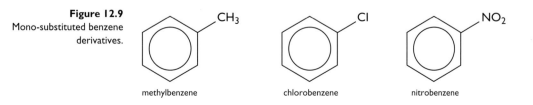

If further substitution occurs then the positioning of the second and third groups becomes more important. In these cases it is necessary to number the position of each of the attached groups. For example, consider the three molecules shown in Figure 12.10. Each of them is a dimethylbenzene, they are isomers of one another, but they are different molecules.

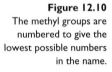

Figure 12.10
The methyl groups are numbered to give the lowest possible numbers in the name.

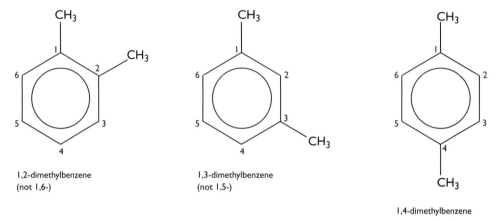

1,2-dimethylbenzene
(not 1,6-)

1,3-dimethylbenzene
(not 1,5-)

1,4-dimethylbenzene

To name the molecules, one of the methyl groups is assumed to be at position 1. The other methyl group is given a number as shown in Figure 12.10. It is possible to number the ring in either direction, but the number obtained should always be the lowest.

The naming of some aromatic compounds is slightly different. For example, the presence of an —OH group on the ring produces the molecule **phenol**, an aromatic alcohol (Figure 12.11). If a carboxylic acid group is present on a benzene ring the compound is called benzoic acid (Figure 12.12).

Figure 12.11
The molecule phenol – an aromatic alcohol.

OH

phenol

Figure 12.12
The molecule benzoic acid – an aromatic carboxylic acid.

benzoic acid

If two or more carboxylic acid groups are present the name changes to indicate the position of the groups on the ring. For example the molecule in Figure 12.13 is known as benzene-1,2-dicarboxylic acid.

If the benzene ring is present in a molecule as a side-chain then it is given the name **phenyl** (C_6H_5—).

Figure 12.13
Numbers are used to indicate the position of the carboxylic acid groups on the ring.

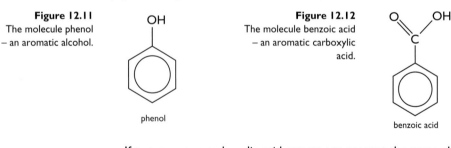

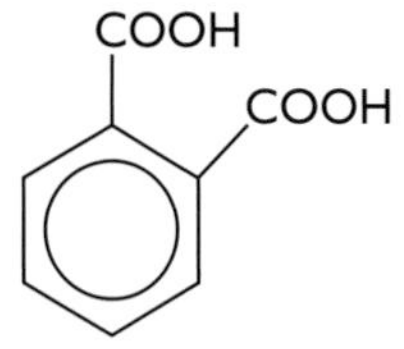

COOH
COOH

benzene-1,2-dicarboxylic acid

2 Draw structures for the following aromatic substances:
a 1,2,3-trimethylbenzene
b 1,3-dichlorobenzene
c 1,4-diethylbenzene.

Reactions of benzene

Because of the high electron density above and below the benzene ring, benzene is attacked by positively charged species or electrophiles. To maintain the inherent stability due to the presence of the delocalised electrons, benzene undergoes substitution reactions. Addition reactions would cause the π bonding to be broken, losing the stability of the delocalised ring system. Benzene will only undergo addition reactions with free radicals.

Halogenation of the benzene ring

Benzene itself is very unreactive because of its delocalised electron structure. It can be made to undergo halogenation in the presence of a suitable catalyst. This can be compared to the more reactive nature of cyclohexene, where the use of a catalyst is not required and the chlorine molecules will *add* across C=C.

For example, chlorine can be substituted into the benzene ring, replacing one of its hydrogen atoms, to give chlorobenzene. The reaction takes place with heating under reflux, with a catalyst of aluminium chloride (or iron(III) chloride), in anhydrous conditions.

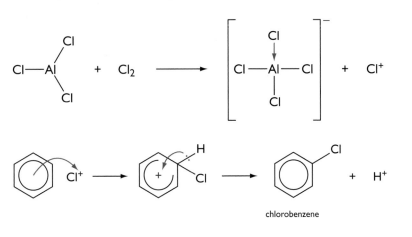

The catalyst forms the electrophile for the reaction, Cl^+.

Figure 12.14
The formation of the electrophile in the chlorination of benzene.

Figure 12.15
The Cl^+ ion is attracted to the benzene ring, forming a positively charged intermediate. Loss of a hydrogen ion gives chlorobenzene.

chlorobenzene

Aluminium chloride is able to catalyse this reaction due to its ability to form a dative covalent bond with a chlorine atom of the chlorine molecule. It can do this because it is an electron-deficient molecule, with only six electrons in its outer shell. It accepts the chlorine molecule bonding pair of electrons to form the tetrachloroaluminate ion $[AlCl_4]^-$ (Figure 12.14). Catalysts which act in this way are sometimes known as **halogen carriers**. The positively charged chlorine ion formed, an electrophile, attacks the benzene ring (Figure 12.15). The hydrogen ion formed then reacts with the tetrachloroaluminate ion, forming hydrogen chloride and regenerating the catalyst:

$$[AlCl_4]^- + H^+ \rightarrow AlCl_3 + HCl$$

This whole process is known as an electrophilic substitution reaction.

Bromination can be achieved in a similar way by using bromine, $Br_2(l)$, and either aluminium bromide or iron(III) bromide as the catalyst/halogen carrier.

Nitration

The nitration of benzene involves the substitution of a nitro group, $-NO_2$, into the benzene ring, replacing one of the hydrogen atoms. This process involves the use of a **nitrating mixture**, a mixture of concentrated sulphuric and nitric acids. The reaction of these two substances produces the electrophile for the subsequent substitution reaction, the **nitronium** or **nitryl ion, NO_2^+**:

$$2H_2SO_4 + HNO_3 \rightarrow NO_2^+ + 2HSO_4^- + H_3O^+$$

3 Produce a mechanism similar to that for the chlorination of benzene but this time using $Br_2(l)/FeBr_3$.

4 In the nitrating mixture which of the two acids is behaving as the stronger acid? Explain your answer.

Figure 12.16
The warming under reflux of benzene with the nitrating mixture to give nitrobenzene.

The reaction to form nitrobenzene (a yellow oily substance) from benzene is carried out at a temperature between 45 °C and 55 °C, but under reflux (Figure 12.16). Restricting the temperature to below 55 °C ensures that no di- or tri-substituted nitrobenzenes are formed. The NO_2^+ electrophile formed bonds with the benzene ring by accepting a pair of delocalised electrons from the benzene ring, forming a positively charged intermediate (Figure 12.17). This intermediate then loses a proton, H^+, which reacts with the hydrogensulphate ions, HSO_4^-, to regenerate the sulphuric acid:

$$H^+ + HSO_4^- \rightarrow H_2SO_4$$

Figure 12.17
The formation of a positively charged intermediate is followed by the loss of a proton, restoring the delocalised electron system in benzene.

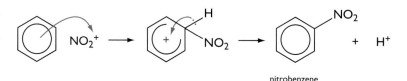

Uses of nitrobenzenes

Nitrobenzenes are important in the manufacture of explosives. TNT, trinitrotoluene (2,4,6-trinitromethylbenzene), is manufactured from methylbenzene by the method described above, but using a higher temperature to allow for a tri-substituted molecule (Figure 12.18).

Figure 12.18
The manufacture of TNT – trinitrotoluene.

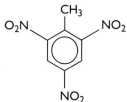

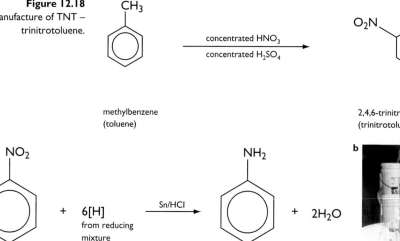

Figure 12.19
(above and right)
a The reduction of nitrobenzene to phenylamine.
b Phenylamine is used in the production of azo dyes.

The formation of nitrobenzene is an important synthetic process as it is an intermediate step in the synthesis of other aromatic compounds containing nitrogen atoms. For example, heating nitrobenzene under reflux with tin and concentrated hydrochloric acid produces phenylamine, an important raw material in the production of dyes (Figure 12.19).

Sulphonation

sulphur trioxide

Sulphonation of a benzene ring involves the electrophilic substitution of one of the hydrogen atoms in the ring with a —SO_3H group. The mechanism of this process is not, as yet, fully understood, although it is widely accepted that sulphur trioxide, SO_3, is the electrophile for the process. The sulphur trioxide molecule is a good electrophile (Figure 12.20).

Figure 12.21
The sulphonation of
benzene.

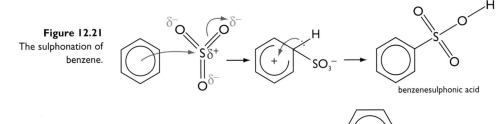

benzenesulphonic acid

Figure 12.22
a A synthetic detergent.
b These cleaning materials
contain benzenesulphonate
derivatives.

a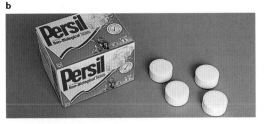

b

Sulphur trioxide attacks the benzene ring, forming a positively charged intermediate (Figure 12.21). Subsequent rearrangement of the positively charged intermediate produces benzenesulphonic acid. Most solid soapless detergents contain benzenesulphonate derivatives (Figure 12.22).

Friedel–Crafts alkylation and acylation

Alkylation

The alkylation of a benzene ring involves the electrophilic substitution of a hydrogen atom on the benzene ring by an alkyl group, for example a methyl or ethyl group. This reaction is important synthetically as it increases the size of the carbon skeleton of the molecule by the formation of a carbon–carbon bond.

The electrophile is the positively charged alkyl ion (a **carbocation**), formed by reaction of a halogenalkane with the appropriate aluminium halide. The example below shows how benzene can be converted into ethylbenzene by this method.

The first step is the production of the necessary carbocation, the ethyl ion ($CH_3CH_2^+$) in this case, from the reaction between chloroethane and aluminium chloride:

$$CH_3CH_2Cl + AlCl_3 \rightarrow [AlCl_4]^- + CH_3CH_2^+$$

5 Write a mechanism for the production of methylbenzene using bromomethane and a suitable aluminium halide.

The $CH_3CH_2^+$ ion then bonds with the benzene ring to form a positively charged intermediate. This then loses a proton to $AlCl_4^-$ to re-form the stable delocalised electron structure of the ring system (Figure 12.23). The proton eliminated reacts with the tetrachloroaluminate ion to re-form the aluminium chloride catalyst.

Ethylbenzene is used in the production of polystyrene.

Figure 12.23
The alkylation of the
benzene ring to form
ethylbenzene.

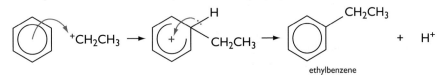

ethylbenzene

Manufacture of polystyrene

6 Why do you think that, in the industrial production of ethylbenzene, ethene is used instead of chloroethane?

Industrially, ethylbenzene is manufactured by a Friedel–Crafts reaction involving benzene and ethene in the presence of hydrogen chloride and aluminium chloride:

$$\langle\bigcirc\rangle(g) + CH_2{=}CH_2(g) \xrightarrow[\substack{HCl \\ 95°C}]{AlCl_3} \langle\bigcirc\rangle\!-CH_2CH_3(g)$$

Figure 12.24
The dehydrogenation of ethylbenzene.

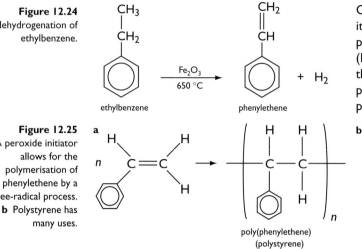

ethylbenzene   phenylethene

Once the ethylbenzene has been produced, it undergoes a dehydrogenation reaction to produce the monomer **phenylethene** (Figure 12.24). Subsequent polymerisation of this monomer produces the polymer polystyrene, the modern name of which is poly(phenylethene) (Figure 12.25).

Figure 12.25
a A peroxide initiator allows for the polymerisation of phenylethene by a free-radical process.
b Polystyrene has many uses.

poly(phenylethene)
(polystyrene)

Acylation

Figure 12.26
An acyl group, RCO$^+$.

the acyl group

Acylation of the benzene ring is the electrophilic substitution of a hydrogen atom on the benzene ring by an acyl group, RCO$^+$ (Figure 12.26). Once again the production of an electrophile is achieved using a Friedel–Crafts catalyst. In order to substitute a CH$_3$CO— group onto the benzene ring the starting materials would be ethanoyl chloride and aluminium chloride:

$$CH_3COCl + AlCl_3 \rightarrow [AlCl_4]^- + CH_3CO^+$$

The electrophile produced would then, on heating under reflux, go on to attack the benzene ring to form a positively charged intermediate (Figure 12.27).

7 Write a balanced chemical equation and describe the mechanism for the reaction of propanoyl chloride with benzene in the presence of aluminium chloride.

Figure 12.27
The formation of phenylethanone.

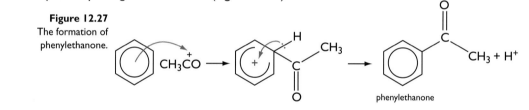

phenylethanone

Oxidation of carbon-containing side-chains

Figure 12.28
The oxidation of the methyl side-chain to give benzoic acid.

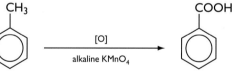

methylbenzene   benzoic acid

Alkanes are generally inert but the alkyl group in aromatic compounds can undergo oxidation reactions with solutions of alkaline potassium manganate(VII). Methylbenzene can be oxidised using this method to form benzoic acid (Figure 12.28). In this reaction the purple manganate(VII) ion is reduced to the green manganate(VI) ion and then on to manganese(IV) oxide, MnO$_2$ (Figure 12.29). The benzene ring has modified the behaviour of the alkyl group in this case.

Figure 12.29
During the oxidation of methylbenzene, the colour changes from the purple manganate(VII) ion to the green manganate(VI). However, the reduction of the manganese compound may go further to produce the dark grey of MnO$_2$.

Reactions of phenol

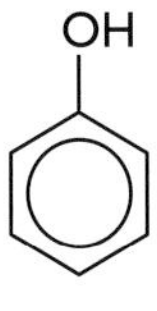

Figure 12.30
A phenol molecule.

Phenol is an aromatic alcohol which consists of an —OH group attached directly to a benzene ring (Figure 12.30). The presence of the —OH group on the benzene ring makes the ring more susceptible to electrophilic attack and therefore phenol is a more reactive molecule than benzene. This is because one of the lone pairs of electrons on the oxygen atom becomes part of the delocalised electron system, increasing the electron density above and below the ring. The —OH group is said to be an **activating group**. It also directs incoming electrophiles to the 2, 4 and 6 positions on the benzene ring.

Salt formation

Figure 12.31
The phenoxide ion is partially stabilised by the delocalised nature of the ring system.

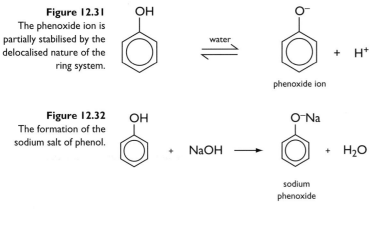

phenoxide ion

Figure 12.32
The formation of the sodium salt of phenol.

sodium phenoxide

The phenol molecule is slightly acidic since the phenoxide ion ($C_6H_5O^-$) produced in aqueous solution is partially stabilised by the delocalisation of the negative charge on the oxygen atom within the benzene ring (Figure 12.31). This acidic nature is shown by the fact that phenol will react with sodium hydroxide to form a salt, sodium phenoxide (Figure 12.32).

Sodium phenoxide can also be produced by the reaction of sodium metal with phenol. Note that this reaction would be very dangerous with a stronger acid such as a carboxylic acid.

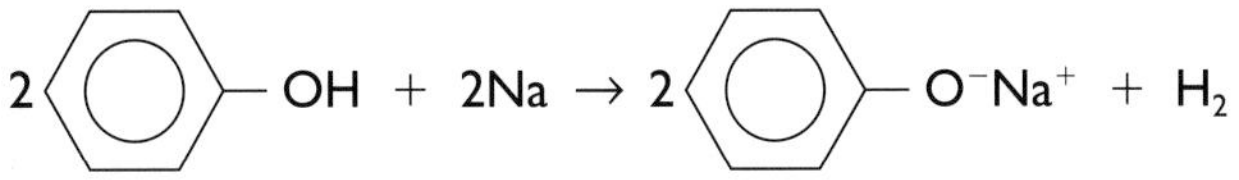

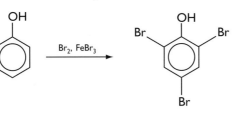

Reaction of phenol with bromine water

Figure 12.33
The formation of 2,4,6-tribromophenol.

2,4,6-tribromophenol

Phenol will react with bromine water to form 2,4,6-tribromophenol. During the reaction the bromine is decolorised and a white precipitate of 2,4,6-tribromophenol is produced (Figure 12.33). This has a familiar antiseptic smell and is an analogue of the much used antiseptic 2,4,6-trichlorophenol, which we know as TCP (Figure 12.34).

Figure 12.34
This antiseptic has a wide variety of uses. 'Always read the label' is the advice given with products like these.

The test for a phenol

The presence of a phenol group can be tested for by the addition of neutral iron(III) chloride solution. The reaction between the phenol group and the iron(III) chloride results in the formation of a complex ion with a purple coloration.

Reactions of phenylamine

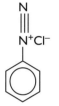

Figure 12.35
A phenylamine molecule.

The $-NH_2$ group of the phenylamine molecule (Figure 12.35) activates the benzene ring to electrophilic attack at the 2, 4 and 6 positions. This is because the lone pair of electrons on the nitrogen atom becomes part of the delocalised electron system of the benzene ring.

One of the major industrial uses of phenylamine is in the production of azo dyes, which are used extensively in the clothing industry. All **azo compounds** contain the following group:

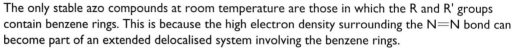

The only stable azo compounds at room temperature are those in which the R and R' groups contain benzene rings. This is because the high electron density surrounding the $N=N$ bond can become part of an extended delocalised system involving the benzene rings.

Azo compounds are formed by the chemical reaction between a **diazonium salt** and a **coupling agent**. This type of reaction is known as a **coupling reaction**.

Diazonium salts

Figure 12.36
Benzenediazonium chloride.

Diazonium salts contain the following group:

$$R-N^+\equiv N$$

The only stable diazonium salts are those in which the R group contains a benzene ring. However, even these are stable only at low temperatures (<5 °C). The most common example of such a salt is **benzenediazonium chloride** (Figure 12.36). Above 5 °C benzenediazonium chloride solution becomes unstable and loses the $-N^+\equiv N$ group as nitrogen gas, with phenol as the other product. Solid benzenediazonium chloride is explosive. Because of this, a solution of benzenediazonium chloride is prepared in an ice bath and is used immediately.

Benzenediazonium chloride can be prepared by a **diazotisation reaction**. In this reaction a cold solution of sodium nitrite, $NaNO_2$ (sodium nitrate(III)), is added to phenylamine in concentrated hydrochloric acid (sulphuric acid can be used) below 5 °C. In the first stage of the reaction the sodium nitrite reacts with the hydrochloric acid to form unstable **nitrous acid, HNO_2** (nitric(III) acid):

$$NaNO_2(aq) + HCl(aq) \rightarrow HNO_2(aq) + NaCl(aq)$$

Figure 12.37 (below)
The formation of benzenediazonium chloride.

The nitrous acid produced then reacts with phenylamine to produce benzenediazonium chloride (Figure 12.37).

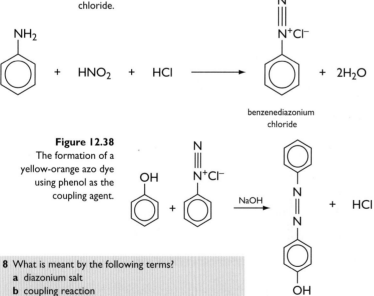

benzenediazonium chloride

Figure 12.38
The formation of a yellow-orange azo dye using phenol as the coupling agent.

Coupling reactions

Coupling reactions result in the production of azo compounds, many of which are used commercially as dyes. In a coupling reaction a diazonium salt reacts with another aromatic compound, called a **coupling agent**. The diazonium salt, due to its positive charge, acts as an electrophile and attacks the benzene ring of the coupling agent. This reaction must be carried out at temperatures below 5 °C and it is usual that the coupling agent is present in alkaline solution.

Changing the coupling agent alters the colour of the azo dye formed. The formation of a yellow-orange azo dye can be achieved by reacting benzenediazonium chloride reacts with phenol as the coupling agent (Figure 12.38). The part of the azo dye molecule responsible for the colour of the dye is the $-N=N-$ group and is known as a **chromophore**. By using different coupling agents, the nature of the groups surrounding this chromophore changes, leading to the production of different coloured dyes.

8 What is meant by the following terms?
 a diazonium salt
 b coupling reaction
 c diazotisation
9 Draw the structure of the products produced when benzenediazonium chloride with phenylamine at a temperature of <5 °C.

a

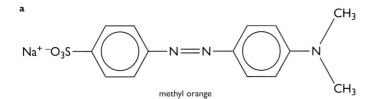

methyl orange

The indicator methyl orange is a familiar azo compound (Figure 12.39). It is red in acid solution and yellow in alkaline solution. The extended delocalised system is affected to a minor extent by the loss or gain of a proton. This affects the wavelength of the light the molecule absorbs. Indicators are dealt with in more detail in Chapter 5, page 75.

Figure 12.39
a Methyl orange is an azo compound, and is used as an acid–base indicator.
b Methyl orange shows these colour changes in acid, neutral and alkaline solutions.

b

● **Key skills** ICT
• Use a desktop publishing program to produce a summary to show the reactions of benzene.

● **Skills task** Use the information in this chapter and the Internet to produce a flow diagram, with chemical equations, to show how aspirin is produced.

CHECKLIST   After studying Chapter 12 you should know and understand the following terms.

● **Arenes:** Aromatic hydrocarbons based on unsaturated rings of carbon atoms.
● **Benzene:** The simplest arene hydrocarbon, formula C_6H_6.
● **Resonance:** A description of a molecule whose structure cannot be represented by a single structure but only by a 'mixture' of two or more differing structures.
● **Electrophilic substitution:** The substitution of hydrogen atoms on a benzene ring by other atoms or groups of atoms. The process involves the attack of an electrophile onto the delocalised electrons in the benzene ring followed by subsequent loss of a hydrogen ion.
● **Halogen carrier:** A catalyst, such as aluminium chloride, used in electrophilic substitution reactions which involve halogen atoms. The halogen atom forms a dative covalent bond to the catalyst and so the electrophile for the process is formed.
● **Alkylation:** The electrophilic substitution of a hydrogen atom on a benzene ring by an alkyl group such as methyl or ethyl.
● **Acylation:** The electrophilic substitution of a hydrogen atom on a benzene ring by an acyl group, RC^+O.
● **Activating group:** An atom or a group of atoms bonded to a benzene ring which specifically directs incoming electrophiles to particular sites on the ring and also increases the rate of reaction.
● **Azo compound:** Contains the group R—N=N—R'.
● **Diazonium salt:** Contains the group R—N$^+$≡N.
● **Coupling agent:** A reagent that reacts with a diazonium salt below 5 °C, in alkaline conditions, to produce an azo dye.
● **Diazotisation reaction:** Reaction in which a diazonium salt is formed by the reaction of an aromatic amine with nitrous acid below 5 °C.

Examination questions

1 Cumene, $C_6H_5CH(CH_3)_2$, is the major organic product obtained when benzene and propene react together in the presence of aluminium chloride and hydrogen chloride.

a **i** Write an equation showing how a reactive species is formed from propene, aluminium chloride and hydrogen chloride.

ii Name the type of substitution reaction which follows the formation of the reactive species above and outline a mechanism for this substitution. (6)

b Explain why propylbenzene, $C_6H_5CH_2CH_2CH_3$, is obtained only as a minor by-product in the above reaction. (3)

c Give the structure of a compound other than propene which could be used to make cumene from benzene in the presence of aluminium chloride. (1)

AQA, A level, Specimen Paper 6421, 2001/2

2 The compound *Mecoprop* is used as a herbicide. It is structurally similar to a number of other compounds which are also used as herbicides. One structural feature which these molecules share with *Mecoprop* is the presence of a chlorinated benzene ring.

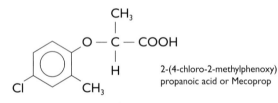

2-(4-chloro-2-methylphenoxy)
propanoic acid or Mecoprop

a **i** State the reagent(s) and conditions which can be used to substitute a chlorine atom into a benzene ring. (3)

ii It is usually necessary to use a catalyst in order to substitute a chlorine atom into a benzene ring. Explain the function of the catalyst in this process. (3)

Mecoprop and related herbicides are solids with very low solubilities in water. They are considerably more soluble in organic solvents such as ethyl ethanoate. However, drinking water in certain areas of the UK, has been found to be contaminated with traces of *Mecoprop* and other herbicides. The identities of the herbicides in a sample of water can be found by first extracting them into a solvent such as ethyl ethanoate, concentrating the solution formed and then analysing this solution.

b Describe, in outline, how you could take a sample of water containing dissolved herbicides and extract the herbicides into a solvent such as ethyl ethanoate (which is immiscible with, and less dense than water), and then concentrate the resulting solution. (5)

c One way to analyse the resulting solution is by gas–liquid chromatography (g.l.c.). Before a sample of this solution is injected into the column, it is treated with diazomethane to convert any carboxylic acid groups present into methyl esters.

i Suggest why the sample is treated in this way. (2)

ii Draw out the full structural formula of the methyl ester produced from *Mecoprop*. (2)

OCR, A level, Specimen Paper A7887, Sept 2000

3 **a** Describe the bonding in benzene and explain why benzene reacts less readily than alkenes with electrophiles. (In this question, 1 mark is available for the quality of written communication.) (7)

b Describe, including a mechanism, the nitration of benzene by electrophilic substitution. (6)

OCR, A level, Specimen Paper A7882, Sept 2000

4 Ethylbenzene is prepared by the reaction between bromoethane and benzene, using anhydrous aluminium bromide as a catalyst in a solution of ethoxyethane (ether). After the reaction is complete the ether and any unreacted bromoethane are distilled off. Finally the ethylbenzene and unreacted benzene are separated by fractional distillation.

a Suggest the mechanism for the reaction between benzene and bromoethane. (4)

b Benzene (boiling temperature 80 °C) and ethylbenzene (boiling temperature 136 °C) dissolve in each other and form solutions which obey Raoult's Law.

Draw a boiling temperature/composition diagram, labelling the liquid and vapour lines. Use your diagram to explain what happens when a mixture containing 40% benzene and 60% ethylbenzene is fractionally distilled. (6)

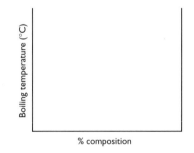

% composition

c Give the reagents and conditions necessary for the conversion of ethylbenzene, $C_6H_5CH_2CH_3$, into potassium benzoate C_6H_5COOK. (3)

Edexcel, A level, Module 4, June 2000

5 Bulletproof vests are made from Kevlar®. The first step in its manufacture is the polymerisation of the two monomers:

benzene-1,4-diamine   benzene-1,4-dicarboxylic acid

186

a Draw the structural formula of the polymer showing **one** repeat unit. (2)

b The infrared spectrum of **one** of the two monomers is given below, together with data for some common IR absorption wavenumber ranges.

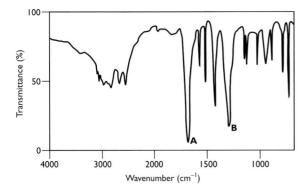

Bond	Wavenumber (cm⁻¹)	Bond	Wavenumber (cm⁻¹)
C—O	1300–1100	C=C (arenes)	1600–1450
C=O	1750–1680	C—H (arenes)	3100–3000
O—H	3500–2500	N—H	3500–3300

Use the data to assign the bonds which cause the peaks **A** and **B** and hence identify which of the monomers produces this spectrum. (3)

c Benzene-1,4-dicarboxylic acid can be made in the laboratory by heating 1,4-dimethylbenzene (boiling temperature 138 °C) under reflux for 1½ hours with an alkaline solution of potassium manganate(VII). The potassium manganate(VII) is reduced to a brown precipitate of manganese(IV) oxide. Concentrated hydrochloric acid is carefully added until the brown precipitate disappears. The mixture is cooled, and the benzene-1,4-dicarboxylic acid is filtered off and recrystallised from boiling water.

i Why is it necessary to have a reflux condenser in this preparation? (1)

ii Why must the solution be cooled before filtering? (1)

iii The purity of the product was tested by observing its melting temperature. What would be noticed if the sample were still impure? (1)

London, A level, Module 4, Jan 2000

6 Phenylamine, $C_6H_5NH_2$, was dissolved in hydrochloric acid and then reacted with excess sodium nitrite solution with the temperature maintained between 0 °C and 5 °C. The product **X** was reacted with a solution of 2-naphthol in aqueous sodium hydroxide and a red precipitate **Y** was obtained.

The structure of 2-naphthol is:

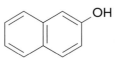

a Write an equation, using structural formulae, to represent the reaction of phenylamine to form **X**. (2)

b Explain why the temperature has to be maintained between 0 °C and 5 °C. (2)

c Draw the structural formula of **Y**. (2)

London, A level, Module 4, Jan 2000

7 Methylbenzene reacts with chlorine in the presence of UV light in a similar way to methane:

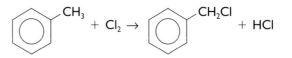

In order for the reaction to take place, chlorine must be passed into boiling methylbenzene in the presence of UV light over a period of time.

a Draw an apparatus which could be used for this reaction. (3)

b What are the hazards **specific** to this experiment? (2)

London, A level, Module 4, June 1999

8 A mixture of concentrated nitric and sulphuric acids is used in the nitration of benzene to form nitrobenzene.

a The first stage of the reaction is the production of the NO_2^+ ion. Write an equation to show its formation. (1)

b i The formation of the NO_2^+ ion occurs in two steps, the first of which is:

$$H_2SO_4 + HNO_3 \rightarrow H_2NO_3^+ + HSO_4^-$$

This is an acid/base reaction. Give the formulae of the acid/base conjugate pairs involved. (2)

ii Hence comment on the relative strengths of nitric and sulphuric acid. (1)

c Give the mechanism for the reaction of the NO_2^+ ion with benzene. (3)

d What do curly arrows show when used in a mechanism? (1)

London, AS/A level, Module 2, Jan 2000

Further polymers

● Addition polymers

Figure 13.1
a Poly(1,1-difluoroethene) is piezoelectric. This makes it useful in the construction of modern stereo loudspeakers.
b Poly(vinyl carbazole) is the photoconducting polymer at the heart of this photocopier.
c Poly(ethenol) has revolutionised the handling of hospital-soiled laundry materials. It dissolves in water.
d PEEK (poly(ether-ether-ketone)) is used for making kettles because it is strong and has a high heat resistance.

a

b

c

d
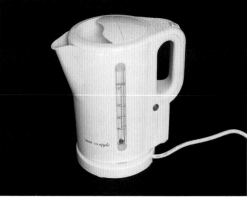

Poly(ethene)

The modern plastics industry was really born with the accidental discovery of poly(ethene) by Reginald Gibson and Eric Fawcett in 1933. Poly(ethene) was the first manufactured example of an addition polymer. The polymer is made by the joining together, or addition, of many small molecules of ethene. The small molecules from which an addition polymer is formed are called **monomers**.

Figure 13.2
The formation of
poly(ethene) from ethene.

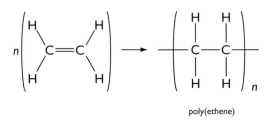

poly(ethene)

Poly(ethene) is now made by heating ethene to a relatively high temperature of 100–300 °C, under a high pressure of 1000–2000 atmospheres in the presence of a trace amount of oxygen as a catalyst. The process by which poly(ethene) is formed is called **addition polymerisation** and it can be represented by the equation shown in Figure 13.2.

In Figure 13.2 the poly(ethene) structure is shown within brackets. This structure is repeated many thousands of times in order to make even a small chain of poly(ethene). Following more recent developments in catalysts, different forms of poly(ethene) have been produced. The poly(ethene) produced by Fawcett and Gibson is now known as **low density poly(ethene)** or **ldpe**. In this form of the polymer, the chains are extensively branched and so take up a lot of space. This therefore forms a lower density material and one which is weaker in strength. The process by which poly(ethene) is made is now understood. The addition polymerisation occurs by a free-radical process.

In 1953, the German scientists Karl Ziegler, Erhard Holzkamp and H. Breil discovered that passing ethene at atmospheric pressure through a liquid alkane solution containing small amounts of titanium tetrachloride and triethylaluminium resulted in the formation of poly(ethene). This polymer contained polymer chains with very little branching and so the chains could pack more closely. It was therefore more dense and more crystalline. This **high density poly(ethene)** or **hdpe** had a greater strength than the ldpe (Figure 13.3).

Figure 13.3
The items shown are made from hdpe. Hdpe is strong and can easily be moulded into complicated shapes.

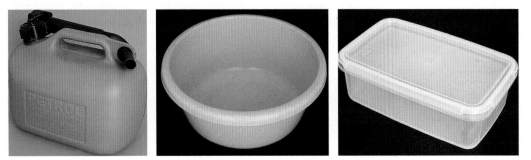

Poly(propene)

Many other addition polymers have been produced following on from the manufacture of poly(ethene). They are all formed from alkene-type monomers. For example, poly(propene) is a polymer made from propene. Poly(propene) has a wide variety of uses, such as car parts, carpets, clothing and sealants (Figure 13.4).

Figure 13.4
These car parts and carpets are manufactured from poly(propene).

There are two forms of poly(propene), which differ in the arrangement of the methyl groups along the carbon chain. If the polymer is manufactured with all the methyl groups on the same side of the carbon chain the polymer is known as **isotactic** (Figure 13.5). The orientation of the methyl groups is determined using special catalysts known as Ziegler–Natta catalysts. Due to the ordered nature of the chains in isotactic poly(propene), they are able to pack closer to one another resulting in greater strength and a higher melting point than poly(propene) manufactured from chains which are unordered. Poly(propene) manufactured with the methyl groups in random orientations is called **atactic** (or **syndiotactic**) poly(propene) (Figure 13.6).

Figure 13.5
In isotactic poly(propene) all the methyl groups are on the same side of the carbon chain.

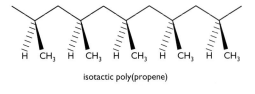

isotactic poly(propene)

Figure 13.6
In atactic poly(propene) the methyl groups are randomly arranged.

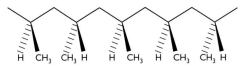

atactic poly(propene)

Other addition polymers

There are many other addition polymers. Poly(vinyl chloride) (PVC) was first made in 1872, but not patented until 1913. The monomer for the manufacture of this polymer is chloroethene, or vinyl chloride as it was known (Figure 13.7). PVC is still used as the name of the polymer, although using modern systematic nomenclature it should be called poly(chloroethene).

Figure 13.7
The formation of PVC.

PVC has different properties to both poly(ethene) and poly(propene). These result from the presence of the chlorine atom in the structure. Chlorine has a high electronegativity and will, therefore, create a polar $C^{\delta+}—Cl^{\delta-}$ bond. This gives rise to permanent dipole–permanent dipole intermolecular forces that are much stronger than the instantaneous dipole–induced dipole forces in poly(ethene) and poly(propene). These stronger intermolecular forces make PVC a harder polymer. In addition, the large size of the chlorine atom means that the chains cannot easily be moved over one another. This results in the polymer being brittle.

Initially PVC had very few uses, because it was hard and brittle. Later it was found that if other substances were added during the polymerisation process, the PVC that was formed had very different properties. These substances became known as **plasticisers**. A plasticiser used in the manufacture of PVC today is di-(2-ethylhexyl)hexanedioate. The plasticiser molecules fit in between the polymer chains, keeping them away from one another (Figure 13.8). This results in the polymer becoming more flexible and softer.

Figure 13.8
a The plasticiser used in the manufacture of PVC.
b The plasticiser molecules separate the polymer chains, allowing them to move freely past one another.

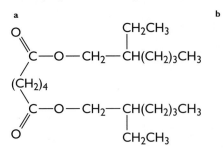

di-(2-ethylhexyl)hexanedioate

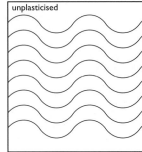

Figure 13.9
The formation of PTFE.

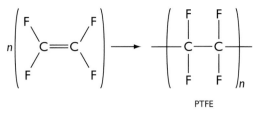

PTFE

Poly(tetrafluoroethene), or **PTFE**, was first marketed under the trade name Teflon (Figure 13.9). PTFE has some unusual properties:

* it will withstand very high temperatures
* it forms a very slippery surface.

These properties make PTFE an ideal 'non-stick' coating for frying pans, along with other uses such as the coating on hip joint replacements where friction-free movement is a requirement (Figure 13.10).

Figure 13.10
Because PTFE forms a slippery surface it is used to coat replacement hip joints.

Figure 13.11
The formation of polystyrene.

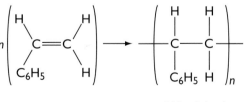

poly(phenylethene)

Polystyrene, or poly(phenylethene), has a wide variety of uses including packaging and insulation (Figure 13.11).

1 The structures of the monomer units used to make two of the polymers shown in Figure 13.1 are:
a $F_2C{=}CH_2$
b $C_{12}H_8NC(H){=}CH_2$
Write equations similar to those shown for other addition polymerisation processes in this section to show the formation of the polymers which can be formed from the monomers **a** and **b**.

Thermosoftening and thermosetting polymers

Polymers can be put into one of two major categories. If they melt or soften when heated (such as the addition polymers poly(ethene), PVC and polystyrene) then they are called **thermoplastics** or **thermosoftening** polymers. If they do not soften on heating but only char and decompose on further heating they are known as **thermosetting** polymers.

Figure 13.12 (below left) Blow moulding/vacuum forming in progress.

Figure 13.13 (below right) Compression moulding in progress.

- Thermosoftening polymers are easily moulded or formed into useful articles. Once they are molten they can be injected or blown into moulds and so produce a variety of different shaped items (Figure 13.12).
- Thermosetting polymers can be heated and moulded only once, usually by compression moulding during their manufacture (Figure 13.13).

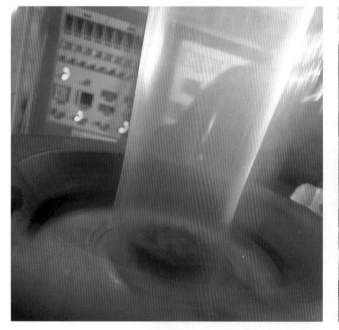

Figure 13.14
a In this type of polymer the chains are cross-linked.
b In this type of polymer there is no cross-linking.

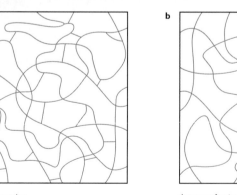

thermosetting

thermosoftening

Figure 13.14 shows the different molecular structures for thermosetting and thermosoftening polymers.

- Thermosetting polymers have polymer chains which are linked or covalently bonded to each other to give a cross-linked structure and so the chains are held firmly in place and no softening takes place on heating.
- Thermosoftening polymers do not have their polymer chains joined in this way. When thermosoftening polymers are heated their polymer chains flow over one another and the polymer softens.

Disposal of addition polymers

Over the last 30 years synthetic polymers have taken over as replacement materials for metals, glass, paper and wood, as well as for natural fibres such as cotton and wool. This is not surprising since synthetic polymers are light, cheap, relatively inert, colourful and can be easily moulded. However, this situation has created a waste problem of enormous proportions. Figure 13.15 shows some average EU figures for solid household waste.

Figure 13.15
Amount of plastic waste compared with other household waste in EU countries.

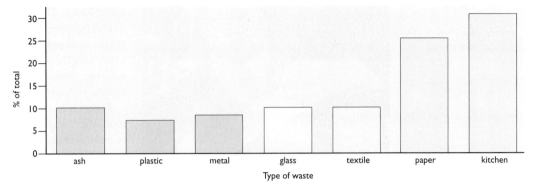

In the recent past much of our polymer waste has been used to landfill disused quarries. These sites are getting harder to find and it is becoming more and more expensive to dispose of the polymer waste. The alternatives to dumping polymer waste are more economical and more satisfactory.

- Incineration – schemes have been developed to use the heat generated for heating purposes. However, care has to be taken because poisonous fumes are produced when some plastics burn.
- Recycling – large quantities of black plastic bags and sheeting are produced for resale from recycled addition polymer.
- Photodegradable – these are polymers that degrade in sunlight. The C=O group is incorporated into the structure of polymer chains: carbonyl groups absorb radiation in the 270–360 nm region and so act as 'energy trappers'. This trapped energy causes the break up of the polymer chains into shorter units that are more easily biodegradable.
- Biodegradable plastics – as well as those polymers which degrade in sunlight others have been developed. These include synthetic biodegradable plastics that are broken down by bacteria (Figure 13.16). Starch units are incorporated into the structure of the polymer. These are attacked and digested by microorganisms such as bacteria.
- Dissolving plastics – poly(ethenol) was developed to be water-soluble. It therefore does not remain for very long in the environment.

Figure 13.16
This polymer is biodegraded by bacteria.

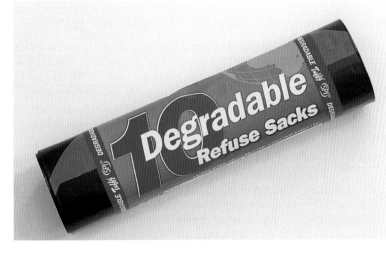

Condensation polymers

Figure 13.17
Wallace Carothers
(1896–1937).

Polyamides

Wallace Carothers discovered a different sort of plastic when he developed nylon in 1935 whilst working for DuPont in the US (Figure 13.17). The first commercial use of nylon was in 1938. Unlike the chance discovery of poly(ethene), Wallace Carothers set out to create new polymers. The development of nylon was through a desire to build up a polymer chain similar to wool and silk – natural polyamide polymers.

Nylon is made by reacting two different monomers together, unlike the manufacture of poly(ethene) which is made from only one monomer unit. Poly(ethene), formed by addition polymerisation, can be represented by:

$$-A-A-A-A-A-A-A-A-A-A-A-A-A-A-A-$$

During addition polymerisation, no molecules are lost.

Nylon was originally made by the reaction of a diamine with a dicarboxylic acid; for example, 1,6-diaminohexane and hexane-1,6-dioic acid. The polymer chain is made up from the two monomers reacting alternately and results in the chain type:

$$-A-B-A-B-A-B-A-B-A-B-A-B-A-B-A-$$

Each time a reaction takes place between the two monomers a molecule of water is lost; because of this it is called **condensation polymerisation** (Figure 13.18). Because an amide link is created during polymerisation, nylon is known as a **polyamide**. The $-CONH-$ group created in the production of nylon is found in proteins, where it is often referred to as a **peptide group** (Chapter 11, page 164).

Figure 13.18
a 1,6-Diaminohexane and hexane-1,6-dioic acid react together to form the $-A-B-$ part of a nylon chain, eliminating a water molecule in the process.
b This is the basic repeating unit of nylon-6,6. It is a polyamide.

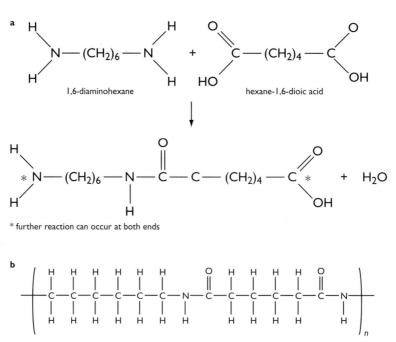

nylon-6,6

In the preparation of nylon the reaction between the diamine and the carboxylic acid is slow. This can be speeded up by reacting the diamine with the diacid dichloride. In this case, the condensation reaction produces hydrogen chloride instead of water (Figure 13.19).

Figure 13.19
1,6-Diaminohexane and hexane-1,6-dioyl dichloride react together to form the —A—B— part of a nylon chain, eliminating a hydrogen chloride molecule in the process.

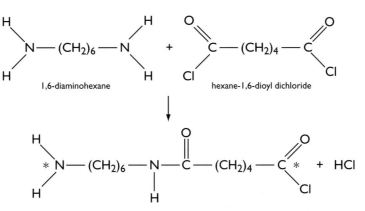

* further reaction can occur at both ends

Figure 13.20
This cyclic amide caprolactam is used to make nylon-6.

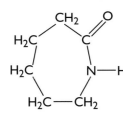

caprolactam

2 Write out the structure of the repeating unit for the condensation polymer, nylon-6.

There are different forms of nylon, for example, nylon-6,6 and nylon-6,10. The type of nylon depends on the number of carbon atoms in the monomers used. So if the diamine contains six carbon atoms and the dicarboxylic acid contains ten carbon atoms then the resulting nylon is called nylon-6,10. It is also possible to make nylon from a single monomer. For this to be done it is necessary for the monomer to contain an amine group at one end of the molecule and an acid group at the other end. So nylon-6 can be made from the monomer $H_2N(CH_2)_5COOH$. Nylon-6 is made nowadays by heating the cyclic amide, caprolactam to 260°C (Figure 13.20).

When nylon is made in industry, it forms a solid, which is melted and it is then forced through fine jets and extruded (Figure 13.21). The long filaments cool and solid nylon fibres are produced, which are stretched to align the polymer molecules and then dried. The resulting yarn can be woven into fabric to make shirts, ties, sheets and parachutes or turned into ropes or racket strings for tennis and badminton rackets.

In nylon it is hydrogen bonding between the H atoms of the N—H groups of the amide link of one polymer chain with the C=O groups on adjacent polymer chains which helps to give this polymer increased strength compared to addition polymers such as poly(ethene) (Figure 13.22).

Figure 13.21
Nylon fibres formed by forcing molten nylon through hundreds of tiny holes.

Figure 13.22 (right)
It is hydrogen bonding that helps to give nylon its increased strength.

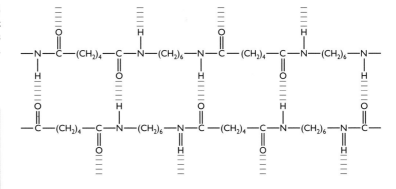

3 Nylon-6,10 can be formed from the reaction of a diamine and a diacid dichloride.
a Write down and name the structures of the two monomer molecules used.
b What is the name and formula of the small molecule produced during this condensation reaction?
c Draw the structure of the repeating unit found in nylon-6,10.
4 Explain why the intermolecular bonding in nylon-6,10 is weaker than that found in nylon-6,6.

5 Would the presence of hydrogen bonding in a fibre help it absorb water? Explain your answer.

A recent development has been to create a polyamide in which the aliphatic hydrocarbon unit within the polymer chain has been replaced by an aromatic unit. This type of polymer is known as an **aramid**. The first aramid was made from 3-aminobenzoic acid. It was not particularly strong even though it had exceptional fire resistance and could be made into fibres. The starting material was modified to create straighter chains in the polymer and a polyaramid was produced with exceptional properties; it is called Kevlar (Figure 13.23).

Figure 13.23
Kevlar is a polyaramid. The polymer chains have the structure shown here.

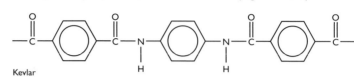

Kevlar

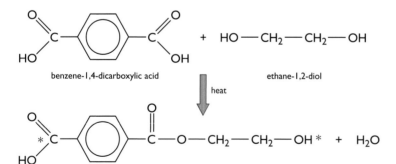

Figure 13.24
Kevlar is an unusual polymer with fire-resistant properties and also great strength. It is found in the crash helmets of Formula 1 racing drivers as well as in the suits of racing motorcyclists.

In nylon the single covalent bonds within the polymer chain are free to rotate and this tends to make the polymer quite flexible. However, in the case of Kevlar the aliphatic hydrocarbon chain parts of the polyamide are replaced by benzene rings. These parts of the polymer chain make the chains inflexible, due to delocalised bonding. Some of this delocalisation extends beyond the benzene rings and onto part of the amide link. This delocalisation also leads to enhanced intermolecular bonding between the Kevlar polymer chains. This in turn makes the bonding throughout Kevlar stronger and so the polymer is far more rigid than nylon. Kevlar is exceptionally strong, being five times the strength of steel on a weight for weight basis. Also it is very fire resistant. These properties have lead to a variety of uses in the aircraft and aerospace industry as well as for the manufacture of cables and ropes, and for protective clothing (Figure 13.24).

6 Draw a section of the structure of Kevlar showing the hydrogen bonding which is present between the polymer chains.

Polyesters

Figure 13.25 (below)
If we react ethane-1,2-diol with benzene-1,4-dicarboxylic acid, then we produce a polyester called Terylene.

Different polymers can be obtained, with different properties, if condensation polymerisation reactions are carried out between other monomer molecules. For example, if a diol such as ethane-1,2-diol is reacted with a dicarboxylic acid such as benzene-1,4-dicarboxylic acid then a **polyester** is produced – in this case the polymer produced is called Terylene (Figure 13.25).

benzene-1,4-dicarboxylic acid

ethane-1,2-diol

heat

* further reaction can occur at both ends

The Terylene chains are linked not by hydrogen bonding but by weaker dipole–dipole interactions. These take place between the polar carbonyl group (C=O), which are found on each of the polymer chains (Figure 13.26). Hence, Terylene is not as strong a polymer as nylon. If the molten polymer is extruded, it forms fibres that are used for materials for making clothes. Since it has the useful property of being able to form permanent creases, it has been used extensively in the production of trousers and skirts.

Due to the fact that polyesters and polyamides are broken down easily by hydrolysis (see Chapter 10, page 153 and Chapter 11, page 165), then these polymer types are biodegradable.

Figure 13.26
The Terylene chains are linked by dipole–dipole interactions between the C=O groups on adjacent polymer chains.

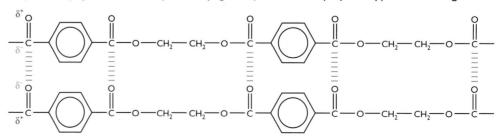

● **Key skills** ICT
  • Use of modelling software to visualise monomers and their addition and condensation polymers.

  Number
  • Production of graphs from data, extracted from a variety of sources, concerning the use of the major addition and condensation polymers.

● **Skills task** Use the Internet and other sources to investigate the polyaramids. Produce an A4 sheet summarising your findings.

CHECKLIST After studying Chapter 13 you should know and understand the following terms.

● **Polymer:** A substance containing very large molecules, that are made up of repeating units or monomers. Polymers therefore have a very large relative molecular mass.
● **Addition polymerisation:** The process by which an addition polymer is formed. The mechanism for this process involves free radicals.
● **Isotactic polymer:** A polymer that possesses a regular structure. This allows close packing to take place and leads to a high degree of crystallinity.
● **Atactic (or syndiotactic) polymer:** A polymer that possesses an irregular structure. This prevents close packing of the polymer chains and results in the polymer being soft and flexible.
● **Condensation polymerisation:** The process that involves the formation of a polymer by a condensation reaction, that is, a reaction in which small molecules, such as water or hydrogen chloride, are produced.
● **Polyamide:** A type of condensation polymer produced by the reaction of an amino group of one molecule and a carboxylic acid group of another molecule to produce a protein-like structure containing the amide or peptide group, —CONH—.
● **Aramid:** This is a type of polyamide in which the aliphatic hydrocarbon unit within the polymer chain has been replaced by an aromatic one.
● **Polyester:** A type of condensation polymer produced by the reaction of an alcohol group of one molecule and a carboxylic acid group of another molecule to produce a plastic containing the ester group, —COO—.

● Examination questions

1 Describe, using an example in each case, the formation of the following polymers:
  a **one** synthetic polymer by addition polymerisation (2)
  b **one** *synthetic* polymer and **one** *natural* polymer by condensation polymerisation. (In this question, 1 mark is available for the quality of written communication.) (11)
(For revision of organic nitrogen-containing molecules, see Chapter 11.)

OCR, A level, Specimen Paper A7882, Sept 2000

2 A large proportion of the nylon manufactured in the UK is nylon-6,6 made from 1,6-diaminohexane and hexanedioic acid. In the industrial process currently in use, both 1,6-diaminohexane and hexanedioic acid are made from benzene.

  a Draw the full structural formula of 1,6-diaminohexane. (2)
  b 1,6-Diaminohexane is not very soluble in water but dissolves readily in dilute hydrochloric acid. Write a balanced equation for this reaction. (You do not need to draw again the full structural formula for 1,6-diaminohexane, but your formula should show the functional groups clearly.) (2)
  c The repeating unit in nylon-6,6 is shown below:

$$-HN-(CH_2)_6-NH-CO-(CH_2)_4-CO-$$

  repeating unit

  Suggest why nylon made in this way is called nylon-6,6. (1)

d Chemists have recently developed a new process for making hexanedioic acid from glucose. Enzymes in some organisms can convert glucose into muconic acid. The chemists modified the bacterium *E. coli* by genetic engineering to produce these enzymes. The modified bacteria are then fed glucose and produce muconic acid. This is collected and converted into hexanedioic acid.

$$HOOC-CH=CH-CH=CH-COOH$$
muconic acid

i Briefly explain, in outline, how a bacterium such as *E. coli* can be modified by genetic engineering. (It is not necessary to give practical details.) (3)

ii The nuclear magnetic spectrum of muconic acid contains two signals. Suggest, giving a reason, a chemical shift for each signal. (2)

(For details on nuclear magnetic resonance (NMR) spectroscopy, see Chapter 15, pages 230–4.)

OCR, A level, Specimen Paper A7887, Sept 2000

3 a Nylon 6:6, a polyamide, has a structure containing the following repeat unit:

$$\left[CO-(CH_2)_4-CONH-(CH_2)_6-NH \right]$$

i Give the structures of the monomers from which this polymer could be made. (2)
ii What type of polymer is this? (1)
b i Show the structure of poly(tetrafluoroethene). (2)
ii State one use of this polymer. (1)

London, A level, Module Test 4, June 1998

4 Nomex is a du Pont fibre, used for flame-retardant clothing, which can resist temperatures of 1000 °C for 12 seconds, enough to have enabled the Benetton Formula 1 team to have survived a serious fire during the 1994 Grand Prix season. Nomex is a polymer which could in principle be made from 1,3-diaminobenzene and benzene-1,3-dicarboxylic acid:

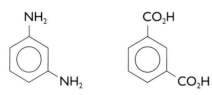

a Draw a representative length of the Nomex polymer chain. (2)
b What **type** of polymer is Nomex? (1)

c Dicarboxylic acids are not usually used in making this type of polymer. They are generally made from acid chlorides. Suggest why this is so. (2)
b Suggest how benzene-1,3-dicarboxylic acid could be converted to its diacid chloride, and draw its structure. (2)

London, A level, Module Test 4, Jan 1998

5 Propenal, CH_2=CHCHO, is one of the materials that gives crispy bacon its sharp odour. In the following question assume that the carbon–carbon double bond and the aldehyde group in propenal behave independently.

a Give the structural formulae of the compounds formed when propenal reacts with:
i hydrogen bromide (2)
ii hydrogen cyanide (1)
iii 2,4-dinitrophenylhydrazine. (2)

b i Give the mechanism for the reaction between hydrogen cyanide and the aldehyde group. You may represent the aldehyde group as

$$\begin{array}{c} R \\ \diagdown \\ C=O \\ \diagup \\ H \end{array}$$

(3)

ii The reaction in part **b i** occurs best in slightly acidic conditions. It is slower if the pH is high or low. Suggest reasons why this is so at high pH and low pH. (3)

c Explain why lithium tetrahydridoaluminate(III) (lithium aluminium hydride), $LiAlH_4$, reacts only with the

$C=O$ bond and not with the $C=C$ bond, even though these bonds have the same electronic structure. (2)

d Suggest reactions, giving equations and conditions, which would convert propenal into a compound which would react with iodine in the presence of sodium hydroxide solution. (4)

London, A level, Module Test 4, Jan 1998

6 a Poly(1,1-difluoroethene) or PDFE is a piezoelectric substance. This means that when it is physically deformed, it produces an electric current.
   Draw the structural formula of this polymer, showing **two** repeat units. (2)
b Name the two different **types** of polymerisation involved in the manufacture of Kevlar® and PDFE. (2)

London, A level, Module Test 4, Jan 2000

Organic synthesis and analysis

Grignard reagents

Grignard reagents were developed by Victor Grignard, who was born in France in 1871 (Figure 14.1). He described them in his PhD thesis of 1901 and received the Nobel Prize in 1912 for his work on Grignard reagents.

Grignard reagents are important and versatile compounds, much used in organic synthesis. They can be used to increase the length of a carbon skeleton, as they can be used to add more than one carbon atom to the skeleton. This is unlike the nucleophilic substitution reaction of halogenoalkanes with cyanide ions, which only increases the carbon skeleton by one atom.

The general formula of these reagents is RMgX, where the R group is an alkyl or phenyl group and X is a halogen. These compounds are classed as organometallic compounds because they contain the metal magnesium covalently bonded to one of the carbon atoms of the alkyl group.

Figure 14.1
Victor Grignard (1871–1935), who discovered Grignard reagents in 1901.

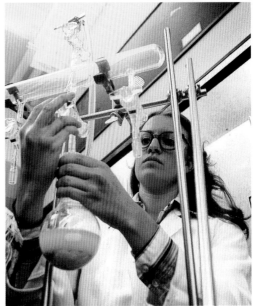

Figure 14.2 (right) Grignard reagents are organometallic compounds used throughout organic synthesis to extend the length of a carbon chain by more than one carbon atom.

Preparation of Grignard reagents

A Grignard reagent is prepared by adding magnesium turnings to a solution of a halogenoalkane, in dry ether, under anhydrous conditions, and heated under reflux. During the preparation of these reagents it is best to exclude air because carbon dioxide and oxygen also react with them. As such, these reagents are not very stable and so must be freshly prepared and used as soon as possible after preparation. For example:

$$CH_3CH_2I + Mg \xrightarrow{ether} CH_3CH_2MgI$$
$$\text{ethylmagnesium iodide}$$

Figure 14.3 (below) Grignard reagents have a polar bond.

$$\overset{\delta^-}{C} - \overset{\delta^+}{Mg}$$

In their reactions, Grignard reagents behave as nucleophiles due to the polar nature of the C—Mg bond. The carbon atom has a higher electronegativity than the magnesium so a polar bond forms (Figure 14.3). During the reaction the nucleophilic carbon atom of the Grignard reagent attacks carbon atoms with a positive charge in other molecules. The reactions always occur in two stages (Figure 14.4). The first step involves the formation of an intermediate by the addition of the Grignard reagent. This intermediate then needs to be decomposed in a second step, by the addition of dilute acid.

Figure 14.4 Grignard reagents react in two stages. The first stage is an addition reaction to form an intermediate. The second stage is the acid hydrolysis of this intermediate.

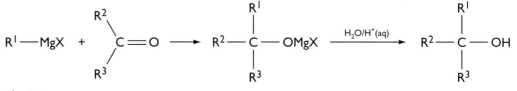

R^1 = alkyl group
$R^2 = R^3$ = alkyl group (except for the reaction with CO_2, when R^2/R^3 = O)

Figure 14.5 Common examples of Grignard reactions.

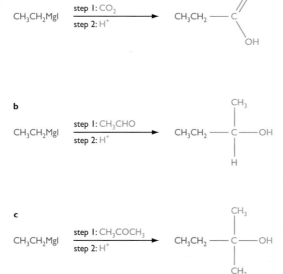

Reactions involving Grignard reagents often produce high yields, in excess of 90%. As a result, they are extensively used in industrial organic synthesis (Figure 14.5).

The number of carbon atoms added to the carbon chain depends on the structure of the Grignard reagent used.

Reaction of a Grignard reagent with an acid produces the parent alkane (Figure 14.6).

Figure 14.6 The acid hydrolysis of a Grignard reagent results in the formation of the parent alkane.

$$R - Mg - X \xrightarrow{H_2O/H+} R - H$$

I Write chemical equations for the reactions of ethylmagnesium bromide with:
a butanal
b butanone.

There are other organometallic compounds that are used in a similar way, for example, organolithium compounds such as $(CH_3)_3CLi$. These compounds are more reactive than the organomagnesium reagents but are more difficult to handle.

Organic synthesis

In this section we will summarise the organic reactions that you have been introduced to both at AS and at A2. We will then explore the use of these reactions in the production of more complex organic compounds such as, for example, ibuprofen (Figure 14.7).

Figure 14.7
Ibuprofen is an anti-inflammatory drug, which is a mild painkiller. It was originally developed to relieve the symptoms of rheumatoid arthritis.

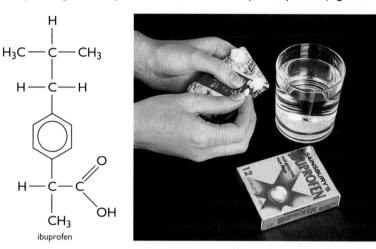

ibuprofen

Reactions of alkenes

Alkenes undergo electrophilic addition reactions. These reactions were introduced in *Introduction to Advanced Chemistry*, Chapter 16, and they are summarised in Figure 14.8.

Figure 14.8
Reactions of alkenes.

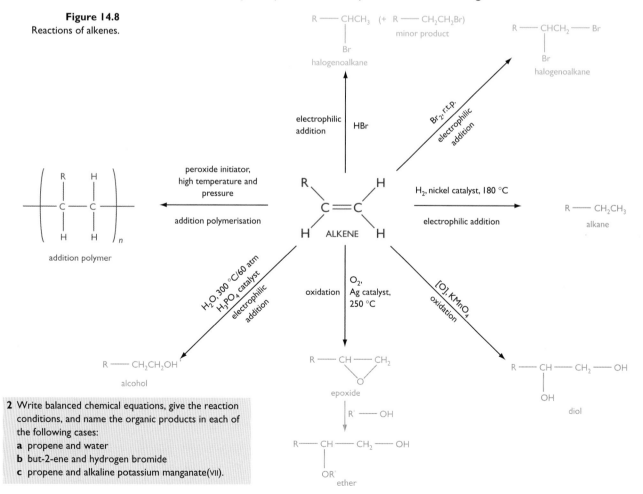

2 Write balanced chemical equations, give the reaction conditions, and name the organic products in each of the following cases:
 a propene and water
 b but-2-ene and hydrogen bromide
 c propene and alkaline potassium manganate(VII).

Reactions of halogenoalkanes

The reactions of halogenoalkanes are essential to preparative organic chemistry. They undergo two main types of reaction: nucleophilic substitution reactions and elimination reactions. These reactions were introduced in *Introduction to Advanced Chemistry*, Chapter 18, and they are summarised in Figure 14.9.

Figure 14.9
Reactions of halogenoalkanes.

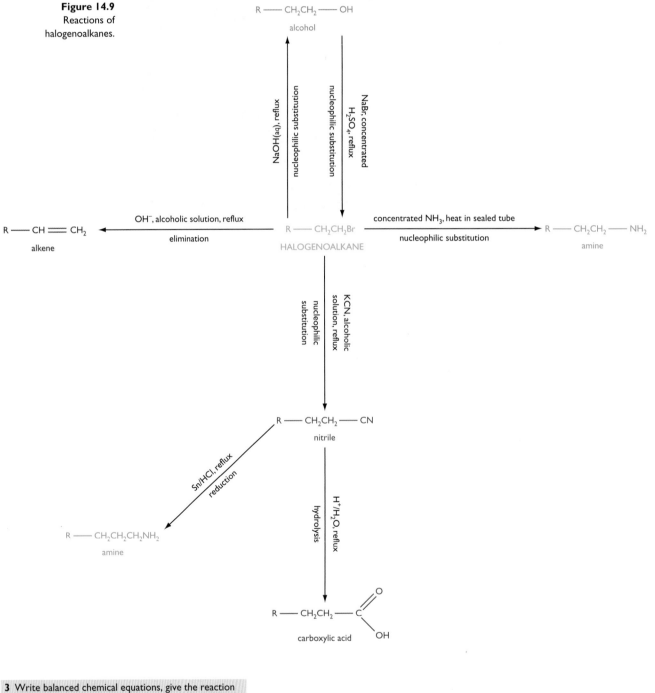

3 Write balanced chemical equations, give the reaction conditions, and name the organic products in each of the following cases:
  a 1-bromopropane and sodium hydroxide solution
  b 2-bromopropane and potassium cyanide
  c 2-bromobutane and potassium hydroxide in ethanol
  d 1-bromobutane and ammonia.

Reactions of alcohols

Alcohols undergo a wide range of reactions, such as oxidation, dehydration and condensation, that are useful in organic synthesis. The important reactions of alcohols were introduced in *Introduction to Advanced Chemistry*, Chapter 19, and they are summarised in Figure 14.10.

Figure 14.10
Reactions of alcohols.

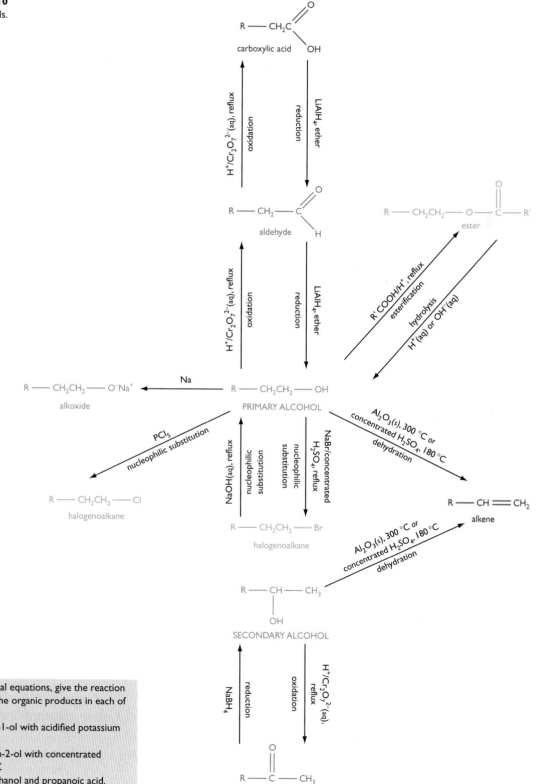

4 Write balanced chemical equations, give the reaction conditions, and name the organic products in each of the following cases:

a oxidation of propan-1-ol with acidified potassium manganate(VII)

b dehydration of butan-2-ol with concentrated sulphuric acid at 180 °C

c reaction between ethanol and propanoic acid.

Reactions of aldehydes and ketones

Aldehydes and ketones are commonly known as carbonyl compounds. Their reactions were introduced briefly in *Introduction to Advanced Chemistry*, Chapter 20, and developed in this book, Chapter 9. They undergo a series of reactions such as reduction, nucleophilic addition and condensation reactions (Figure 14.11).

Figure 14.11
Reactions of aldehydes and ketones. (NAE = nucleophilic addition–elimination)

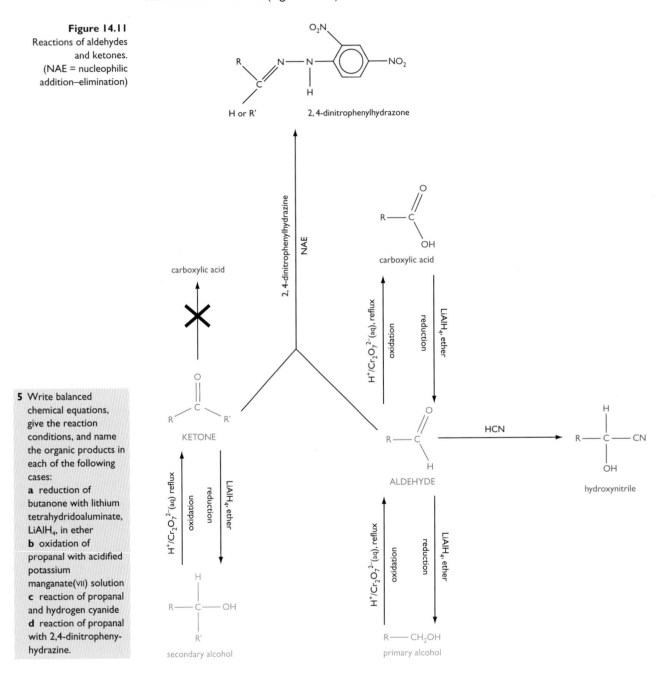

5 Write balanced chemical equations, give the reaction conditions, and name the organic products in each of the following cases:
a reduction of butanone with lithium tetrahydridoaluminate, LiAlH$_4$, in ether
b oxidation of propanal with acidified potassium manganate(VII) solution
c reaction of propanal and hydrogen cyanide
d reaction of propanal with 2,4-dinitrophenylhydrazine.

Reactions of carboxylic acids and their derivatives

Carboxylic acids themselves have few synthetic uses due to their unreactive nature, relative to their derivatives such as acid chlorides (acyl chlorides) and acid anhydrides. These reactions are summarised in this book, Chapter 10, and in Figure 14.12 for carboxylic acids, Figure 14.13 for acid chlorides and Figure 14.14 for acid anhydrides.

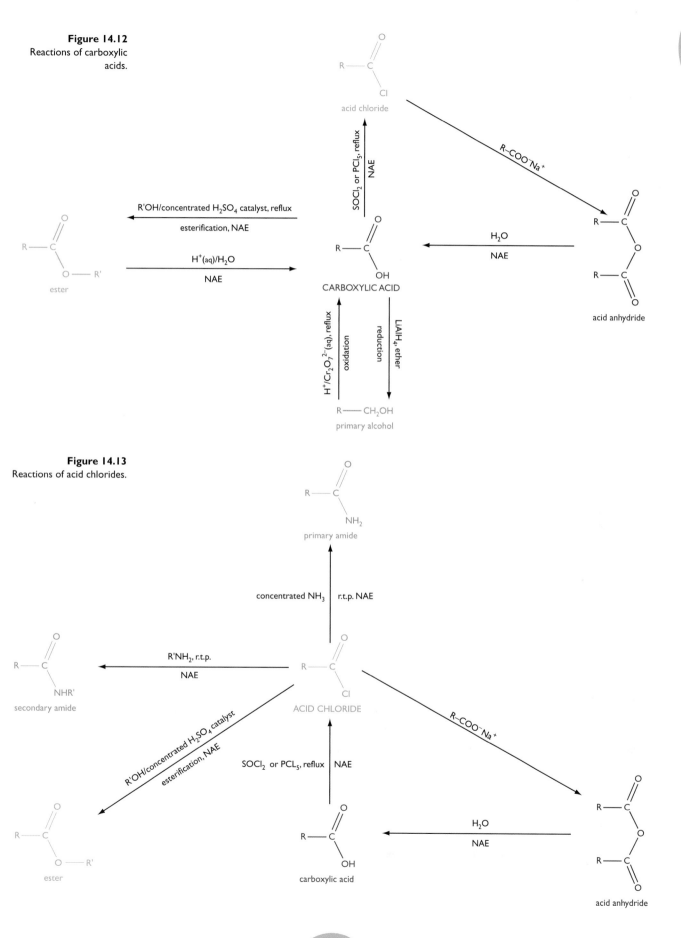

Figure 14.12
Reactions of carboxylic acids.

Figure 14.13
Reactions of acid chlorides.

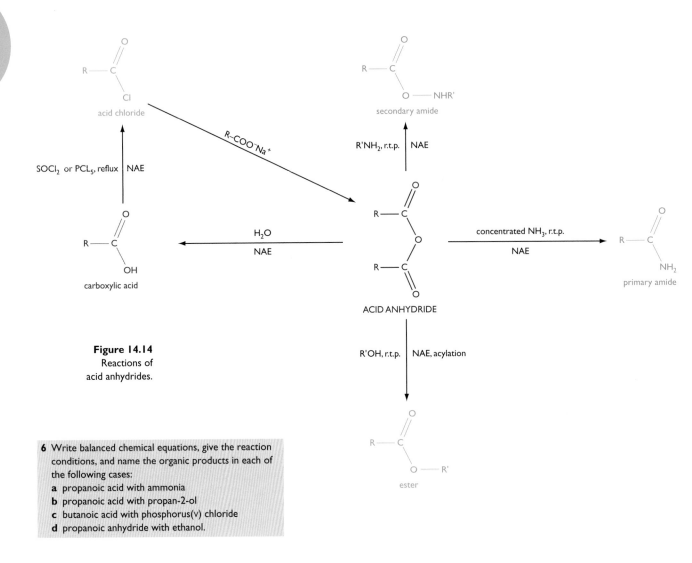

Figure 14.14
Reactions of
acid anhydrides.

6 Write balanced chemical equations, give the reaction
conditions, and name the organic products in each of
the following cases:
a propanoic acid with ammonia
b propanoic acid with propan-2-ol
c butanoic acid with phosphorus(V) chloride
d propanoic anhydride with ethanol.

Reactions of the benzene ring

Many synthetic routes involve an initial step of introducing a side-group containing another
functional group, onto a benzene ring. The resulting side-group can then undergo further
reactions to produce more complex organic molecules. For example, nitrobenzene can be
reduced to phenylamine, which can then be converted to a diazo dye. Owing to the delocalised
electrons in the benzene ring (Figure 14.15) it undergoes electrophilic substitution reactions.
These reactions (see Chapter 12) are summarised in Figure 14.16. The reaction sequence
starting with nitrobenzene is shown in Figure 14.17.

Figure 14.15
The delocalised nature of
the benzene ring results in
it undergoing electrophilic
substitution reactions.

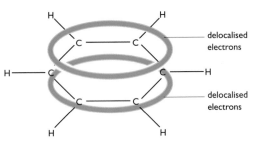

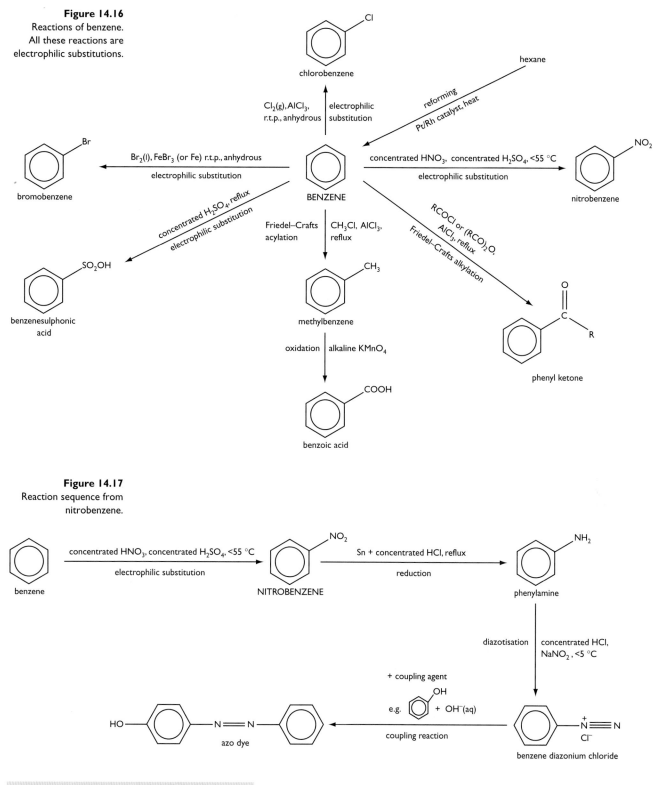

Figure 14.16
Reactions of benzene.
All these reactions are
electrophilic substitutions.

chlorobenzene

hexane

$Cl_2(g), AlCl_3$,
r.t.p., anhydrous | electrophilic
substitution

reforming
Pt/Rh catalyst, heat

$Br_2(l)$, $FeBr_3$ (or Fe) r.t.p., anhydrous
electrophilic substitution

concentrated HNO_3, concentrated H_2SO_4, <55 °C
electrophilic substitution

bromobenzene

BENZENE

nitrobenzene

concentrated H_2SO_4, reflux
electrophilic substitution

Friedel–Crafts
acylation | CH_3Cl, $AlCl_3$,
reflux

$RCOCl$ or $(RCO)_2O$,
$AlCl_3$, reflux
Friedel–Crafts alkylation

SO_2OH

benzenesulphonic
acid

methylbenzene

phenyl ketone

oxidation | alkaline $KMnO_4$

COOH

benzoic acid

Figure 14.17
Reaction sequence from
nitrobenzene.

benzene

concentrated HNO_3, concentrated H_2SO_4, <55 °C
electrophilic substitution

NITROBENZENE

Sn + concentrated HCl, reflux
reduction

phenylamine

diazotisation | concentrated HCl,
$NaNO_2$, <5 °C

+ coupling agent

e.g. OH + OH^-(aq)

coupling reaction

azo dye

benzene diazonium chloride

7 Write balanced chemical equations, give the reaction
conditions, and name the organic products in each of
the following cases:
a nitration of methylbenzene
b reaction of 2-chloropropane with benzene
c reaction of butanoyl chloride with benzene
d reduction of benzene with hydrogen.

Synthetic routes

A synthetic route is the series of steps carried out to convert a starting molecule to the final product of an organic synthesis. This final product is known as the target molecule.

Synthesis of ibuprofen

Figure 14.18
The ibuprofen molecule.

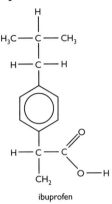

ibuprofen

Ibuprofen was developed by the Boots company in response to a need. The need was to help people suffering from rheumatoid arthritis who had very painful and swollen joints. After many years of testing and molecular structural modifications, a suitable molecule was synthesised. The structural formula of ibuprofen is shown in Figure 14.18. In order to manufacture the ibuprofen molecule in sufficient quantities to meet the demand, a suitable synthetic route had to be found. In deciding on a suitable route several factors needed to be considered:

- the availability of a suitable starting material
- the number of steps involved in producing the target molecule
- the yield of each of the steps involved, and hence the overall yield
- the cost of the overall process.

Looking at the ibuprofen molecule it is possible to see that benzene is a suitable starting molecule. To the benzene ring two carbon–carbon bond-forming reactions can be used to introduce the required side-chains. The six-step synthesis of ibuprofen is shown in Figure 14.19.

Figure 14.19
The six-step synthesis of ibuprofen from benzene (continued opposite).

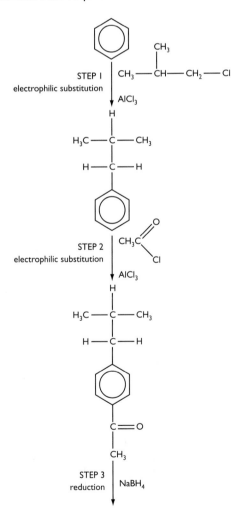

STEP 1
electrophilic substitution

STEP 2
electrophilic substitution

STEP 3
reduction | $NaBH_4$

Step 1: starting with the benzene molecule, the alkyl group can be introduced using a Friedel–Crafts alkylation reaction (see Chapter 12, page 181). A halogenoalkane with the required carbon skeleton is used, along with an aluminium chloride catalyst, under anhydrous conditions and heated under reflux. The halogenoalkane is 1-chloro-2-methylpropane.

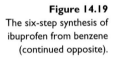

1-chloro-2-methylpropane

Step 2: the introduction of a second group into the 4-position of the benzene ring. This is also a carbon–carbon bond-forming reaction but in order to produce the required functional group at this point it is necessary to introduce a group which can undergo further reactions. A suitable group is an acyl group, and in this case ethanoyl chloride can be used:

ethanoyl chloride

The acyl group is introduced in a similar manner to the alkyl group, but by a Friedel–Crafts acylation reaction, using aluminium chloride as a catalyst, under anhydrous conditions and heated under reflux.

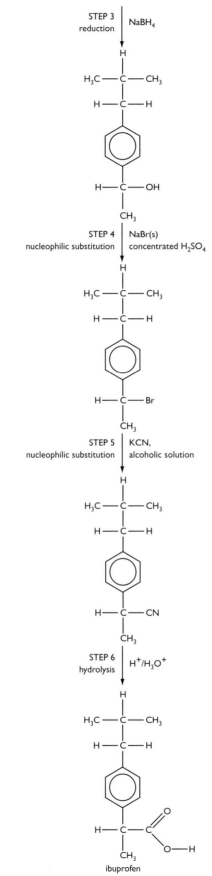

Figure 14.19 (continued)
The six-step synthesis of
ibuprofen from benzene.

The group we have just introduced is one carbon short of that required in the target molecule. So we now have to carry out a series of reactions on this side-chain to first extend the carbon chain and then produce the final functional group of the target molecule.

The extra carbon atom can be introduced by the use of a cyanide group. You will have seen in your study of organic chemistry that the introduction of a cyanide group ($-C\equiv N$) requires a nucleophilic substitution reaction of a halogenoalkane. So, how do we obtain a halogenoalkane from the carbonyl group present?

Step 3: the initial reduction of the carbonyl group by warming with sodium tetrahydroborate, $NaBH_4$ (with water as a solvent). This is a reduction reaction as hydrogen is added across the double bond (Chapter 9, page 140). The result is the formation of a secondary alcohol group.

Step 4: the formation of the halogenoalkane from the secondary alcohol group. This is done by reacting the alcohol group with phosphorus(v) chloride, PCl_5, at room temperature.

Step 5: we are now in a position to increase the length of the carbon chain by a nucleophilic reaction between the halogenalkane group and a cyanide ion, CN^-. This reaction is carried out by heating the molecule under reflux with potassium cyanide in alcoholic solution.

Step 6: the cyanide group can be converted into a carboxylic acid group by acid hydrolysis (Chapter 10, page 149). The hydrolysis is carried out by heating under reflux using dilute hydrochloric acid.

The overall yield of these six steps is high because of the small number and relative simplicity of each of the steps. However, the starting material for each of the steps needs to be separated and purified from the previous step. This involves distillation, recrystallisation and solvent extraction (see pages 211–13).

---

8 Describe a two-stage synthesis of 2-hydroxypropanoic acid (lactic acid), starting from ethanal.

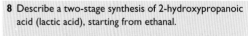

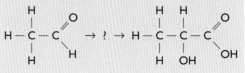

Drug manufacture and chirality

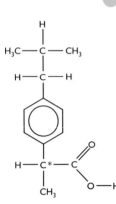

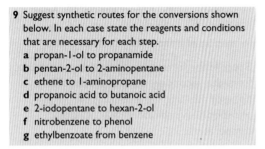

Figure 14.20
The chiral carbon in the ibuprofen molecule is labelled (*). It gives rise to two enantiomeric forms.

The ibuprofen molecule contains a chiral carbon atom (Figure 14.20). When ibuprofen is manufactured both optical isomers, or enantiomers, are produced. It has been found that one enantiomer is a more effective painkiller than the other and so routes which result in the production of only this enantiomer are being developed.

In the mid-1950s a drug called thalidomide was developed as a sedative. Further testing showed that it was very effective in the treatment of morning sickness in pregnant women. It was marketed for this use from 1958. As with ibuprofen, when it was manufactured equal amounts of each enantiomer were produced (Figure 14.21). One of these enantiomers is a safe and effective drug; the other was found to cause birth defects. It was withdrawn during 1961/62, after between 5000 and 10 000 children had been born with severe deformities.

Since this catastrophe any new pharmaceutical compound which contains a chiral carbon has had to have the two enantiomers tested separately. This requires the enantiomeric forms to be separated from one another, which is expensive due to their similar chemical properties. However, even the separation of the enantiomeric forms does not always prevent problems. The separation would not have helped in the case of thalidomide, because the 'safe' enantiomer of thalidomide is converted (racemised) by the body into a mixture of both enantiomers.

9 Suggest synthetic routes for the conversions shown below. In each case state the reagents and conditions that are necessary for each step.
 a propan-1-ol to propanamide
 b pentan-2-ol to 2-aminopentane
 c ethene to 1-aminopropane
 d propanoic acid to butanoic acid
 e 2-iodopentane to hexan-2-ol
 f nitrobenzene to phenol
 g ethylbenzoate from benzene

Figure 14.21
The structure of the thalidomide molecule showing the chiral carbon atom.

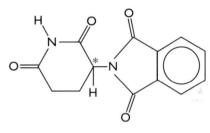

thalidomide

Calculating the overall yield of a synthetic route

During any organic preparation it is unlikely that the reaction will give a 100% yield. There are many reasons for less than 100% yield, including:

- reactant(s) may not be pure
- the reaction may involve an equilibrium
- some of the products may be left behind in the reaction vessel.
- the product may be difficult to purify
- the possibility of side reactions occurring

In a series of reactions the yield of one step has an effect on the yield of the next. Consider the two-step process in the production of phenylamine from benzene (Figure 14.22). If the first step in the process gave a yield of 90% and the second step a yield of 85%, what would be the overall yield for this two-step process?

The overall yield is found by the following expression:

overall % yield $= (0.90 \times 0.85) \times 100 = 76.5\%$

Figure 14.22
Phenylamine can be made in two steps from benzene.

benzene $\xrightarrow[<55\ °C]{85\%}$ nitrobenzene $\xrightarrow{90\%}$ phenylamine

If the phenylamine was subsequently converted to the amide and that reaction had a yield of 85%, what would now be the overall yield from the starting material benzene?

overall % yield $= (0.90 \times 0.85 \times 0.85) \times 100$
$= 65\%$

You should now see how even a three-step process, in which each individual step has a high yield, results overall in a yield of only 65%. In the manufacture of ibuprofen, therefore, it is very important that in order to have a high overall yield each step must have an excellent yield.

10 A compound K can be prepared by three different routes, as shown in Figure 14.23. The yield for each step is shown under the arrows. Which route gives the best yield of compound K?

Figure 14.23

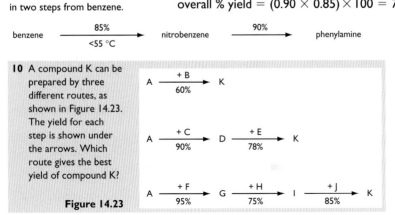

Separation and analysis

At some point in an organic synthesis it will be necessary to use separation techniques to isolate a particular substance from a mixture of products. Separation is then followed by the use of analytical techniques to identify the organic substance (see Chapter 15).

The four main methods of separation used are:

- solvent extraction
- filtration under reduced pressure
- distillation
- recrystallisation
- chromatography.

Solvent extraction

Solvent extraction is a very useful technique for separating the organic products of a synthesis from an aqueous reaction mixture. To do this an organic solvent, such as ethoxyethane, is chosen which is immiscible with water but in which the required organic product is soluble. The aqueous reaction mixture and the organic solvent are shaken together in a separating funnel (Figure 14.24). The organic product dissolves preferentially in the organic solvent (it is partitioned between the two solvents, see Chapter 2, page 23). The layers are allowed to separate and then the lower aqueous layer is run off. The aqueous layer should now be re-extracted using fresh samples of the organic solvent, so that more of the organic product is obtained. The final step of this process involves collecting together the different samples of the organic solution and removing the final traces of water using a drying agent such as anhydrous sodium sulphate. The dried organic solution is then filtered and the product obtained by evaporation using a rotary evaporator (Figure 14.25).

Figure 14.24 (left)
A separating funnel can be used to separate immiscible solvents. Separation occurs if the required organic product is more soluble in one solvent than the other. For example, aspirin is more soluble in ethoxyethane than in water. Hence, after shaking, the upper organic layer will contain a higher concentration of aspirin than the lower aqueous layer.

Figure 14.25 (right)
A rotary evaporator is used to remove the organic solvent.

Figure 14.26
Reducing the pressure in the Buchner flask produces a partial vacuum, which draws the liquid through the filter paper much more quickly.

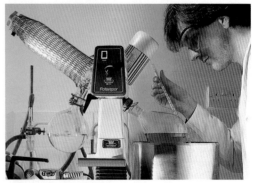

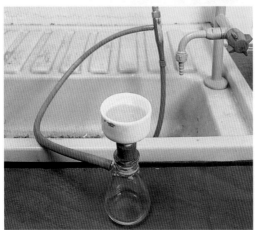

Filtration under reduced pressure

If a solid needs to be separated quickly from a liquid then filtration under reduced pressure is used. This technique involves the use of a Buchner flask and funnel (Figure 14.26). A partial vacuum is created inside the Buchner flask by a water or vacuum pump. This causes the liquid component of the mixture to be drawn through the filter paper. Because of the reduced pressure the Buchner flask is made of thick glass.

Distillation

There are three main distillation techniques:

- simple distillation
- fractional distillation
- steam distillation.

Figure 14.27
The apparatus used to carry out a simple distillation.

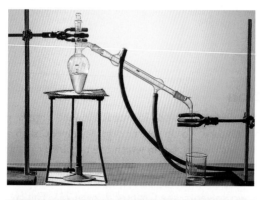

Figure 14.27
The apparatus used to carry out a simple distillation.

Figure 14.28 (above)
For the most efficient and effective separation a tall fractionating column packed with glass beads or rings is used.

Figure 14.29 (right)
A large fractionating column is used in an oil refinery to ensure efficient separation of the individual fractions from crude oil.

Figure 14.30
Steam distillation apparatus used for separating an organic liquid with a high boiling point which has a low solubility in water.

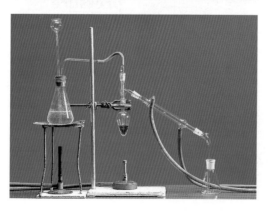

Simple distillation is used to separate a solvent from a non-volatile solute (Figure 14.27). For example, this technique is used to recover ethanol in the final stages of the preparation of aspirin.

Fractional distillation is the technique used to separate mixtures of miscible liquids. For the most effective separation, a fractionating column is packed with glass beads or rings to increase the surface area. A long column also improves the efficiency of the process (Figure 14.28). This allows for the most efficient and effective separation of the miscible liquid mixture. This technique is the way in which fractions are obtained from crude oil. Industrial fractionating columns designed for this process can be tens of metres high (Figure 14.29).

Steam distillation is a technique which is particularly useful for separating liquids that decompose significantly at their boiling points (Figure 14.30). A high boiling point organic liquid which has a very low solubility in water can be removed from a mixture by steam distillation. In this process steam is passed through the heated mixture and the distillate is collected. This distillate will be present as two layers, one being the aqueous layer and the other an immiscible organic layer. The two layers can be separated using a separating funnel.

An example of the use of this technique is in the extraction of limonene from the peel of oranges and lemons (Figure 14.31). Limonene is used in many household fragrances and herbal medicines.

Figure 14.31
Limonene is extracted from orange and lemon peel by steam distillation.

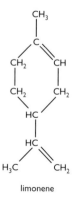

limonene

Recrystallisation

Recrystallisation is used where the solid product required readily dissolves in a hot solvent but is much less soluble in the cold solvent. In this technique the impurities must either be insoluble, or remain in solution in both the cold and hot solvent. The process to purify the solid product is carried out by the following steps:

- a suitable solvent is chosen which should not react with the compound
- the minimum amount of *hot* solvent should be used to dissolve the product from the reaction
- the mixture is then filtered to remove the insoluble impurities
- the filtrate is allowed to cool slowly allowing the crystals to re-form, with soluble impurities staying in solution in the cold solvent
- the crystals can be obtained by further filtration.

Chromatography

Chromatography is a general term used to cover the separation of complex mixtures that contain only very small quantities of different substances. There are many different types of chromatography, but each is based on the same fundamental principles.

There are two phases involved:

- a **stationary phase**, which does not move during the process
- a **mobile phase**, which moves over or through the stationary phase.

The separation occurs because the different components of the mixture have different attractions for the stationary and mobile phases. The result is that the mobile phase carries the components of the mixture through or over the stationary phase at different rates, separating the mixture.

Paper chromatography

Paper chromatography is the simplest of the chromatographic techniques, and is often used to separate dyes from a mixture (Figure 14.32). In this technique the stationary phase is made up of water molecules that are trapped in the cellulose fibres of paper. The mobile phase is the

Figure 14.32
The separation of the dyes found in black ink.

aqueous or organic solvent that moves up the paper by capillary action. This capillary action is caused by the forces between the cellulose fibres of the paper and the solvent. The dyes that are more soluble in the solvent than they are in the water molecules of the stationary phase move rapidly up the paper, whilst those which are more soluble in water are not carried as far up the paper (Figure 14.33).

Figure 14.33
The black ink separates into three different dyes.

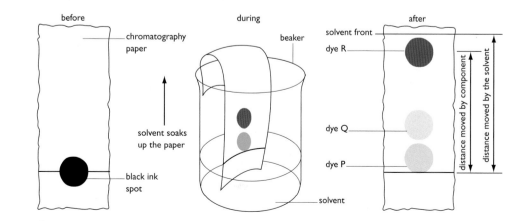

When the solvent front has almost reached the top, the paper is removed and left to dry in a fume cupboard.

For each of the different components a **R_f** value (or **retention factor**) can be calculated:

$$R_f = \frac{\text{distance moved by the component}}{\text{distance moved by the solvent}}$$

Standard R_f values are tabulated for a wide variety of substances using particular solvents under standard conditions. These allow the identification of unknown components by comparison with these standard R_f values. It should be noted that the R_f value is not dependent on the distance travelled by the solvent. It is dependent on the nature of the mobile and stationary phases as well as on the components of the substance being separated.

This type of chromatography is used extensively in hospital and forensic science laboratories to separate a variety of intriguing mixtures. It is sometimes necessary in these cases to carry out a variation of simple chromatography, which requires the paper to be rotated by 90°. After the initial separation process the paper is allowed to dry and then a different solvent is used after the paper has been rotated. This technique is known as **two-way chromatography** and allows the separation of complex mixtures (see Figure 11.21, page 170).

The substances to be separated do not have to be coloured. Colourless substances can be made visible by spraying or treating the chromatogram with a locating agent. For example, ninhydrin can be used as a locating agent for amino acids formed by the hydrolysis of proteins (Figure 14.34). The locating agent reacts with the colourless substance either to form a coloured product or to allow the spots to be located, for example, using ultraviolet light.

If required, the individual components can be extracted from the paper by cutting out the individual spots using scissors and then extracting the component using a suitable solvent.

The rate at which an individual component moves up the stationary phase depends on the partition coefficient of that component between the different phases (see Chapter 2, page 23).

Figure 14.34
Ninhydrin can be used to locate the spots formed by amino acids following the hydrolysis of proteins.

Thin-layer chromatography (TLC)

The basis of thin-layer chromatography is similar to that of paper chromatography. The paper stationary phase is replaced by a TLC plate, which has a thin layer of a substance such as silica (SiO_2) or alumina (Al_2O_3) coated onto a glass, aluminium foil or plastic plate. The components may be recovered by scraping the areas containing the spots into a suitable solvent.

This chromatographic technique is used in hospitals and forensic laboratories to identify amino acids from blood samples.

Column chromatography

Column chromatography is a convenient variation of the chromatography technique that allows for the large scale separation of mixtures. A glass column is packed with an inert substance such as silica or alumina, which acts as an inert stationary phase. The mixture to be separated is introduced to the top of the column and then the solvent is allowed to run through under the force of gravity. As the separation takes place the different components will arrive at the bottom of the column at different times and may be collected in different flasks for further use (Figure 14.35).

Larger, more efficient, columns are used in industry where larger quantities of complex mixtures are separated.

Figure 14.35
Column chromatography can be used to separate the components of a mixture for further use.

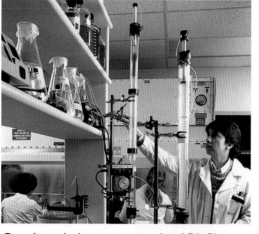

Gas–liquid chromatography (GLC)

GLC is one of the most important techniques used in analytical laboratories (Figure 14.36). The stationary phase is an inert solid onto which a high boiling point liquid has been coated. The solid is packed into a thin glass column, up to 100 m in length, which is coiled and set into an oven. The mobile phase, which passes through the coiled tube, is a chemically inert gas of high purity, such as nitrogen.

Figure 14.36
The main components of a gas–liquid chromatograph.

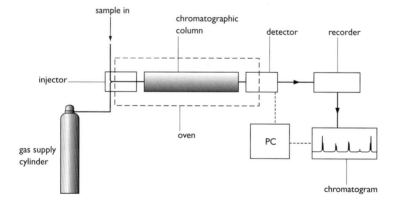

A sample of the mixture is injected into the coiled tube and the carrier gas moves the mixture through the tube. The individual components have different affinities for the mobile and stationary phases and therefore separation takes place. It is found that the most volatile components usually emerge first and those which favour the stationary phase emerge last from the tube. The time taken for each component to emerge from the tube after injection is called the **retention time**, which is dependent upon:

- the carrier gas being used
- the flow of the carrier gas
- the temperature at which the column is maintained
- the nature of the stationary phase
- the length of the column.

Because of these variables the instrument has to be calibrated on a regular basis using pure samples of known substances. Once calibrated, the conditions mentioned above have to remain constant.

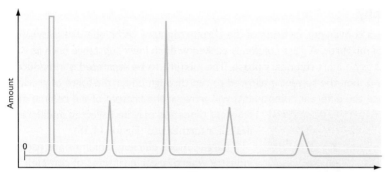

Figure 14.37 (above)
The chromatogram is a plot of peak height against retention time.

Figure 14.38 (right)
The use of the two techniques, gas–liquid chromatography and mass spectrometry, is a very powerful tool for analytical chemists.

As the components emerge they are detected by various forms of detector, for example one which determines the thermal conductivity of the components, and a chromatogram is produced (Figure 14.37). The area under each peak is proportional to the amount of a substance present in the mixture. If the peaks are very sharp then the height of the peak can be taken as being proportional to the amount of substance.

It is more usual today to have a mass spectrometer linked to the GLC machine. As the components emerge from the tube they are passed into the mass spectrometer (see Chapter 15) and the resulting mass spectrum can be used to identify the component (Figure 14.38).

The technique of GLC is used, for example, to estimate the amount of alcohol in blood and to identify the presence of drugs in urine.

High-performance liquid chromatography (HPLC)

HPLC is a development from column chromatography. Instead of using the force of gravity to move the mobile phase down a column, high pressure is used to force the mobile phase through the stationary phase. Because of this the column containing the stationary phase needs to be made of stainless steel to withstand the pressures used (Figure 14.39). Separation is much faster as a result of the increased pressure. Because of the faster movement through the column, substances with similar R_f values can be separated before they begin to diffuse or interact with one another.

Figure 14.39
The HPLC system.

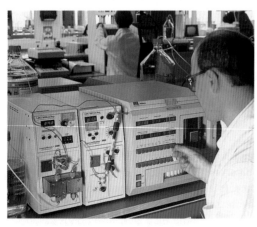

11 a Explain why fractional distillation is used in preference to simple distillation to separate crude oil into its fractions.
b In GLC, why does the mobile phase have to be an *inert* carrier gas?
c Why is the baseline in paper chromatography drawn with a pencil and not a pen?
d Why do consistent conditions have to be used when carrying out GLC?

The chromatogram obtained is similar to that obtained from a GLC machine. For example, a plot can be of the absorption of ultraviolet light against retention time. An example of where HPLC is used is in the quality control stage for aspirin tablets.

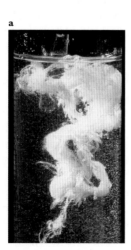

Electrophoresis

Electrophoresis is a variation on paper chromatography, based on how easily molecules can form ions. Once formed, the ions migrate in a buffered solution under the influence of an electric field to electrodes. The negative ions move towards the anode whilst the positive ions move towards the cathode (Figure 14.40).

Electrophoresis has been used to identify the amino acids from which protein molecules are made. The protein molecule would first be hydrolysed in the presence of an enzyme, which breaks the peptide links in the protein. The products of the hydrolysis would then be placed on a suitable medium, for example, paper or a polyacrylamide gel.

If the buffer used has a low pH then the amine groups on the amino acid molecules will be protonated to form the $-NH_3^+$ group (Chapter 11, page 168), giving an overall positive charge. Those amino acids with increased numbers of ionisable amine groups will gain a higher positive charge.

When the electric field is applied, the positive ions will move towards the cathode at different rates depending on the size of their positive charge and the mass of the ion.

Buffers of a high pH would cause negative ions to be formed by ionisation of the carboxyl group. In this case the ions would migrate to the anode at different rates.

The distance which each amino acid moves, under controlled conditions of pH and electric field, can be measured and compared with standard values.

This technique was used by Frederick Sanger to determine the sequence of amino acids in the insulin molecule (see Figure 11.23, page 171), and aided the subsequent synthesis of the molecule for the improved treatment of diabetes. Sanger was awarded the Nobel Prize for Chemistry in 1958 for his work in this area.

Figure 14.40
In electrophoresis, separation is caused by the movement of ions, under the influence of an electric field, to the anode or cathode.

DNA fingerprinting

Electrophoresis has an increasing use in forensic work to help in the identification of criminals from minute traces of flakes of skin, hair and blood found at the scene of a crime. No two individuals have exactly the same DNA sequence, except for identical twins. Therefore, using electrophoresis on the fragments of DNA produced by the use of restriction endonucleases, leads to a unique pattern that relates to a particular individual (Figure 14.41). The pattern obtained from the forensic samples can be compared with a DNA sample from a suspect taken, for example, in the form of a blood test (Figure 14.42). This technique is also established as a method of resolving paternity cases, as the identity of the parents and their children can be identified due to similarities in their DNA make-up.

Figure 14.41
a In DNA fingerprinting the DNA chains are cut up in a predictable way using enzymes called restriction endonucleases. These act as 'molecular scalpels' in breaking up the DNA molecule.
b The double helix structure of DNA.

a
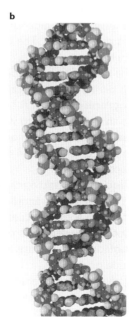

b

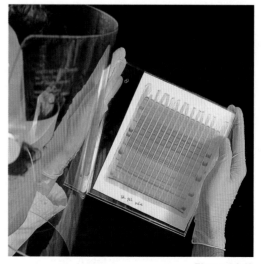

Figure 14.42
DNA fingerprinting is used in the detection of criminals.

12 During electrophoresis, the amino acid alanine, $H_2NCH(CH_3)COOH$, migrates towards the cathode at low pH and towards the anode at high pH. Explain why.

● **Key skills** ICT

  • Produce a Powerpoint presentation to describe the necessary steps required to make paracetamol. (For structure of paracetamol, see page 164.)

● **Skills task** Use the information in this chapter to summarise the possible reactions of the anaesthetic Novocaine.

Novocaine

CHECKLIST After studying Chapter 14 you should know and understand the following terms.

● **Grignard reagents:** Have the general formula RMgX where R is an alkyl or phenyl group and X is a halogen. They are classed as organometallic compounds and are used extensively in organic synthesis.

● **Synthetic route:** The series of steps that are carried out to convert a starting molecule to the target molecule of an organic synthesis.

● **Chirality:** Used in reference to a molecule that contains a chiral carbon atom. A molecule with a chiral carbon atom will be present as two optically active isomers or enantiomers. The enantiomers are non-superimposable mirror images of one another.

● **Solvent extraction:** A method used to separate and purify substances using a solvent which dissolves the desired product but leaves other compounds dissolved in the original solvent. It is essential that the two solvents do not mix.

● **Filtration under reduced pressure:** The process of separating a solid from a liquid rapidly. A partial vacuum is created inside a Buchner flask using a water pump. This causes the liquid component of the mixture to be drawn rapidly through the filter paper.

● **Steam distillation:** Particularly useful for separating liquids that decompose significantly at their boiling points. Steam is passed through the heated mixture and the organic substance is carried over and collected. The distillate will be present as two layers and the aqueous layer may be run off in a separating funnel.

● **Recrystallisation:** A technique used to separate a solid product. The required product must readily dissolve in the hot solvent but must be insoluble in the cold solvent. The impurities must either be insoluble or remain in solution in both the cold and hot solvents.

● **Chromatography:** A general term used to cover the separation of complex mixtures that contain only very small quantities of different substances. All types of chromatography have a stationary phase and a mobile phase. The mixture is separated as the mobile phase moves through the stationary phase.

● **Retention factor:** The ratio of the distance travelled by the solute to the distance travelled by the solvent during chromatography. Symbol: R_f.

● **Gas liquid chromatography (GLC):** A technique used to analyse mixtures of liquids. The stationary phase is an inert solid onto which a high boiling point liquid has been coated, packed into a coiled heated column. The mobile phase is a chemically inert gas of high purity.

● **Retention time:** The time taken for each component of the mixture to emerge from the GLC machine after injection.

● **High-performance liquid chromatography (HPLC):** Used to separate components of a mixture that are very similar to one another. A high pressure is used to drive the mobile phase, a pure solvent, through a column containing the stationary phase, a solid such as silica.

● **Electrophoresis:** A technique used to separate and identify organic compounds. It involves the formation of ions that migrate in a buffered solution under the influence of an electric field to the anode or cathode.

● **DNA fingerprinting:** A powerful forensic science tool in the investigation of crimes, based on electrophoresis analysis of DNA.

Examination questions

1 You are required to plan an experiment to determine the percentage by mass of bromine in a bromoalkane. The bromoalkane, which boils at 75 °C, can be hydrolysed completely by heating with an appropriate amount of boiling, aqueous sodium hydroxide for about 40 minutes. The bromide ion released can be estimated by converting it into a silver bromide precipitate which is subsequently weighed.

a Write equations for the reactions which occur. (2)

b Describe how you would carry out the hydrolysis, giving details of the apparatus and the conditions which you would use. (5)

c Describe, giving details of the apparatus and reagents, how you would obtain a silver bromide precipitate from the hydrolysis solution and how you would determine the mass of the silver bromide. (6)

d Show how the percentage by mass of bromide ion, in the original haloalkane, can be calculated. (2)

AQA, A level, Specimen Paper 6421, 2001/2

2 a State the reagent(s) used for the conversion of ethanol to
i iodoethane (1)
ii ethene. (1)

b Suggest a series of reactions by which ethanol can be converted to 2-hydroxypropanoic acid, $CH_3CH(OH)COOH$. For each reaction specify the reagents and conditions necessary. (6)

c Explain whether a solution of 2-hydroxypropanoic acid, **made in this way**, would have any effect on a beam of plane polarised monochromatic light. (2)

d 2-hydroxypropanoic acid reacts with lithium tetrahydridoaluminate(III) (lithium aluminium hydride).
State the conditions necessary for this reaction and give the structural formula of the organic product. (2)

Edexcel, A level, Module 4, June 2000

3 Grignard reagents are useful in organic syntheses where it is necessary to form a new carbon–carbon σ bond.

a Draw a diagram of suitable apparatus, state the conditions and **one** specific safety precaution for the preparation of the Grignard reagent 1-propyl-magnesium bromide, $CH_3CH_2CH_2MgBr$, from 1-bromopropane. The following data will be useful.

Substance	Boiling temperature (°C)	Flammability
1-Bromopropane	70.8	low
Ethoxyethane (ether)	34.5	very high

(7)

b Give the names and structural formulae of the organic products obtained when 1-propylmagnesium bromide reacts with:
i ethanal, followed by dilute hydrochloric acid (2)
ii carbon dioxide followed by dilute hydrochloric acid. (2)

Edexcel, A level, Module 4, June 2000

4 Zingerone is a constituent of oil of ginger which makes ginger hot. Zingerone could be obtained from coniferyl alcohol.

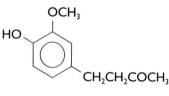

zingerone

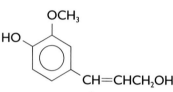

coniferyl alcohol

For the purposes of this question these substances can be represented as $RCH_2CH_2COCH_3$ and $RCH=CHCH_2OH$. Assume that the R group is unreactive.

a Coniferyl alcohol could be converted into zingerone by the series of reactions given below:

$$RCH=CHCH_2OH \xrightarrow[\text{step 1}]{\text{oxidation}} RCH=CHCHO$$

$$\xrightarrow[\text{step 2}]{} RCH=CHCH(OH)CH_3$$

$$\xrightarrow[\text{step 3}]{\text{oxidation}} RCH=CHCOCH_3$$

$$\xrightarrow[\text{step 4}]{H_2/Pt} RCH_2CH_2COCH_3$$

i Suggest suitable reagents which could be used in step 1. (1)

ii The conditions need to be carefully controlled so as to avoid further oxidation of the aldehyde. Outline how you would carry out the experiment so as to minimise further oxidation. (2)

iii Step 2 involves the use of a Grignard reagent. Give the formula of an appropriate Grignard reagent and outline how you could make this Grignard reagent from a halogenoalkane. (3)

b Step 4 is:

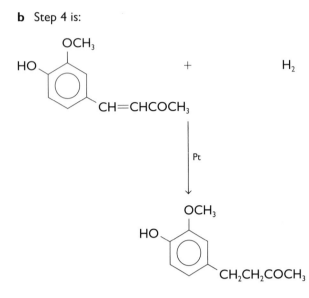

The reaction takes place at room temperature.

i The hydrogenation reaction uses platinum, a transition metal, as a catalyst. On what property of transition metals does their catalytic activity often depend? (1)

ii The benzene ring is not reduced under the conditions used. Explain this in terms of structure and bonding in the benzene ring. (3)

Edexcel, A level, Synoptic Paper (CH6), June 2000

5 An oily liquid **Z** can be produced from propene by the following scheme:

$$CH_3CH{=}CH_2 \xrightarrow[\text{step 1}]{HBr} CH_3CHBrCH_3 \xrightarrow{\text{step 2}} (CH_3)_2CHMgBr$$

step 3   step 4

$$CH_3CH(OH)CH_3 \quad (CH_3)_2CHCOOH$$

X   +   **Y**

warm/reflux | concentrated sulphuric acid

Z

a Write the mechanism for the reaction between propene and hydrogen bromide. (3)

b Give the names of the reagents and the conditions for:

i step 2 (3)

ii step 3. (2)

c Step 4 takes place in two stages. Name the reagents for each of the stages 1 and 2. (2)

d Give the full structural formula of the liquid **Z** produced by the reaction of **X** with **Y**. (2)

London, A level, Module 4, Jan 2000

6 a Consider the following reaction scheme:

$$C_3H_7OCl \xrightarrow[\text{H}_2\text{SO}_4\text{(aq)/K}_2\text{Cr}_2\text{O}_7\text{(aq)}]{\text{heat under reflux}} C_3H_5O_2Cl \xrightarrow{\text{NaOH(aq)}}$$

A   **B**

$$C_3H_5O_3Na \xrightarrow[\text{2. H}_2\text{SO}_4\text{(aq)}]{\text{1. I}_2\text{/NaOH(aq)}} H_2C_2O_4 + CH_3I$$

C   **D**

Substance **A** gives steamy fumes with phosphorus pentachloride, but not with water.

Substances **B** and **D** give carbon dioxide with sodium hydrogen carbonate solution.

i Deduce the structural formulae of **A**, **B**, **C** and **D**. (4)

ii State the name of substance **A**. (1)

b Outline a series of steps by which you could synthesise 1-aminopropane, $CH_3CH_2CH_2NH_2$, from propanoyl chloride, C_2H_5COCl. (5)

London, A level, Module 4, Jan 2000

7 Salicylic acid, shown below, has been used as a painkiller.

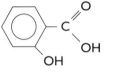

a Name the functional groups present in salicylic acid. (2)

b Deduce the molecular formula of salicylic acid. (1)

c Show a displayed formula of a likely organic product formed when salicylic acid reacts with

i ethanol and concentrated sulphuric acid under reflux

ii bromine

iii aqueous sodium hydroxide. (4)

d Salbutamol, shown below, is used in inhalers to relieve asthma.

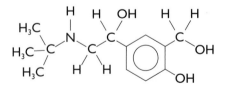

i Salbutamol is a chiral compound. Copy the structure above and mark the chiral centre with an asterisk (*).

ii List **two** reasons why salbutamol may be used as a pharmaceutical as a single optical isomer.

iii Salbutamol is reacted with $K_2Cr_2O_7/H_2SO_4$ under reflux. Predict the likely organic product of this reaction and draw a displayed formula of this product. (5)

OCR, A level, Specimen Paper A7882, Sept 2000

Determination of the structure of molecules

Figure 15.1
The differences in chemical structure of these molecules give them different chemical properties.
a Aspirin's structure means that it acts as an analgesic.
b Ibuprofen's structure means that it reacts differently within our bodies and so acts as an anti-inflammatory drug.

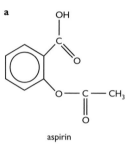

a

aspirin

b

ibuprofen

The properties of a compound depend upon the arrangement of the atoms in that substance as well as on the type of chemical bonds that hold the atoms together. Because of this, the determination of the chemical structure of substances is very important. We can then build 'scale models' using molecular modelling computer programs. Because atoms are incredibly small and cannot be seen with the naked eye, scientists have had to develop instrumental techniques which allow them to investigate substances and so discover how the atoms are arranged within the molecules. Many of the instrumental methods that have been developed are quite sophisticated and they work by analysing the effect of different types of electromagnetic radiation on matter (Figure 15.2). These techniques can be used to obtain not only qualitative information about molecular structure but also quantitative information about the molecules, such as bond length, bond angles and the concentration of the substance present.

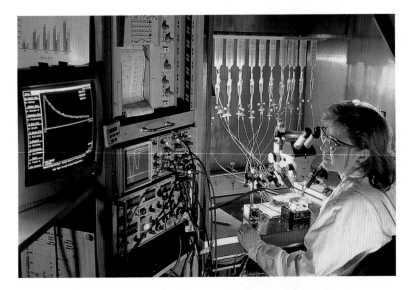

Some of the techniques that can be used to determine the arrangement of atoms in a substance are:

- mass spectrometry (or mass spectroscopy) (MS)
- infrared spectroscopy (IR)
- nuclear magnetic resonance spectroscopy (NMR)
- visible and ultraviolet spectroscopy (vis–UV).

Figure 15.2
This laboratory at Glaxo is dedicated to the determination of the chemical structure of new substances that may have a pharmacological activity.

Introduction to spectroscopy

The energy possessed by electrons in atoms and molecules, the energy of vibrating atoms within molecules as well as the energy of rotating and translating molecules cannot be of any random value. In all these cases we say that the energy is **quantised**, with fixed levels. If there are fixed energy levels, there will be energy gaps, or differences, between these levels (Figure 15.3).

Figure 15.3
The electronic, vibrational and rotational energy levels in a molecule. If enough energy is put into the system then an electron, for example, may be promoted to a higher energy level.

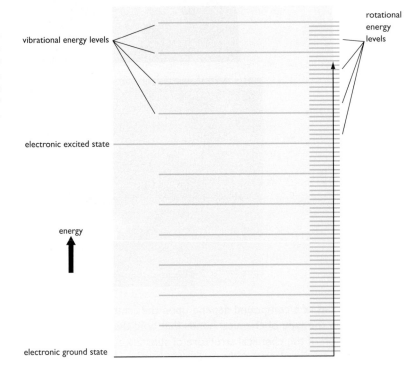

If a particle is in its **ground state** (lowest energy state) and gains energy equivalent to the energy difference between two energy levels (electronic, vibrational, and so on) then it is promoted to a higher energy level. It is then said to be in an **excited state**. Spectroscopy generally deals with this absorption of energy as the particle undergoes the various changes in energy level. The changes in energy can be analysed and information about the chemical structure of the molecules can be derived.

Mass spectrometry

Francis Aston built the first mass spectrometer in 1919 (Figure 15.4). This enabled scientists to determine the relative atomic masses of atoms (A_r) and relative molecular masses of molecules (M_r) accurately for the first time. Typical mass spectra are shown in Figure 15.5.

Figure 15.4
a Francis Aston (1877–1945).
b This is the first mass spectrometer, built by Francis Aston.

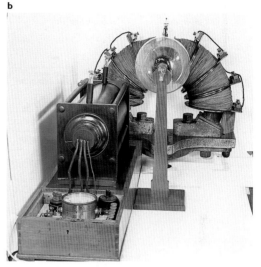

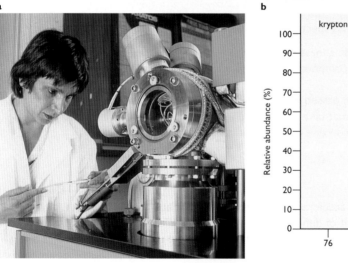

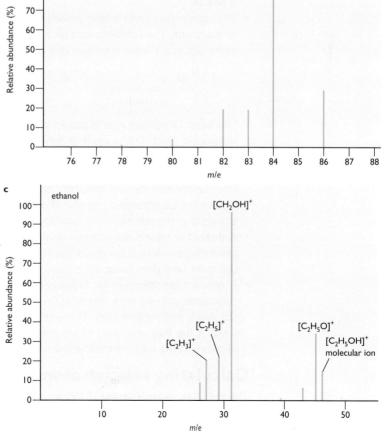

Figure 15.5
a A modern mass spectrometer allows very accurate values of the relative atomic masses of elements and relative molecular masses of most compounds to be obtained.
b Mass spectrum of krypton.
c Mass spectrum of ethanol.

Figure 15.6
A diagram of a modern
mass spectrometer.

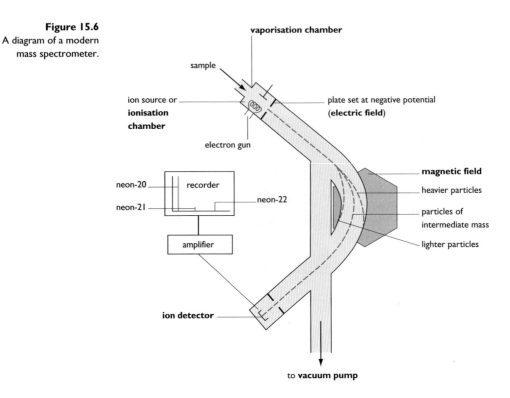

The technical details of a mass spectrometer are as follows (Figure 15.6):

- A vacuum is maintained inside a mass spectrometer.
- A sample of the element or compound is injected into the **vaporisation chamber** where it is heated.
- The vapour produced is then passed into the **ionisation chamber** where it is bombarded by electrons. The collisions that take place in the ionisation chamber cause the atoms or molecules in the vapour to lose one of their electrons and form positive ions:

$$X(g) + \underset{\substack{\text{bombarding}\\\text{electron}}}{e^-} \rightarrow X^+(g) + 2e^-$$

- The beam of positive ions is accelerated by an **electric field** and then is deflected by a **magnetic field**. The amount of deflection depends on the different charge to mass ratio of the positive ions. Because the charge is the same (one positive charge from the loss of one electron), the deflection depends on the mass of the positive ions. Lighter ions that are formed from lighter particles are deflected more than heavier ones. In this way particles with different masses can be separated and identified. The magnetic field is varied during the process to ensure that the different ions actually hit the detector.
- A **detector** counts the number of each of the ions that fall upon it. The detector does this by producing an electric current when hit by a positive ion. If two ions arrive at the detector at the same time then twice the current is produced.
- A typical mass spectrum for krypton is shown in Figure 15.5b. This type of diagram is sometimes called a stick diagram. You will notice that the relative abundance is recorded on the y-axis (the vertical axis). The x-axis (the horizontal axis) displays the mass to charge ratio (m/e) of the fragment ions. As we only consider ions with one positive charge, this is a direct measure of the mass of the fragment.

Calculating relative atomic masses

The **relative atomic mass** (A_r) of an element is the average mass of a large number of atoms related to an agreed scale. This quantity takes into account the percentage abundance of all the isotopes of an element.

In 1961 the International Union of Pure and Applied Chemistry (IUPAC) recommended that the standard used for the A_r scale be carbon-12. An atom of carbon-12 was taken to have a mass of 12 amu (atomic mass units) exactly. The A_r of an element is now defined as the average mass of its isotopes compared to $\frac{1}{12}$th the mass of one atom of carbon-12:

$$A_r = \frac{\text{average mass of isotopes of the element}}{\frac{1}{12}\text{th of the mass of one atom of carbon-12}}$$

Note: $\frac{1}{12}$th of the mass of one carbon-12 atom = 1 amu. Therefore, the A_r of an element is effectively its average isotopic mass divided by 1.

Information about the isotopes of krypton can be found from Figure 15.5b. The data are 0.35% ^{78}Kr, 2.25% ^{80}Kr, 11.6% ^{82}Kr, 11.5% ^{83}Kr, 57.0% ^{84}Kr and 17.3% ^{86}Kr. The relative atomic mass of krypton can be calculated as shown below:

$$A_r(Kr) = \frac{(78 \times 0.35)}{100} + \frac{(80 \times 2.25)}{100} + \frac{(82 \times 11.6)}{100} +$$

$$\frac{(83 \times 11.5)}{100} + \frac{(84 \times 57.0)}{100} + \frac{(86 \times 17.3)}{100}$$

$$= 83.89$$

Mass spectra of molecules

Molecules will ionise and produce what is known as a **molecular ion**. For example, Figure 15.5c shows the mass spectrum of ethanol. The molecular ion is produced as follows:

$$CH_3CH_2OH(g) + e^- \rightarrow CH_3CH_2OH^+(g) + 2e^-$$
$$\text{molecular ion}$$

In addition to this, molecules will undergo **fragmentation**. In this process the molecule splits up under the influence of further electron bombardment in the ionisation chamber. This creates many smaller positively charged fragments. For ethanol, $CH_3CH_2OH^+$ fragments to give positive ions such as $CH_3CH_2^+$, CH_3^+, CH_2OH^+ and $CH_3CH_2O^+$. These peaks can be seen in Figure 15.5c.

In the mass spectrum of ethanol there is a small peak at m/e 47, to the right of the molecular ion peak at m/e 46. This peak at m/e 47 corresponds to the $(M + 1)^+$ molecular ion. This peak usually occurs in the mass spectrum of organic compounds because of the presence of some molecules containing ^{13}C, the carbon-13 isotope. This isotope of carbon has a natural abundance of 1.10% of that of ^{12}C. The height of this $(M + 1)^+$ peak depends upon the number of carbon atoms present in the organic molecule. If a molecule contains four carbon atoms, for example, then the $(M + 1)^+$ peak will be approximately 4.40% of the height of the actual molecular ion peak.

The mass spectra of organic compounds are recorded by showing the relative percentage abundance. The most abundant (stable) fragment is called the **base peak** and by convention is given a relative abundance of 100%. The abundance of each fragment ion is measured relative to this base peak.

All the mass spectra of organic compounds obtained by a particular instrument are obtained under the same conditions of temperature and ionising voltage. Therefore, as fingerprints are unique to a person, mass spectra are unique to the particular molecule being studied. Extensive databases are maintained of the mass spectra of the vast majority of organic molecules. This makes it relatively quick and easy to identify a substance whose mass spectrum has just been obtained. For this reason:

- pharmaceutical companies and oil refineries use mass spectrometers
- space probes that have been sent to Mars have used mass spectrometry to analyse samples from the surface of the planet
- the International Amateur Athletic Federation (IAAF) uses mass spectrometry to help in the detection of minute traces of illegal substances that athletes may have taken (Figure 15.7).

1 Calculate the A_r value for Ni. The percentage abundance of its isotopes are:
^{58}Ni 67.76
^{60}Ni 26.16
^{61}Ni 1.25
^{62}Ni 3.66
^{64}Ni 1.16

2 A mass spectrum of an organic molecule shows an $(M + 1)$ peak of intensity 11% of the molecular ion peak. How many carbon atoms does this molecule contain? Explain your answer.

Figure 15.7
Mass spectrometers are also used in drug detection in athletes. They can be used to identify extremely small quantities of illegal substances taken by athletes. The sprinter, Ben Johnson, was banned from competing for life by the IAAF in 1993 after testing positive for steroid use.

Isotope peaks

Most elements have more than one naturally occurring, stable isotope. For example, the element chlorine has two isotopes, ^{35}Cl and ^{37}Cl. These are present in the ratio of:

$$^{35}Cl : {}^{37}Cl = 3 : 1$$

Therefore, any molecule or fragment that contains chlorine atoms will give rise to two peaks, separated by 2 mass units and in a ratio of $3 : 1$ in height.

If you see a pair of peaks like this in a mass spectrum then it is almost certain that chlorine is present in the molecule. For example, Figure 15.8 shows the mass spectrum of chloroethane. There are two molecular ion peaks, at masses of 64 and 66. These are due to $CH_3CH_2{}^{35}Cl^+$ and $CH_3CH_2{}^{37}Cl^+$, respectively. There are also peaks at masses of 49 and 51. These also are in the ratio of $3 : 1$. These peaks represent the loss of 15 units of mass, or a CH_3 group, from the molecule leaving $CH_2{}^{35}Cl^+$ and $CH_2{}^{37}Cl^+$ ions.

Figure 15.8
The mass spectrum of chloroethane shows peaks 2 mass units apart at 49/51 and 64/66 whose heights are in the ratio $3 : 1$. This is a clear indication of the presence of chlorine in this molecule.

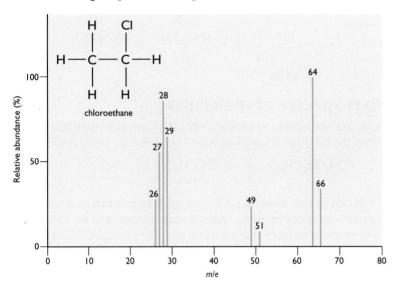

3 The approximate abundance of the two isotopes of bromine, ^{79}Br and ^{81}Br, are 50% each. By considering the mass spectrum of chloroethane and using the idea of 'isotope peaks' construct the mass spectrum for bromoethane.

4 Obtain the molecular formula of an organic compound X which contains 40.0% C, 6.7% H and the rest O. Its mass spectrum shows that it has an M_r of 60.
(A_r: O, 16; C, 12; H, 1)

Infrared spectroscopy

Figure 15.9
The typical vibrations shown by a molecule of carbon dioxide.

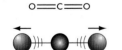

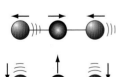

5 Look at Figure 15.9, which shows the typical vibrations shown by a molecule of carbon dioxide. Draw diagrams to represent the likely vibrations shown by a molecule of water.

Covalent bonds in molecules are not rigid; they can vibrate, bend and even twist (Figure 15.9). However, they will only do this with certain frequencies. In infrared spectroscopy we make use of this property. When infrared radiation is passed through a molecule, the molecules *absorb* radiation with just the right energy to cause the bonds to vibrate, twist or bend more energetically. The frequency of this vibration depends upon the nature of the atoms in the bond and the type of bond. The greater the mass of the atoms in the bond then the slower will be the vibration that is taking place and vice versa. Also, multiple bonds vibrate at a higher frequency than single bonds. It is also true that the other atoms in the molecule being studied influence the vibration of a bond even though they do not contribute to the particular bond. The frequency of such vibrations may be quoted in Hz; however, in infrared spectroscopy it is more usual to give the values in cm^{-1} (reciprocal wavelength in cm). This is called the **wavenumber** and is the number of waves per cm.

The infrared spectrometer

Figure 15.10
a The essential parts of a double beam infrared spectrometer.
b This is a modern infrared spectrometer. It is used in analysis to obtain the so-called 'fingerprint' spectrum of a substance that will allow it to be identified. The more modern spectrometers are called Fourier transform spectrometers. This method is much quicker than that used in conventional instruments.

A schematic diagram of the basic parts of a double beam infrared spectrometer is shown in Figure 15.10a.

a

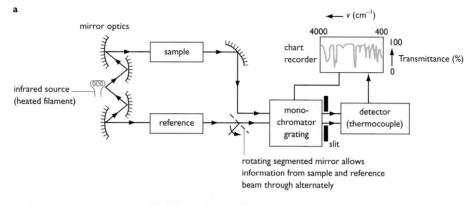

b

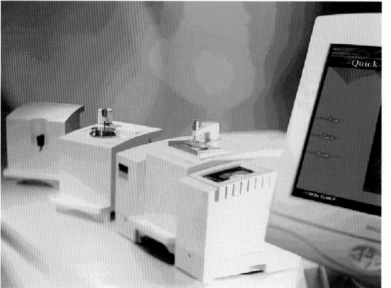

- A **heated filament** acts as a source of infrared radiation.
- The beam of radiation is split into two parallel beams. One of these passes through the sample being studied whilst the other is used as a **reference beam**. This ensures that allowance can be made for any absorption by carbon dioxide and water vapour present in the air.
- **Mirrors** direct the two beams so that they follow parallel paths.
- The **diffraction grating** or sodium chloride prism then splits up the infrared radiation into separate frequencies. (Sodium chloride is used for the prism since it is transparent to infrared radiation.)
- By rotating the grating or prism, infrared radiation of a particular frequency is focused on the **detector**.
- The spectrum is obtained by rotating the grating or prism so that the detector scans the individual frequencies and records their intensities on a chart recorder. The detector continuously compares the two beams; the spectrum shows the difference between them. The spectrum produced is of the **transmission**, that is, the percentage of the infrared radiation that passes through the sample, against the frequency of the infrared radiation. (In practice the frequency is measured in **wavenumbers**, as discussed above, and is measured in cm^{-1}.)

Interpretation of infrared spectra

The infrared spectrum of a substance can give you an idea of the structural formula of its constituent molecules. Consider, for example, the IR spectra of ethanol, ethanal and ethanoic acid, shown in Figure 15.11, overleaf. The spectra show absorptions characteristic of the particular compounds. For example, the carbonyl group, $C{=}O$, absorbs very strongly at about $1680-1750\,cm^{-1}$. Hence, ethanal and ethanoic acid absorb infrared radiation in this region of the infrared spectrum but ethanol does not. The alcohol group, $O{-}H$, in ethanol absorbs strongly at $3230-3550\,cm^{-1}$ (broad) whilst the $O{-}H$ of the carboxyl group of ethanoic acid absorbs to give an infrared band of medium intensity at $2500-3300\,cm^{-1}$. Using this information it is possible to identify which substance is which. Figure 15.12 shows further characteristic infrared absorptions found in organic molecules.

Figure 15.11
The infrared spectra of
a ethanol,
b ethanal and
c ethanoic acid. These
spectra have been labelled
with the important
absorptions shown by
these molecules.

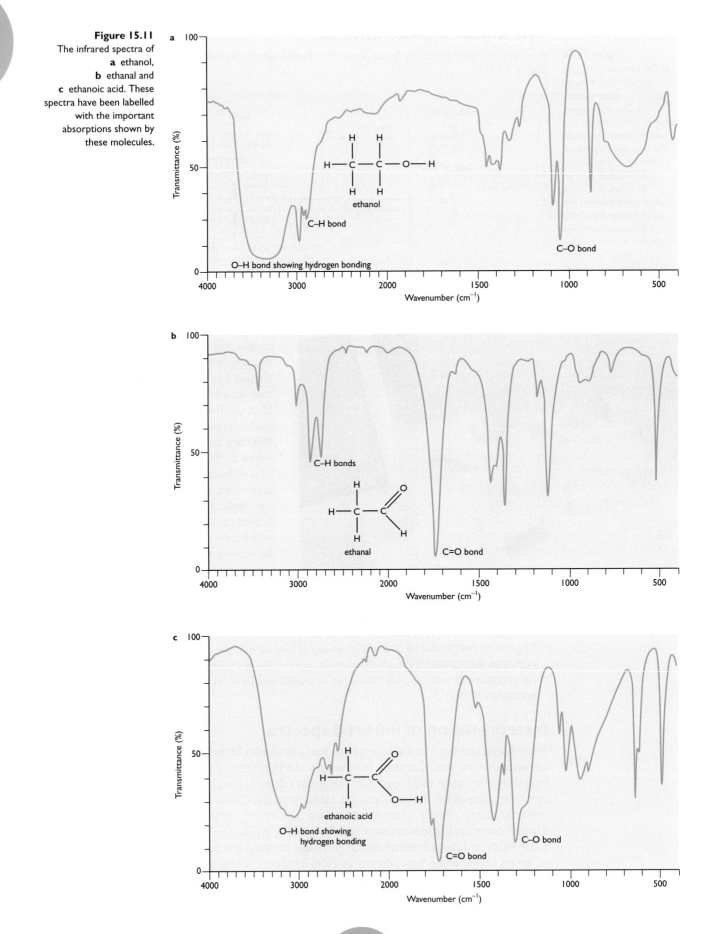

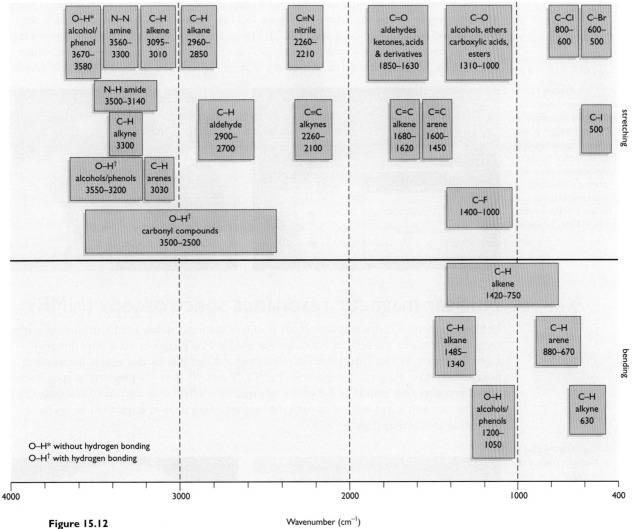

Figure 15.12
Characteristic infrared
absorptions.

6 Use the data in Figure 15.12 to identify further characteristic infrared absorptions which could aid the identification of ethanol, ethanal and ethanoic acid.

7 3-Methylhex-2-enoic acid (MHA) has been identified as the compound that gives rise to the smell associated with underarm sweat. The structure of this substance is shown below.

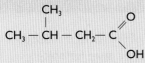

MHA

At one time, iso-valeric acid (3-methylbutanoic acid) was thought to cause the smell. Its structure is:

Use Figure 15.12 to identify the common absorption bands that may be found both in MHA and iso-valeric acid. Also identify an infrared band that would not be found in iso-valeric acid but would be found in MHA.

Infrared absorption bands can be used in quantitative analysis, although this use is not very common. The breathalyser used by the police at the roadside measures the concentration of alcohol found in a person's breath; this depends on the amount of alcohol in a person's blood. This evidence, however, is not admissible in court. To obtain legally admissible evidence, the police have to take the accused back to the police station and test a sample of breath using the Intoximeter (Figure 15.13, overleaf).

The Intoximeter is an infrared spectrometer and it measures the infrared absorption in the 3000 cm^{-1} region of the spectrum (see Figure 15.11a). By doing this the police obtain a measure of the **B**lood **A**lcohol **C**oncentration (BAC). The police can then ask for a confirmatory blood sample to be taken, which is analysed using gas–liquid chromatography. A similar instrument to that used by the police is used to analyse the alcohol content of beers and wines by brewers.

Figure 15.13
The Intoximeter measures the infrared absorption in the 3000 cm^{-1} region of the infrared spectrum. It is used by the police to obtain an accurate value of a person's BAC, also by brewers to obtain a measure of the alcohol content of beers or wines.

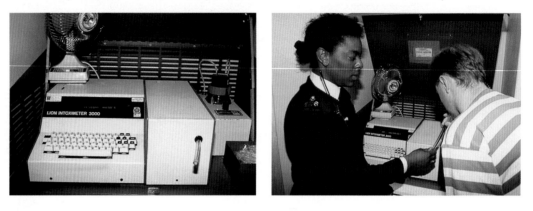

Nuclear magnetic resonance spectroscopy (NMR)

Nuclear magnetic resonance spectroscopy is one of the most widely used instrumental analytical techniques. It is also the basis of the technique used in MRI (magnetic resonance imaging) scanners (Figure 15.14). This technique relies upon the fact that certain atomic nuclei that possess odd mass numbers, for example ^1H, ^{13}C, ^{19}F and ^{31}P, have the property of spin. Because of this property they behave as if they are tiny magnets. When they are placed in a magnetic field they interact with it and will line up with the magnetic field (known as parallel) or against it (known as antiparallel) (Figure 15.15).

Figure 15.14
a Magnetic resonance imaging, which is used throughout the medical world, uses the principles of NMR as the basis of its operation.
b An MRI scan of a patient's brain.

a
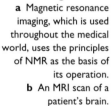
b

Figure 15.15
There are two different energies associated with a small magnet placed in a strong magnetic field. The alignment parallel to the magnetic field is at a lower energy than that anti-parallel to the magnetic field.

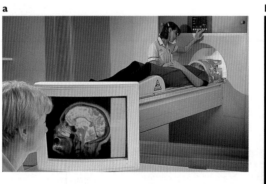
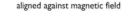
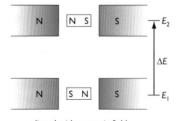

Chemists tend to examine protons, ^1H, because it is the simplest of the nuclei mentioned above, because it is very common in organic compounds and it is present at an isotopic abundance of 100%. The two different orientations shown in Figure 15.15 give rise to energy levels with slightly different energy, ΔE. If the protons are subject to a frequency of electromagnetic radiation equivalent to the energy gap between the two levels then some of the protons that are aligned with the magnetic field will interact with it and move into the higher energy position; that is, they 'flip over' into the higher energy orientation. In the case of NMR it is low-energy radio waves that are used (they have a frequency of 10–100 MHz). Fortunately, there is a slight majority of protons aligned with the applied magnetic field at room temperature. As a result, there is a net absorption of energy when the sample is subjected to electromagnetic radiation because of this excess of proton nuclei in the lower energy level.

An NMR spectrometer is shown in Figure 15.16.

Figure 15.16
a NMR is used as one of the main analytical techniques in research chemistry to help in the elucidation of chemical structure. It is also used in quality control on a routine basis. For example, the fat content of chocolate is found by using proton NMR spectroscopy.
b A diagram of the main features of an NMR spectrometer.

a

b

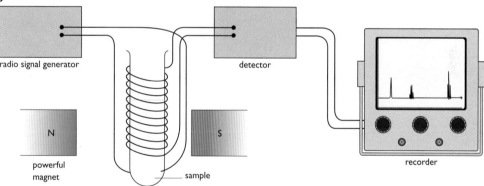

radio signal generator   detector

N   S

powerful magnet   sample   recorder

- The sample used may be in the liquid or solid state. However, it is more usual for it to be in a liquid state, either as the pure substance or dissolved in a proton-free solvent, that is one that does not contain any H atoms, such as trichlorodeuteromethane ($CDCl_3$) or tetrachloromethane (CCl_4). Deuterium, ^2H (sometimes shown as D), an isotope of ^1H, is introduced into the solvents in place of the usual ^1H atoms.
- The sample being studied is placed in a narrow glass tube, which is transparent to the radiation being used.
- It is then placed in the magnetic field.
- The sample is spun on a cushion of nitrogen to ensure all variations in the magnetic field are averaged out and so each of the proton nuclei experience the same magnetic field.
- The sample is now given pulses of radio frequencies. These pulses contain a range of radio frequencies.
- When the radio frequency corresponds to the amount of energy (quanta) equivalent to ΔE then the sample absorbs some of the pulsed radio waves.
- The detector records the differences in the signal that it receives compared to the original signal. An absorption peak results on the chart recorder.

If all the protons in a compound were exactly the same then you would expect to only get one absorption in the spectrum. However, because the nuclei of the protons are shielded by other electrons in the compound to a greater or lesser extent, then the protons in different chemical environments, or different chemical structural arrangements, will give rise to a slightly different magnetic field. This creates different energy gaps, ΔE. Hence, they will absorb different radio frequencies and give rise to different NMR absorptions. The spectrum produced is a plot of the radio frequency energy absorbed against the particular radio frequency. Therefore, if we analyse a proton NMR spectrum we can identify hydrogen atoms in different structural environments within the molecule being studied.

To ensure reproducibility it is essential to ensure that the positions of the NMR absorption peaks are measured relative to the signal of a given standard. The reference compound used is tetramethylsilane (TMS for short). It has 12 protons in identical environments (Figure 15.17) and it therefore produces a sharp single peak which is well separated from the majority of the proton peaks found in other molecules. A small amount of this substance is added to a sample before it is put in the spectrometer. The signal produced by TMS is given a value of 0.0 by convention and the extent to which the other proton signals differ from the TMS signal position is called the **chemical shift** and given the symbol δ. The chemical shifts of some of the protons you are likely to find are shown in Table 15.1.

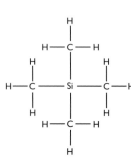

tetramethylsilane

Figure 15.17
The tetramethylsilane (TMS) molecule has 12 equivalent protons.

Table 15.1
Average chemical shift data for protons in different environments.

Type of proton	Chemical shift, δ (compared to TMS)
R—CH$_3$	0.9
R—CH$_2$—R	1.3
R$_3$CH	2.0
CH$_3$—COOR	2.0
CH$_3$—COR	2.1
[benzene ring]—CH$_3$	2.3
R—C≡C—H	2.6
R—CH$_2$—X (X = halogen or O†)	3.2–3.7
R—O—CH$_3$	3.8
R—O—H*	4.6
RHC=CH$_2$	4.9
RHC=CH$_2$	5.9
[benzene ring]—OH	7.0
[benzene ring]—H	7.3
R—CO—H	9.7
R—COOH	11.5

H indicates the particular proton the chemical shift refers to.
† RCH$_2$—O can be as low as $\delta = 2.2$
* R—O—H can be in the range $\delta = 3.0$–5.0
Note: many of the values of δ are sensitive to the nature of the solvent used and temperature.

The area under each absorption peak is proportional to the number of protons in a particular environment within the molecule being studied.

Interpretation of NMR spectra

Figure 15.18 shows the NMR spectrum of ethanol.

Figure 15.18
The low-resolution NMR
spectrum of pure ethanol
shows three peaks in
addition to the TMS peak.
This indicates that the
hydrogen atoms in this
molecule are located in
three distinct structural
environments.

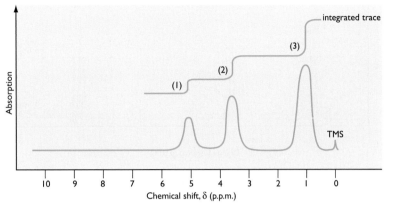

Figure 15.18
The low-resolution NMR spectrum of pure ethanol shows three peaks in addition to the TMS peak. This indicates that the hydrogen atoms in this molecule are located in three distinct structural environments.

- It has three peaks at different chemical shift positions relative to TMS, showing that the hydrogen atoms in this molecule are located in three distinct structural environments. By comparing the chemical shift data for the three peaks with the values given in Table 15.1 it is possible to identify the structural units that are in the molecule.
- The spectrometer, in addition to the NMR spectrum, has drawn what is called 'an integrated spectrum trace'. This trace is drawn at the same time as the spectrum and the height of the steps is a measure of the area under the peak. It is therefore proportional to the number of protons absorbing at this point in the spectrum. In this case there are three steps. The heights of the steps are in the ratio of $1:2:3$ (from left to right).

Table 15.2 summarises the chemical shifts, integration ratios and structural features of ethanol. Note that the presence of the oxygen atom has a 'deshielding' effect on the protons either attached directly to it or attached to carbon atoms it is attached to. This is due to the electronegative nature of the oxygen atom and has a marked effect on the chemical shift of these protons. The hydrogen nuclei are shielded by their electrons from the external magnetic field. The greater the electron density around the hydrogen atom then the higher is the magnetic field that is going to cause the 'flip' of the hydrogen nuclei. However, if the hydrogen atom is attached to a highly electronegative atom, such as oxygen, then the electron density around that hydrogen is reduced. The atom is said to be 'deshielded' and therefore requires a smaller magnetic field to 'flip'.

8 Using the data given in Table 15.1 construct the NMR spectra of:
a the isomer of ethanol, that is, ethoxyethane (CH_3OCH_3)
b ethanal (CH_3CHO)
c methanol (CH_3OH).

Table 15.2
NMR of ethanol.

δ	Integration ratio	Structural feature
5.0	1	—OH
3.6	2	—OCH$_2$R
1.0	3	CH$_3$—R

High-resolution NMR spectroscopy and spin–spin splitting

Modern NMR machines can be run in one of two ways: either low-resolution or high-resolution. So far we have only considered low-resolution NMR spectroscopy, in which a series of single peaks is obtained. Under the conditions of high-resolution the single peaks are often split into a number of peaks (Figure 15.19, overleaf). You will notice that two of the peaks are split. The splitting occurs because the spins of neighbouring protons can interact, or 'couple', with each other. To further understand this splitting in the high-resolution spectrum, we need to consider the $(n + 1)$ rule. This states that a single peak will be split into $(n + 1)$ peaks by the protons on an adjacent group containing n equivalent protons. There is no interaction between protons on the same carbon atom, since they are all equivalent, and generally there is no interaction to cause splitting of peaks over more than one carbon–carbon bond.

Figure 15.19
The high-resolution
spectrum of ethanol.
Compare this with the
low-resolution spectrum in
Figure 15.18.

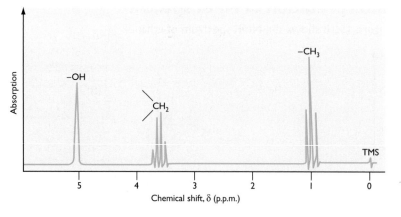

Using the $(n + 1)$ rule, the $-CH_3$ peak is split into three by the interaction with the protons
of the adjacent $-CH_2-$ group. Also the $-CH_2-$ peak is split into four by the interaction with
the protons of the adjacent $-CH_3$ group.

Why is there still only one peak associated with the proton of the $-OH$ group in the high-
resolution NMR spectrum? This is because the protons on this $-OH$ group are hydrogen-
bonded to adjacent $-OH$ groups. These hydrogen atoms rapidly exchange with one another and
therefore experience only an average environment and so do not interact sufficiently with the
protons of their neighbouring $-CH_2-$ group. These protons are said to be **labile**.

9 Use your knowledge of
the $(n + 1)$ rule to:
a account for the
numbers of peaks
shown in the high-
resolution NMR
spectrum of ethanal
(Figure 15.20)
b predict the high-
resolution spectrum of
methanol.

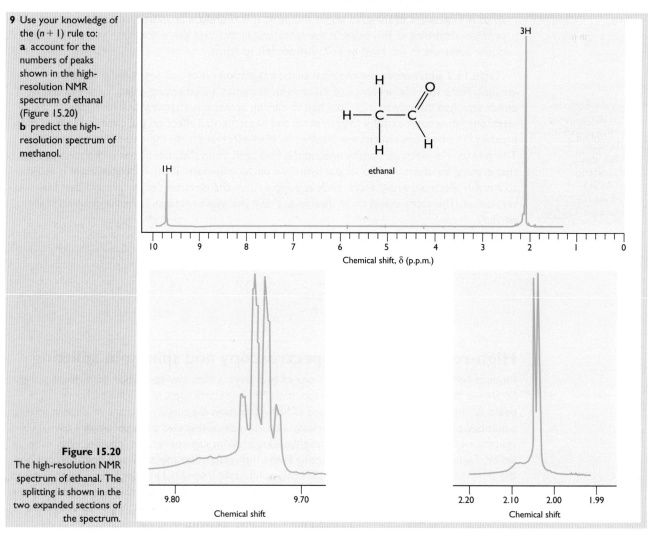

Figure 15.20
The high-resolution NMR
spectrum of ethanal. The
splitting is shown in the
two expanded sections of
the spectrum.

UV/visible spectroscopy

Figure 15.21
Forensic scientists use
UV/visible spectroscopy to
help establish the identity
of an unknown substance.
The inference made, by
comparing the absorption
spectrum of the unknown
substance with that of
known substances, can be
confirmed by using other
techniques – such as mass
spectrometry and infrared
spectroscopy.

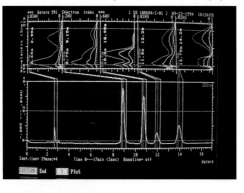

Ultraviolet and visible radiation (wavelength of approximately 10^{-8} to 10^{-6} m) causes the electrons furthest from the nuclei to be promoted to a higher electronic energy level. For this reason UV/visible spectroscopy is referred to as **electronic absorption spectroscopy**. This increase in energy level is particularly likely if the organic molecule contains delocalised electrons. If UV/visible radiation is passed through a compound then an absorption spectrum can be obtained. A UV/visible spectrometer operates on much the same principles as an infrared spectrometer, but with the IR source being replaced by one that produces UV and visible radiation.

UV/visible spectroscopy can be used:

* to suggest the presence of certain functional groups *or*
* to make concentration measurements by using the intensity of certain absorption bands in this region of the spectrum.

Figure 15.22 shows the UV/visible spectrum of propanone, CH_3COCH_3. This substance contains a $-C=O$ group and gives rise to a characteristic absorption centred near 280 nm (although there is a variation covering 275–290 nm, depending on the carbonyl compound being used).

The vast majority of transition element compounds are coloured, both in solution as well as in the solid state. In the case of transition elements ions, the frequency of the radiation absorbed is part of the visible region of the spectrum. The colour of these compounds can be related to incompletely filled 3d sub-shells in the transition element ion present (see Chapter 8, page 119). The colour is also associated with the oxidation state of the central metal ion as well as the nature and geometrical arrangement of the ligands around it (Figure 15.23).

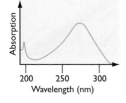

Figure 15.22 (above)
The electronic absorption spectrum of propanone shows a strong absorption centred near 280 nm.

Figure 15.23
The absorption spectra of
a $[Co(H_2O)_6]^{2+}$ and
b $[CoCl_4]^{2-}$ are quite different because their structures are different. $[Co(H_2O)_6]^{2+}$ is octahedral and $[CoCl_4]^{2-}$ is tetrahedral.

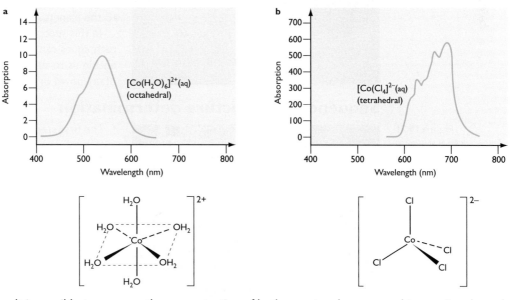

10 An aqueous solution of $[Ti(H_2O)_6]^{3+}$ absorbs most strongly at 500 nm. Explain how a calibration curve can be used to determine the concentration of this solution containing $[Ti(H_2O)_6]^{3+}$.

It is possible to measure the concentration of both organic substances and inorganic coloured substances in solution. To do this we have to have a UV or visible spectrum to obtain a value for the maximum absorption band in that spectrum, λ_{max}. The intensity of this absorption is proportional to the concentration of the substance present (according to what is known as the Beer–Lambert Law). In order to convert the reading given by the UV/visible spectrometer into a concentration, it is necessary to use a '**calibration curve**'. The calibration curve in this case can be plotted by obtaining readings for solutions of known concentration for the substance being studied.

Structure determination using combined instrumental techniques

Earlier in this chapter you studied a variety of instrumental techniques that have been developed to allow us to study substances. We can then discover how the atoms are arranged within the molecules that make up those substances (Figure 15.24).

Figure 15.24
The different instrumental techniques discussed in the previous sections of this chapter are used extensively 'in combination' to allow the determination of the structure of substances such as morphine and heroin.

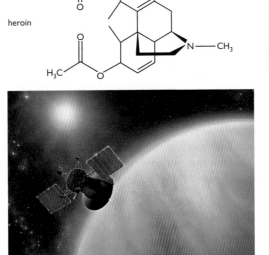

morphine

heroin

11 What are the structural differences between the molecules of morphine and heroin, as shown in Figure 15.24?

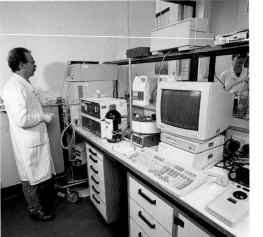

Figure 15.25
The spacecraft sent to other planets, for example Venus, contain a combination of instruments including UV spectrometers and mass spectrometers.

These instrumental techniques are such a powerful aid to the elucidation of chemical structure and the identification of molecules that they are present in forensic science, research and quality control laboratories throughout the world. NASA scientists even send these analytical instruments on missions to the planets (Figure 15.25).

In this section we will explore how these techniques can be used 'in combination' to enable us to successfully identify the structure of unknown organic compounds.

Sequence for structure determination

Figure 15.26
An elemental analyser can be used to determine the composition by mass of an organic substance in one operation. It can do this for the elements carbon, hydrogen, nitrogen, the halogens and sulphur. These determinations by mass can be converted into the empirical formula for the substance.

- The first step in the sequence for the structural determination of organic compounds is to establish which elements are present in the substance and find the composition by mass of each of these elements in the unknown compound. This involves the use of an **elemental analyser** (Figure 15.26). The composition by mass is determined by burning a known mass of the sample in oxygen and analysing the products. Any gases formed are separated by gas–liquid chromatography (see Chapter 14, page 215). The resultant chromatogram can be analysed. The area under each peak is proportional to the mass of each of the gases produced. The mass ratio of the elements present in the substance can then be determined from this data. The empirical formula can then be determined for the substance.
- From a mass spectrum the molecular ion peak can be identified and the relative molecular mass found. This is then compared with the empirical formula to establish the molecular formula.
- The structural formula is then established by using a combination of infrared, NMR and UV/visible spectroscopy.

Worked example

Elemental analysis of compound X gave the composition by mass of the elements present as follows:

- $C = 66.67\%$ ($A_r = 12$)
- $H = 11.11\%$ ($A_r = 1$)
- $O = 22.22\%$ ($A_r = 16$).

This data allows us to calculate the empirical formula of X.

	C	**H**	**O**
% by mass	66.67	11.11	22.22
number of moles	$\dfrac{66.67}{12}$	$\dfrac{11.11}{1}$	$\dfrac{22.22}{16}$
	$= 5.56$	$= 11.11$	$= 1.39$
ratio of moles	4 :	8 :	1

(obtained by dividing by the smallest number)
Empirical formula of X is C_4H_8O.

Figure 15.27 (overleaf) shows the mass spectrum, IR spectrum and NMR spectrum for the unknown substance X.

- The composition by mass gives an empirical formula of C_4H_8O which has a mass of 72. The mass spectrum shows an M^+ at 72. This means that the empirical and molecular formula are the same, that is, C_4H_8O.
- The mass spectrum of X shows that there is a smaller peak at $m/e\,73$ to the right of the molecular ion peak at $m/e\,72$. The peak at $m/e\,73$ corresponds to the $(M + 1)^+$ molecular ion. The height of this peak is 4.40% of the M^+ peak. This supports the fact that there are four carbon atoms in the molecule.
- The IR spectrum shows a strong band at $1718\,cm^{-1}$. This is due to the presence of a $-C{=}O$ bond from an aldehyde or a ketone. This band cannot be due to $-C{=}O$ found in an acid or ester since these molecules contain two oxygen atoms.
- There are also several medium to strong IR bands in the region of $2900\,cm^{-1}$. These indicate the presence of several $-C-H$ bonds found in alphatic compounds, perhaps a $-CH_3$ group as well as a $-CH_2-$ group in an alkane chain. Perhaps we are dealing with an aliphatic aldehyde or ketone.
- The NMR spectrum shows three sets of peaks.
  - The triplet of peaks centred at approximately $\delta = 1$ are from the protons of a $-CH_3$ group that has been split under the $(n + 1)$ rule by the protons of an adjacent $-CH_2-$ group.
  - The single peak at approximately $\delta = 2.1$ is due to a $-CH_3$ group that is adjacent to a $-C{=}O$ or carbonyl group.
  - The quartet of peaks centred at approximately $\delta = 2.5$ is due to the protons of a $-CH_2-$ group split under the $(n + 1)$ rule by an adjacent $-CH_3$ group.

Taking all this analytical information into consideration the molecule X will have the structure shown below. The structure also shows the correlation between the structure and the NMR spectrum.

$$\delta = 2.1 \qquad \delta = 2.5 \quad \delta = 1$$

$$
\begin{array}{ccccccc}
 & H & & O & & H & & H \\
 & | & & \| & & | & & | \\
H - & C & - & C & - & C & - & C & - H \\
 & | & & & & | & & | \\
 & H & & & & H & & H \\
\end{array}
$$

12 Which chemical test could you use to confirm that the substance X was a ketone and not an aldehyde? Explain how you would carry out this test.

You will have noted that the final elucidation of the structure of compound X required information taken from all the spectra. It is important that you are able to recognise the important features found in each spectrum when trying to establish the identity of unknown structures.

Figure 15.27
a The mass spectrum,
b the IR spectrum and
c the NMR spectrum of
compound X.

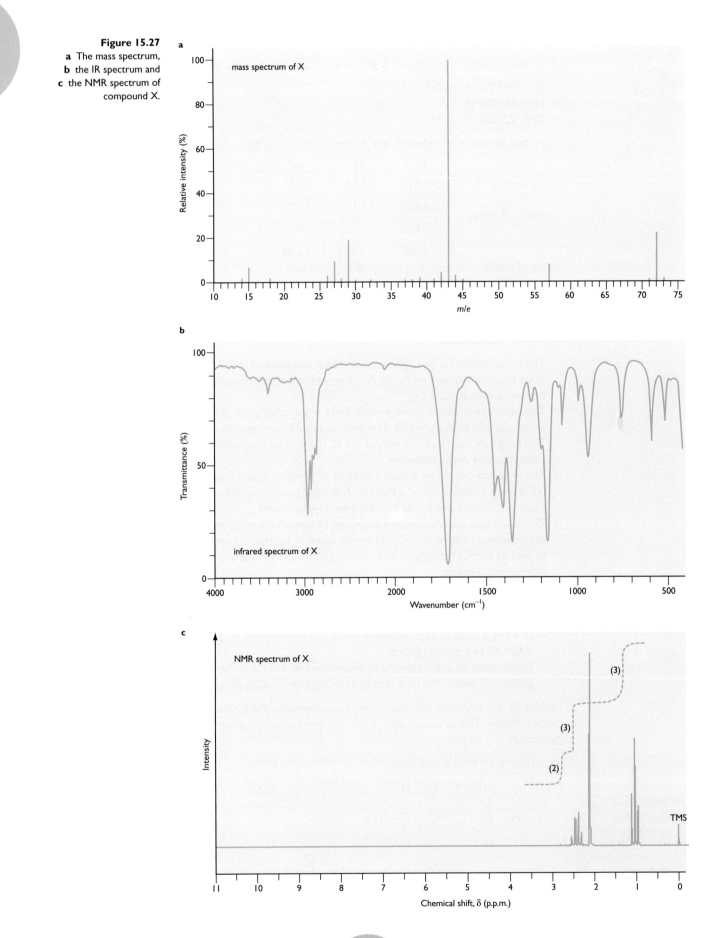

13 Identify compound D by analysis of the following information.

- Figure 15.28 shows the mass spectrum, IR spectrum and NMR spectrum for compound D.
- Elemental analysis:
  C = 64.86%
  H = 13.51%
  O = 21.62%
- The mass spectrum shows an $(M + 1)^+$ peak that has a height which is 4.40% of the molecular ion peak.

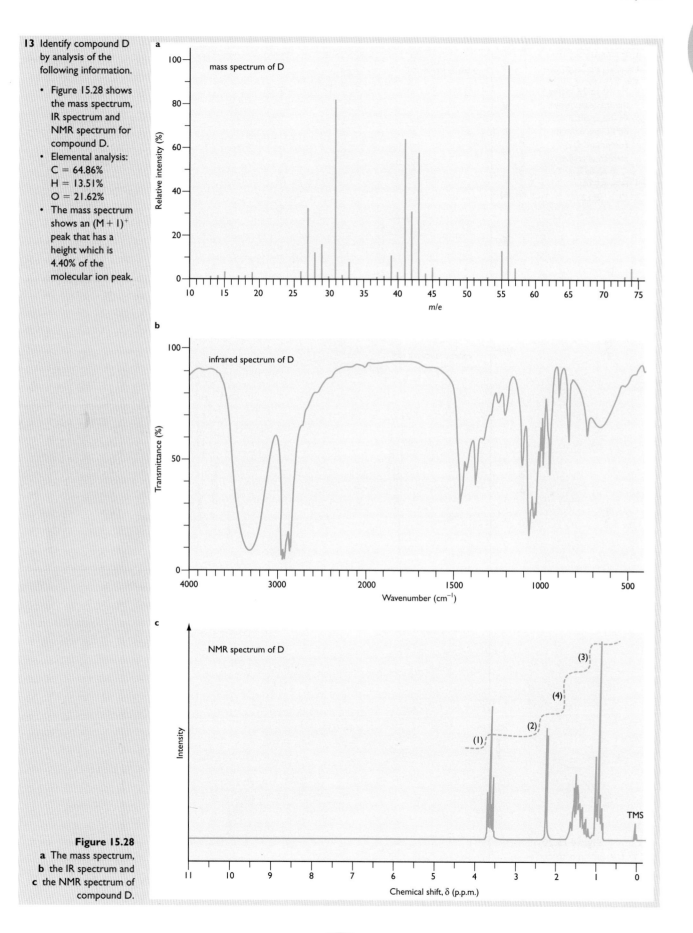

Figure 15.28
a The mass spectrum,
b the IR spectrum and
c the NMR spectrum of compound D.

14 Identify compound E
by analysis of the
following information.

- Figure 15.29 shows
  the mass spectrum,
  IR spectrum and
  NMR spectrum for
  compound E.
- Elemental analysis:
  C = 45.86%
  H = 8.92%
  Cl = 45.22%

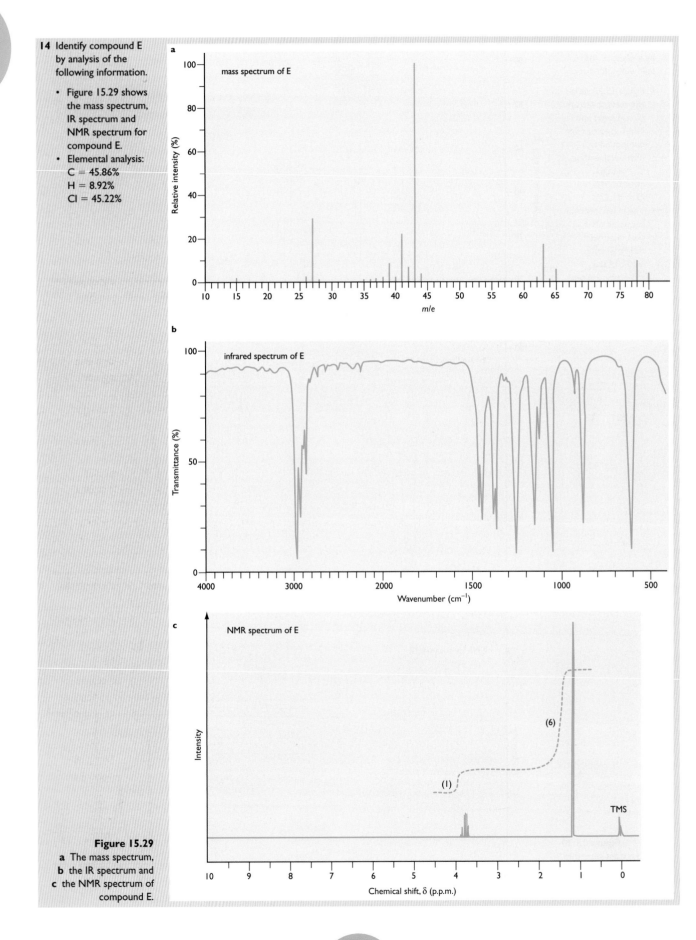

Figure 15.29
a The mass spectrum,
b the IR spectrum and
c the NMR spectrum of
compound E.

⬤ **Key skills**   **ICT**
- Use computer software to obtain the mass spectra of simple organic molecules.

Number
- Use information on the height of the M + 1 peak to determine the number of carbon atoms in a molecule.
- Determination of the molecular formula of an organic compound given its elemental percentage composition and its molecular mass.

⬤ **Skills task**   Produce a summary of the instrumental techniques outlined in this chapter, highlighting the strengths of each technique.

CHECKLIST   After studying Chapter 15 you should know and understand the following terms.

- ⬤ **Mass spectrometry (MS):** In this technique the atoms and molecules are ionised and then sorted according to the relative masses of the ions. Different molecules can be identified by their characteristic pattern of lines on their mass spectra.
- ⬤ **Molecular ion:** The ion formed when a molecule loses an electron in the mass spectrometer, but is otherwise unchanged.
- ⬤ **Fragmentation process:** Caused by the bombarding electrons having sufficient energy to break covalent bonds in the molecules being studied. Most of the molecules fragment during MS into smaller, positively charged ions.
- ⬤ **Base peak:** The most abundant (stable) fragment produced in the mass spectrum gives rise to the tallest peak, known as the base peak.
- ⬤ **Isotope peaks:** The extra peaks in the mass spectra of molecules due to the presence in the molecules of atoms that possess isotopes.
- ⬤ **Infrared spectroscopy (IR):** A technique based upon the principle that molecular vibrations occur in the infrared region of the electromagnetic spectrum and functional groups have characteristic infrared absorption frequencies.
- ⬤ **Nuclear magnetic resonance spectroscopy (NMR):** NMR depends on the fact that certain atoms, for example hydrogen, will absorb different radio frequencies when in different structural environments within a molecule. This gives rise to characteristic NMR spectra.
- ⬤ **Chemical shift:** The position of a NMR peak relative to a standard. The most common standard used is tetramethysilane (TMS). Symbol: δ.
- ⬤ **Labile protons:** Protons attached to the oxygen of an OH group in ethanol, for example, that exchange with one another so rapidly that they experience an average environment and so do not interact with protons on neighbouring parts of the molecule. In a high-resolution NMR spectrum their peaks are not split.
- ⬤ **UV/visible spectroscopy:** A technique based on the principle that electronic transitions in molecules occur in the visible and ultraviolet regions of the electromagnetic spectrum and that a given transition occurs at a characteristic wavelength. For this reason it is often called electronic absorption spectroscopy.
- ⬤ **Elemental analyser:** A technique used in the analysis of organic compounds to determine their composition by mass. The elemental composition by mass is carried out by burning a known mass of the compound in oxygen in the instrument followed by separation of the resulting gases by gas–liquid chromatography.

● Examination questions

1 Butanone can be reduced with NaBH₄ to form an alcohol **G**. Compound **G** has a chiral centre and can display optical isomerism.

   a **i** Explain the meaning of the term **chiral centre**.
      ii Deduce the identity of compound **G** and draw its optical isomers.   (3)

   b Butanone has the infrared spectrum below.

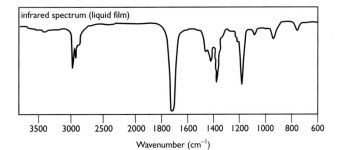

infrared spectrum (liquid film)

Wavenumber (cm⁻¹)

      i How does this infrared spectrum confirm the presence of the functional group present in butanone?
      ii How would you expect the infrared spectrum of compound **G** to differ from that of butanone? Explain your answer clearly.   (4)

   c Butanone reacts with hydrogen cyanide in the presence of potassium cyanide.
      i Describe, with the aid of curly arrows, the mechanism for this reaction.
      ii What type of reaction is this?   (4)

          OCR, A level, Specimen Paper A7882, Sept 2000

2 Compound **E** is an aromatic hydrocarbon with the molecular formula C_8H_{10}.

   a Draw structures for the **four** possible isomers of **E**.   (4)

   b The NMR spectrum of **E** is shown below.

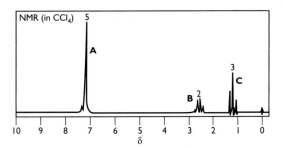

NMR (in CCl₄)

   Suggest the identity of the protons responsible for the groups of peaks **i A**, **ii B** and **iii C**. For each group of peaks, explain your reasoning carefully in terms of both the chemical shift value and the splitting pattern.   (9)

   c Using the evidence from part **b**, identify and show the structure of hydrocarbon **E**.   (1)

          OCR, A level, Specimen Paper A7882, Sept 2000

3 The spectra shown below were obtained from an organic compound **G**. Using data from the three spectra, suggest a structure for **G**, indicating what evidence you have used from the spectra.   (9)

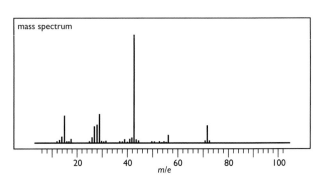

mass spectrum

m/e

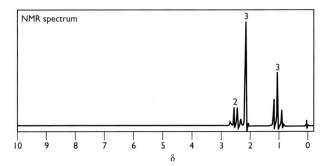

NMR spectrum

δ

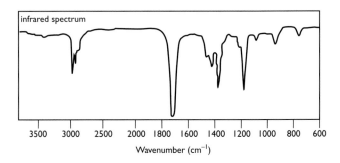

infrared spectrum

Wavenumber (cm⁻¹)

          OCR, A level, Specimen Paper A7882, Sept 2000

4 Compound **A**, $C_5H_{10}O$, reacts with NaBH₄ to give **B**, $C_5H_{12}O$. Treatment of **B** with concentrated sulphuric acid yields compound **C**, C_5H_{10}. Acid-catalysed hydration of **C** gives a mixture of isomers, **B** and **D**.

   Fragmentation of the molecular ion of **A**, $[C_5H_{10}O]^{+\bullet}$, leads to a mass spectrum with a major peak at $m/z\,57$. The infrared spectrum of compound **A** has a strong band at $1715\,cm^{-1}$ and the infrared spectrum of compound **B** has a broad absorption at $3350\,cm^{-1}$ (see the following table). From the spectrum below, the proton NMR spectrum of **A** has two signals at δ 1.06 (triplet) and 2.42 (quartet), respectively.

Bond	Wavenumber (cm^{-1})
C—H	2850–3300
C—C	750–1100
C=C	1620–1680
C=O	1680–1750
C—O	1000–1300
O—H (alcohols)	3230–3550
O—H (acids)	2500–3000

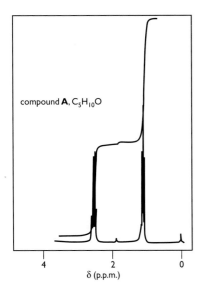

compound **A**, $C_5H_{10}O$

δ (p.p.m.)

Use the analytical and chemical information provided to deduce structures for compounds **A**, **B**, **C** and **D**, respectively. Include in your answer an equation for the fragmentation of the molecular ion of **A** and account for the appearance of the proton NMR spectrum of **A**. Explain why isomers **B** and **D** are formed from compound **C**. (20)

AQA, A level, Specimen Paper 6421, 2001/2

5 This question concerns the compounds linked by the reaction scheme

$$C_4H_8O \rightarrow C_4H_{10}O \rightarrow C_4H_9Br \rightarrow C_4H_8$$
$$\textbf{A} \qquad\quad \textbf{B} \qquad\quad \textbf{C} \qquad\quad \textbf{D}$$

A reacts with 2,4-dinitrophenylhydrazine to give a solid **E** which, when recrystallised, has a melting temperature of 126 °C. The melting temperatures of some 2,4-dinitrophenylhydrazine derivatives are listed below:

Compound	Melting temperature of 2,4-dinitrophenylhydrazine derivative (°C)
Propanone	126
Butanone	116
Propanal	155
Methylpropanal	187
Butanal	126

The infrared spectrum of **A** has a peak at 1720 cm^{-1}, but none at about 3500 cm^{-1} or 1650 cm^{-1}. The spectrum of **B** has a very broad peak at 3500 cm^{-1}, but none at about 1720 cm^{-1} or 1650 cm^{-1}.

Some typical infrared absorption wavenumbers are shown in the table below:

Bond	Wavenumber (cm^{-1})
O—H	3600–3300
C—O	1200–1150
C=C	1680–1620
C=O	1750–1680

a i Why must solid **E** be recrystallised before its melting temperature is measured? (1)
   ii What bond in **A** is responsible for the peak at 1720 cm^{-1}? (1)
   iii Why is the peak at 3500 cm^{-1} in the spectrum of **B** very broad? (2)
   iv Draw the structural formula of **A** and of **B**. (2)
b i Name the reagent and the solvent used for the conversion of **C** to **D**. (2)
   ii Draw the structural formula of **D**. (1)
   iii Draw the structural formula of the major product of the addition of HBr to **D**. (1)
   iv Suggest why the reaction in part **iii** does not produce **C** as the major product. (1)

Edexcel, A level, Module 4, June 2000

Numerical and short answers

● Chapter 1 **Marginal questions**

1 Increase temperature, increase concentration (or increase pressure if gas) of reactant, increase surface area, use a catalyst.

2 a first order in each case   **b** third order overall

3 a first order in each case   **b** second order overall

5 In experiments 1 and 3, where [NO] is constant, doubling [H_2] doubles rate.
∴ first order w.r.t. H_2.
In experiments 2 and 5, where [NO] is constant, doubling [H_2] doubles rate.
∴ first order w.r.t. H_2.
In experiments 2, 3 and 4, where [H_2] is constant, doubling [NO] quadruples rate.
∴ second order w.r.t. NO.

6 Half-lives are not constant but progressively increase, therefore **not** first order but probably second order w.r.t. phenylamine.

7 c slope of graph of ln k versus 1/T has slope $\dfrac{-20}{0.89 \times 10^{-3}\,K^{-1}} = 2.25 \times 10^4\,K$

but slope $= E_a / R$
∴ $E_a = 2.25 \times 10^4\,K \times 8.31\,J\,K^{-1}\,mol^{-1} = 1.87 \times 10^5\,J\,mol^{-1} = 187\,kJ\,mol^{-1}$

Examination questions

1 a i rate $= k[X][Y]^2$
  ii 3
  iii 8

 b i 2; doubling both [A] and [B] leads to rate $\times 4$
  ii 0; doubling [A] at constant [B] also leads to rate $\times 4$
  iii rate $= k[A]^2$

  iv $k = \dfrac{\text{rate}}{[A]^2} = \dfrac{3.5 \times 10^{-4}\,mol\,dm^{-3}\,s^{-1}}{(0.2)^2\,mol^2\,dm^{-6}}$

       $= 8.75 \times 10^{-3}\,dm^3\,mol^{-1}\,s^{-1}$

2 a $pV = nRT$
number of moles of H_2O_2 used $= 0.10$
number of moles O_2 formed $= 0.05$

  $V = \dfrac{nRT}{p} = \dfrac{0.05 \times 8.31 \times 293}{100\,000}$

       $= 1.217 \times 10^{-3}\,m^3$ or $1217\,cm^3$

3 a Compare experiments 1 and 2: [CN^-] is constant, [C_2H_5Br] doubles, rate doubles, so first order w.r.t. C_2H_5Br
Compare experiments 1 and 3: [C_2H_5Br] is constant, [CN^-] doubles, rate doubles, so first order w.r.t. CN^-
rate $= k$ [CN^-] [C_2H_5Br]

4 a ii first order w.r.t. propanone, zero order w.r.t. iodine, first order w.r.t. hydrogen
   iv rate $= k$ [CH_3COCH_3] [H^+]

5 a i NO(g), 2nd order: using experiments 1 and 2, [NO_2] doubled, rate quadrupled
CO(g), zero order: using experiments 2 and 3, [CO] doubled, rate constant
O_2(g), 1st order: using experiments 3 and 4, [O_2] quadrupled, rate quadrupled

    ii rate $= k[NO]^2[O_2]$

      e.g. use data from experiment I

      $0.44 \times 10^{-3} \, mol \, dm^{-3} s^{-1} = k \, (1.00 \times 10^{-3})^2 \, mol^2 \, dm^{-6} \times (1.00 \times 10^{-1}) \, mol \, dm^{-3}$

                 $= 4400$

      units $= dm^6 \, mol^{-2} \, s^{-1}$

    iii temperature of reactants increases/rate increases

       rate constant k increases

 c **D** Na : N : O $= \dfrac{27.1}{23} : \dfrac{16.5}{14} : \dfrac{56.4}{16}$

              $= 1.18 : 1.18 : 3.52$

              $= \;\; 1 \;\; : \;\; 1 \;\; : \;\; 3$

              $= NaNO_3$

  either

        equation: $2NaNO_3 \rightarrow 2NaNO_2 + O_2$

  or

              $0.04 \, mol \;\; \rightarrow$ $0.02 \, mol$ (idea of a 2 : 1 ratio)

  $24 \times 0.0200 \, dm^3 = 0.48 \, dm^3$ (units required)

● Chapter 2 **Marginal questions**

I a $K_c = \dfrac{[PCl_3(g)]_{eqm}[Cl_2(g)]_{eqm}}{[PCl_5(g)]_{eqm}} \; mol \, dm^{-3}$

 b $K_c = \dfrac{[CH_3COOC_2H_5(l)]_{eqm}[H_2O(l)]_{eqm}}{[CH_3COOH(l)]_{eqm}[C_2H_5OH(g)]_{eqm}}$ (no unit)

 c $K_c = \dfrac{[C_2H_5OH(g)]_{eqm}}{[C_2H_4(g)]_{eqm}[H_2O(g)]_{eqm}} \; mol^{-1} \, dm^3$

2

	$H_2(g)$	$+$	$I_2(g)$	$\rightleftharpoons$	$2HI(g)$
initial amount/mol	2.5		2.5		0
eqm. amount/mol	$2.5 - 2.0$		$2.5 - 2.0$		4.0
	$= 0.5$		$= 0.5$		

$K_c = \dfrac{[HI(g)]^2_{eqm}}{[H_2(g)]_{eqm}[I_2(g)]_{eqm}} = \dfrac{(4.0 \, mol \, dm^{-3})^2}{(0.5 \, mol \, dm^{-3})^2} = 64$ (no unit)

3 $K_c = \dfrac{[H_2(g)]^2_{eqm}[S_2(g)]_{eqm}}{[H_2S(g)]^2_{eqm}}$

  $\therefore [H_2(g)]_{eqm} = \sqrt{\dfrac{3.0 \times 10^{-4} \, mol \, dm^{-3} \times (5.0 \times 10^{-3} \, mol \, dm^{-3})^2}{2.4 \times 10^{-3} \, mol \, dm^{-3}}} = 1.8 \times 10^{-3} \, mol \, dm^{-3}$

4 a to the left   **b** no change   **c** to the left   **d** to the right

5 a $K_p = \dfrac{p(N_2(g)_{eqm}) \times p(O_2(g)_{eqm})}{p(NO(g)_{eqm})^2}$ (no unit)

 b $K_p = \dfrac{p(SO_3(g)_{eqm})^2}{p(SO_2(g)_{eqm})^2 \times p(O_2(g)_{eqm})} \; Pa^{-1}$

 c $K_p = \dfrac{p(CO_2(g)_{eqm})^2}{p(CO(g)_{eqm})^2 \times p(O_2(g)_{eqm})} \; Pa^{-1}$

6 total number of moles of gas at eqm. $= (0.30 + 0.70 + 0.70)\,\text{mol} = 1.70\,\text{mol}$

partial pressure of $PCl_5 = \dfrac{0.30}{1.70} \times 1.0 \times 10^6\,\text{Pa} = 1.76 \times 10^5\,\text{Pa}$

partial pressure of $PCl_3 = \dfrac{0.70}{1.70} \times 1.0 \times 10^6\,\text{Pa} = 4.12 \times 10^5\,\text{Pa}$

partial pressure of $Cl_2 = \dfrac{0.70}{1.70} \times 1.0 \times 10^6\,\text{Pa} = 4.12 \times 10^5\,\text{Pa}$

$\therefore K_p = \dfrac{p(PCl_3(g)_{eqm}) \times p(Cl_2(g)_{eqm})}{p(PCl_5(g)_{eqm})} = \dfrac{(4.12 \times 10^5\,\text{Pa})^2}{1.76 \times 10^5\,\text{Pa}} = 9.6 \times 10^5\,\text{Pa (or 960\,kPa)}$

7 total number of moles of gas at eqm. $= (0.95 + 0.060 + 0.060)\,\text{mol} = 1.07\,\text{mol}$

partial pressure of $NO_2 = \dfrac{0.95}{1.07} \times p_T = 0.888\,p_T$

partial pressure of $NO = \dfrac{0.060}{1.07} \times p_T = 0.0561\,p_T$

partial pressure of $O_2 = \dfrac{0.060}{1.07} \times p_T = 0.0561\,p_T$

$\therefore K_p = \dfrac{p(NO(g)_{eqm})^2 \times p(O_2(g)_{eqm})}{p(NO_2(g)_{eqm})^2} = \dfrac{(0.0561 \times p_T)^3}{(0.888 \times p_T)^2} = 2.24 \times 10^{-4} \times p_T$

$\therefore p_T = \dfrac{K_p}{2.24 \times 10^{-4}} = \dfrac{7.0 \times 10^{-6}\,\text{Pa}}{2.24 \times 10^{-4}} = 3.1 \times 10^{-2}\,\text{Pa}$

8 a $AgBr(s) \rightleftharpoons Ag^+(aq) + Br^-(aq)$

$K_{sp} = [Ag^+(aq)][Br^-(aq)] = [Ag^+(aq)]^2 = 5.0 \times 10^{-13}\,\text{mol}^2\,\text{dm}^{-6}$

$\therefore [Ag^+(aq)] = \sqrt{5.0 \times 10^{-13}\,\text{mol}^2\,\text{dm}^{-6}} = 7.1 \times 10^{-7}\,\text{mol}\,\text{dm}^{-3}$

b $Ca(OH)_2(s) \rightleftharpoons Ca^{2+}(aq) + 2OH^-(aq)$

$K_{sp} = [Ca^{2+}(aq)][OH^-(aq)]^2 = 4[Ca^{2+}(aq)]^3$ (since $[OH^-(aq)] = 2[Ca^{2+}(aq)]$)

$\therefore [Ca^{2+}(aq)] = \sqrt[3]{\dfrac{5.5 \times 10^{-6}\,\text{mol}^3\,\text{dm}^{-9}}{4}} = 0.011\,\text{mol}\,\text{dm}^{-3}$ *[Lit. 0.0153\,mol\,dm^{-3}]*

c $PbCl_2(s) \rightleftharpoons Pb^{2+}(aq) + 2Cl^-(aq)$

$K_{sp} = [Pb^{2+}(aq)][Cl^-(aq)]^2 = 4[Pb^{2+}(aq)]^3$ (since $[Cl^-(aq)] = 2[Pb^{2+}(aq)]$)

$\therefore [Pb^{2+}(aq)] = \sqrt[3]{\dfrac{2.0 \times 10^{-5}\,\text{mol}^3\,\text{dm}^{-9}}{4}} = 0.017\,\text{mol}\,\text{dm}^{-3}$ *[Lit. 0.039\,mol\,dm^{-3}]*

9 a $CdCO_3(s) \rightleftharpoons Cd^{2+}(aq) + CO_3^{2-}(aq)$

$K_{sp} = [Cd^{2+}(aq)][CO_3^{2-}(aq)] = [Cd^{2+}(aq)]^2 = (1.6 \times 10^{-7}\,\text{mol}\,\text{dm}^{-3})^2 = 2.6 \times 10^{-14}\,\text{mol}^2\,\text{dm}^{-6}$

b $CaF_2(s) \rightleftharpoons Ca^{2+}(aq) + 2F^-(aq)$

$K_{sp} = [Ca^{2+}(aq)][F^-(aq)]^2 = 4[Ca^{2+}(aq)]^3 = 4(2.2 \times 10^{-4}\,\text{mol}\,\text{dm}^{-3})^3 = 4.3 \times 10^{-11}\,\text{mol}^3\,\text{dm}^{-9}$

c $Cr(OH)_3(s) \rightleftharpoons Cr^{3+}(aq) + 3OH^-(aq)$

$K_{sp} = [Cr^{3+}(aq)][OH^-(aq)]^3 = 27[Cr^{3+}(aq)]^4 = 27(1.4 \times 10^{-8}\,\text{mol}\,\text{dm}^{-3})^4 = 1.0 \times 10^{-30}\,\text{mol}^4\,\text{dm}^{-12}$

10 In the solution in NaBr, $[Br^-(aq)] = 0.0050\,\text{mol}\,\text{dm}^{-3}$ (neglecting the tiny amount from AgBr)

$K_{sp} = [Ag^+(aq)][Br^-(aq)]$

$\therefore [Ag^+(aq)] = \dfrac{K_{sp}}{[Br^-(aq)]} = \dfrac{5.0 \times 10^{-13}\,\text{mol}^2\,\text{dm}^{-6}}{0.0050\,\text{mol}\,\text{dm}^{-3}} = 1.0 \times 10^{-10}\,\text{mol}\,\text{dm}^{-3}$

(1/7100 the value in Question **8a**!)

Examination questions

1 b $C_2H_6(g) \rightleftharpoons C_2H_4(g) + H_2(g)$

   0.64 mol   0.36 mol   0.36 mol   at equilibrium

   total mol   1.36

   partial pressure $=$ mol fraction $\times$ total pressure

$$p(C_2H_4) \equiv p(H_2) = \frac{0.36}{1.36} \times 180 = 47.6 \, kPa$$

$$p(C_2H_6) = \frac{0.64}{1.36} \times 180 = 84.7 \, kPa$$

$$K_p = \frac{p(C_2H_4)p(H_2)}{p(C_2H_6)}$$

$$= \frac{47.6 \times 47.6}{84.7} = 26.8 \, kPa$$

2 b iii $N_2O_4(g) \rightleftharpoons 2NO_2(g)$ $K_p = \dfrac{p(NO_2(g)_{eqm})^2}{p(N_2O_4(g)_{eqm})}$

iv $\therefore p(NO_2(g)_{eqm}) = \sqrt{(K_p \times p(N_2O_4(g)_{eqm}))} = \sqrt{(48 \, atm \times 0.15 \, atm)} = 2.7 \, atm$

4 b   moles of N_2 at eqm. $= 9.0 \times \frac{1}{3}$ $= 3.0 \, mol$ $\therefore [N_2(g)]$ $= 0.30 \, mol \, dm^{-3}$

   moles of H_2 at eqm. $= 27 \times \frac{1}{3}$ $= 9.0 \, mol$ $\therefore [H_2(g)]$ $= 0.90 \, mol \, dm^{-3}$

   moles of NH_3 at eqm. $= 9.0 \times \frac{2}{3} \times 2 = 12 \, mol$ $\therefore [NH_3(g)] = 1.20 \, mol \, dm^{-3}$

c $K_c = \dfrac{[NH_3(g)]^2_{eqm}}{[N_2(g)]_{eqm}[H_2(g)]^3_{eqm}} = \dfrac{(1.20 \, mol \, dm^{-3})^2}{0.30 \, mol \, dm^{-3} \times (0.90 \, mol \, dm^{-3})^3} = 6.6 \, dm^6 \, mol^{-2}$

5 c i $K_p = \dfrac{p(CH_3OCH_3) \times p(H_2O)}{[p(CH_3OH)]^2}$

ii   at equilibrium, partial pressure of CH_3OCH_3 equals that of H_2O

   (partial pressure of $CH_3OCH_3)^2 = 9.00 \times (0.142)^2$

   partial pressure of CH_3OCH_3 $= 0.426 \, atm$

6 c i $C_2H_5OH + 3O_2 \rightarrow 2CO_2 + 3H_2O$

   $C_8H_{18} + 12.5O_2 \rightarrow 8CO_2 + 9H_2O$

   1 g ethanol requires $\dfrac{3}{46} = 0.065 \, mol \, O_2$

   1 g octane requires $\dfrac{12.5}{114} = 0.11 \, mol \, O_2$

● Chapter 3   Marginal questions

1 moles of MnO_4^- ions used $= 0.020 \, mol \, dm^{-3} \times \dfrac{22.80}{1000} \, dm^3 = 4.56 \times 10^{-4} \, mol$

from the equation given, moles of Fe^{2+} in 25 $cm^3 = 5 \times 4.56 \times 10^{-4} \, mol = 2.28 \times 10^{-3} \, mol$

$\therefore [FeSO_4 \cdot (NH_4)_2SO_4 \cdot 6H_2O(aq)] = [Fe^{2+}(aq)] = \dfrac{2.28 \times 10^{-3} \, mol}{0.025 \, dm^3} = 0.0912 \, mol \, dm^{-3}$

2 moles of $Cr_2O_7^{2-}$ used $= 0.00100 \, mol \, dm^{-3} \times 0.0225 \, dm^3 = 2.250 \times 10^{-5} \, mol$

equation shows moles of Fe^{2+} (and Fe) in 25.0 $cm^3 = 6 \times$ moles of $Cr_2O_7^{2-} = 1.35 \times 10^{-4} \, mol$

$\therefore$ moles of Fe in 250 $cm^3 = 1.35 \times 10^{-3} \, mol$

$\therefore$ mass of Fe dissolved $= 1.35 \times 10^{-3} \, mol \times 55.85 \, g \, mol^{-1} = 0.0754 \, g$

and percentage iron $= 7.54\%$

3 moles of Na_2SO_3 used $= 0.00100 \, mol \, dm^{-3} \times 0.01580 \, dm^3 = 1.58 \times 10^{-5} \, mol$

half that amount of I_2 must have been released $= 7.90 \times 10^{-6} \, mol$

equation shows moles of $HOCl$ = moles of $I_2 = 7.90 \times 10^{-6} \, mol$

$$\therefore [HOCl(aq)] = \frac{7.90 \times 10^{-6} \, mol}{0.0500 \, dm^3} = 1.58 \times 10^{-4} \, mol \, dm^{-3}$$

5 a $Mg(s)|Mg^{2+}(aq)||Cu^{2+}(aq)|Cu(s)$

b $Mg(s)|Mg^{2+}(aq)||Zn^{2+}(aq)|Zn(s)$

6 a $E^{\ominus}_{cell} = -0.76 \, V - -2.37 \, V = +1.61 \, V$

b $E^{\ominus}_{cell} = +1.51 \, V - +0.34 \, V = +1.17 \, V$

c $E^{\ominus}_{cell} = +1.09 \, V - -1.66 \, V = +2.75 \, V$

7 a $2Al(s) + 3Pb^{2+}(aq) \rightarrow 2Al^{3+}(aq) + 3Pb(s)$

b $6Br^-(aq) + Cr_2O_7^{2-}(aq) + 14H^+(aq) \rightarrow 2Cr^{3+}(aq) + 7H_2O(l) + 3Br_2(aq)$

c $5Fe^{2+}(aq) + MnO_4^-(aq) + 8H^+(aq) \rightarrow 5Fe^{3+}(aq) + Mn^{2+}(aq) + 4H_2O(l)$

d $Mg(s) + Fe^{2+}(aq) \rightarrow Mg^{2+}(aq) + Fe(s)$

8 a Cl: 0 to -1 and $+1$

b I: $+1$ to -1 and $+5$

c O: -1 to -2 and 0

Examination questions

1 d
$$
\begin{array}{lll}
MnO_4^{2-} + 4H^+ + 2e^- & \rightarrow & MnO_2 \quad + \quad 2H_2O \\
2MnO_4^{2-} & \rightarrow & 2MnO_4^- \quad + \quad 2e^- \\
\hline
3MnO_4^{2-} + 4H^+ & \rightarrow & 2MnO_4^- \quad + \quad MnO_2 + 2H_2O
\end{array}
$$

$E^{\ominus} = +1.55 - (+0.60) = 0.95 \, V$

3 b ii $E^{\ominus}_{cell} = E^{\ominus}[Cu^{2+}(aq)/Cu(s)] - E^{\ominus}[Fe^{2+}(aq)/Fe(s)]$

$\therefore E^{\ominus}[Cu^{2+}(aq)/Cu(s)] = E^{\ominus}_{cell} + E^{\ominus}[Fe^{2+}(aq)/Fe(s)]$

$\qquad = +0.78 \, V + -0.44 \, V = +0.34 \, V$

5 b i $Cu(s) \rightarrow Cu^{2+}(aq) + 2e^-$

ii $Ag^+(aq) + e^- \rightarrow Ag(s)$

c $Cu(s) + 2Ag^+(aq) \rightarrow Cu^{2+}(aq) + 2Ag(s)$

d i $E^{\ominus}_{cell} = +0.80 - (+0.34) = +0.46 \, V$

ii silver: explanation in terms of electron gain/change in oxidation state/ Ag^+/Ag system moves to right/has more positive standard electrode potential

Chapter 4   Marginal questions

6 $\Delta H^{\ominus}_{sol}(NaCl) = -\Delta H^{\ominus}_{LE}(NaCl) + \Delta H^{\ominus}_{hyd}(Na^+) + \Delta H^{\ominus}_{hyd}(Cl^-)$

$\qquad\qquad = +787 \, kJ \, mol^{-1} - 499 \, kJ \, mol^{-1} - 364 \, kJ \, mol^{-1} = -76 \, kJ \, mol^{-1}$

7 a Br_2: more electrons, $\therefore$ more energy levels, $\therefore$ more ways of distributing energy

b Cl_2: more electrons, $\therefore$ more energy levels, $\therefore$ more ways of distributing energy

c NH_3: more atoms, $\therefore$ more energy levels, $\therefore$ more ways of distributing energy

d butane: more atoms, $\therefore$ more energy levels, $\therefore$ more ways of distributing energy

e CO_2: more atoms, $\therefore$ more energy levels, $\therefore$ more ways of distributing energy

8 a 7

b 15

9 $\Delta S_{surr} = \dfrac{-\Delta H}{T} = \dfrac{-(-6.01) \, kJ \, mol^{-1}}{258 \, K} = 0.0233 \, kJ \, K^{-1} \, mol^{-1}$ (or $23.3 \, J \, K^{-1} \, mol^{-1}$)

$\Delta S_{total} = \Delta S_{sys} + \Delta S_{surr} = -22.0 \, J \, K^{-1} \, mol^{-1} + 23.3 \, J \, K^{-1} \, mol^{-1} = +1.3 \, J \, K^{-1} \, mol^{-1}$

10 b $\Delta S_{surr} = \dfrac{-\Delta H}{T} = \dfrac{-118\,kJ\,mol^{-1}}{298\,K} = -0.396\,kJ\,K^{-1}\,mol^{-1}$ (or $-396\,J\,K^{-1}\,mol^{-1}$) at 298 K

$\Delta S_{surr} = \dfrac{-\Delta H}{T} = \dfrac{-118\,kJ\,mol^{-1}}{600\,K} = -0.197\,kJ\,K^{-1}\,mol^{-1}$ (or $-197\,J\,K^{-1}\,mol^{-1}$) at 600 K

$\Delta S_{surr} = \dfrac{-\Delta H}{T} = \dfrac{-118\,kJ\,mol^{-1}}{1300\,K} = -0.0908\,kJ\,K^{-1}\,mol^{-1}$ (or $-90.8\,J\,K^{-1}\,mol^{-1}$) at 1300 K

$\Delta S_{total} = \Delta S_{sys} + \Delta S_{surr} = (160.4 - 396)\,J\,K^{-1}\,mol^{-1}\quad = -235.6\,J\,K^{-1}\,mol^{-1}\quad$ at 298 K

$= (160.4 - 197)\,J\,K^{-1}\,mol^{-1}\quad = -36.6\,J\,K^{-1}\,mol^{-1}\quad$ at 600 K

$= (160.4 - 90.8)\,J\,K^{-1}\,mol^{-1}\quad = +69.6\,J\,K^{-1}\,mol^{-1}\quad$ at 1300 K

Graphically, $\Delta S_{total} = 0$ at eqm. at approximately 720 K, the lowest possible temperature for decomposition.

or, for $\Delta S_{total} = 0$, $\Delta S_{surr} = -\Delta S_{sys} = -160.4\,J\,K^{-1}\,mol^{-1}$ (or $0.1604\,kJ\,K^{-1}\,mol^{-1}$)

$\therefore T = \dfrac{-\Delta H}{\Delta S_{surr}} = \dfrac{-118\,kJ\,mol^{-1}}{-0.1604\,kJ\,K^{-1}\,mol^{-1}} = 736\,K$

11 a $\Delta G = \Delta H - T\Delta S = +118\,kJ\,mol^{-1} - (298\,K \times 0.1604\,kJ\,K^{-1}\,mol^{-1}) = +70.2\,kJ\,mol^{-1}$

b $\Delta G = \Delta H - T\Delta S = +118\,kJ\,mol^{-1} - (1300\,K \times 0.1604\,kJ\,K^{-1}\,mol^{-1}) = -90.5\,kJ\,mol^{-1}$

Examination questions

1 a

b $-\Delta H_f + \Delta H_{sub} + \Delta H_i - \tfrac{1}{2}\Delta H_{vap} + \tfrac{1}{2}\Delta H_{diss} + \Delta H_{ea} - \Delta H_L = 0$

$\Delta H_{vap} = 2(\Delta H_f - \Delta H_{sub} - \Delta H_i - \tfrac{1}{2}\Delta H_{diss} - \Delta H_{ea} + \Delta H_L)$

$= 2(-361 - 107 - 498 - 97 + 325 + 753) = +30\,kJ\,mol^{-1}$

2 a steam condenses to water when $\Delta G \leqslant 0$

$\therefore \Delta H = T\Delta S$

$\Delta S = 189 - 70 = 119\,J\,K^{-1}\,mol^{-1}$

$\therefore \Delta H = 373 \times 119 = 44\,kJ\,mol^{-1}$

b spontaneous reaction when $\Delta G \leqslant 0$

$CH_4(s) + H_2O(g) \rightarrow CO(g) + 3H_2(g)$

$\Delta S = 198 + 3 \times 131 - 189 - 186 = 216\,J\,K^{-1}\,mol^{-1}$

+ve entropy change if $-T\Delta S$ makes ΔG −ve

$\therefore \Delta H = T\Delta S$

$T = \dfrac{\Delta H}{\Delta S} = \dfrac{210\,kJ\,mol^{-1}}{0.216\,kJ\,mol^{-1}} = 972\,K$

c diamond → graphite $\Delta S = +3 \, J K^{-1} mol^{-1}$
since $\Delta H < 0$, then ΔG is always < 0
kinetics: large E_a makes reaction too slow

d $CaO(s) + CO_2(g) \rightarrow CaCO_3(s)$
$\Delta S = 90 - 40 - 214 = -164 \, J K^{-1} mol^{-1} = -0.164 \, kJ K^{-1} mol^{-1}$
at $\Delta G = 0$, $\Delta H = T\Delta S$

$$T_s = \frac{\Delta H}{\Delta S} = \frac{-178 \, kJ mol^{-1}}{-0.164 \, kJ K^{-1} mol^{-1}} = 1085 \, K$$

3 a ii $\Delta H^{\ominus} = 2\Delta H_f^{\ominus}(NaOH) - 2\Delta H_f^{\ominus}(H_2O)$
$= 2(-470 \, kJ mol^{-1}) - 2(-286 \, kJ mol^{-1}) = -368 \, kJ mol^{-1}$

4 a

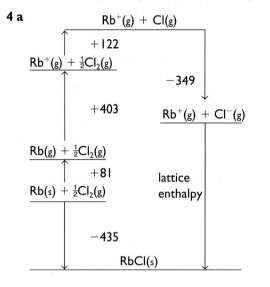

$-(-435) + 81 + 403 + 122 + (-349) +$ lattice enthalpy $= 0$
lattice enthalpy $= -435 - 81 - 403 - 122 + 349$
$= -692 \, kJ \, mol^{-1}$

5 a ii $\Delta H_f^{\ominus} = (+109 + 494 + 121 - 364 - 770) \, kJ mol^{-1} = -410 \, kJ mol^{-1}$

● **Chapter 5** **Marginal questions**

1

	acid	+ base	conjugate acid	conjugate base
a	HBr	+ H_2O →	H_3O^+	+ Br^-
b	CH_3COOH	+ OH^- →	H_2O	+ CH_3COO^-
c	H_2SO_4	+ H_2O →	H_3O^+	+ HSO_4^-

8 a $pH = -\log[H_3O^+(aq)] = -\log(0.01) = 2$
b $= -\log(0.001) = 3$
c $= -\log(0.0001) = 4$

10 $[H_3O^+(aq)] = \sqrt{(1.6 \times 10^{-4} \, mol \, dm^{-3} \times 0.50 \, mol \, dm^{-3})} = 8.94 \times 10^{-3} \, mol \, dm^{-3}$
$pH = -\log[H_3O^+(aq)] = 2.05$

11 $pH = 3.5 \therefore [H_3O^+(aq)] = 10^{-pH} = 3.2 \times 10^{-4} \, mol \, dm^{-3}$

$[H_3O^+(aq)] = \sqrt{(K_a \times \text{concentration of acid})}$

$\therefore$ concentration of acid $= \dfrac{[H_3O^+(aq)]^2}{K_a} = \dfrac{(3.2 \times 10^{-4} \, mol \, dm^{-3})^2}{1.7 \times 10^{-5} \, mol \, dm^{-3}} = 6.0 \times 10^{-3} \, mol \, dm^{-3}$

12 $[OH^-(aq)] = 0.50\,mol\,dm^{-3}$

$$\therefore [H_3O^+(aq)] = \frac{1.0 \times 10^{-14}\,mol^2\,dm^{-6}}{0.50\,mol\,dm^{-3}} = 2.0 \times 10^{-14}\,mol\,dm^{-3}$$

$$\therefore pH = -\log[H_3O^+(aq)] = 13.7$$

13 $[acid] = 0.050\,mol\,dm^{-3}$ $[salt] = \dfrac{2.05\,g}{82.1\,g\,mol^{-1} \times 1.0\,dm^3} = 0.025\,mol\,dm^{-3}$

$$[H_3O^+(aq)] = K_a \times \frac{[acid]}{[salt]} = 1.7 \times 10^{-5}\,mol\,dm^{-3} \times \frac{0.050\,mol\,dm^{-3}}{0.025\,mol\,dm^{-3}} = 3.4 \times 10^{-5}\,mol\,dm^{-3}$$

$$\therefore pH = -\log[H_3O^+(aq)] = 4.5$$

Skills task

moles of HCl in $20.50\,cm^3 = 0.250\,mol\,dm^{-3} \times 0.02050\,dm^3 = 5.125 \times 10^{-3}\,mol$
$\therefore$ moles of NaOH in $25.0\,cm^3$ of diluted solution $\quad = 5.125 \times 10^{-3}\,mol$

(HCl and NaOH react 1:1)

$\therefore$ moles of NaOH in $1.00\,cm^3$ of the original solution $\quad = 5.125 \times 10^{-3}\,mol$
$\therefore$ moles of NaOH in $500\,cm^3$ of the original solution $\quad = 5.125 \times 10^{-3}\,mol \times 500 = 2.56\,mol$
$\therefore$ mass of NaOH in $500\,cm^3$ of the original solution $\quad = 2.56\,mol \times 40.0\,g\,mol^{-1} = 102\,g$

Examination questions

1 a iii

$$K_a = \frac{[H^+]^2}{[CH_3CH_2COOH]}$$

$[CH_3CH_2COOH] = 0.10\,mol\,dm^{-3}$
$$\therefore [H^+] = \sqrt{(0.1\,mol\,dm^{-3} \times 1.35 \times 10^{-5}\,mol\,dm^{-3})}$$
$$= 1.16 \times 10^{-3}\,mol\,dm^{-3}$$
$$\therefore pH = 2.93$$

2 b i $K_a = \dfrac{[H^+][HCO_3^-]}{[CO_2]}$

$$[H^+] = \frac{[CO_2] \times K_a}{[HCO_3^-]}$$

$$= \frac{(1.25 \times 10^{-3}\,mol\,dm^{-3}) \times (4.5 \times 10^{-7}\,mol\,dm^{-3})}{2.5 \times 10^{-2}\,mol\,dm^{-3}}$$

$$= 2.3 \times 10^{-8}\,mol\,dm^{-3}$$

ii pH = 7.6

3 b i (+)5

ii   moles $H_3PO_4 = 0.100\,mol\,dm^{-3} \times \dfrac{20.0}{1000}\,dm^3$

moles NaOH $= 3 \times$ moles H_3PO_4
moles NaOH $= 3 \times 0.00200 = 0.00600\,mol$

$$\text{volume NaOH} = \frac{0.00600\,mol}{0.250\,mol\,dm^{-3}} \times 1000 = 24.0\,cm^3$$

c $K_w = [H^+(aq)][OH^-(aq)]$

$$\therefore [H^+(aq)] = \frac{1.00 \times 10^{-14}\,mol^2\,dm^{-6}}{0.250\,mol\,dm^{-3}} = 4 \times 10^{-14}\,mol\,dm^{-3}$$

$$pH = -\log[H^+] = -\log(4 \times 10^{-14}) = 13.4$$

4 a i $K_a = \dfrac{[H^+][CH_3COO^-]}{}$

    ii $[H^+] = [CH_3COO^-]$
        $[H^+]^2 = K_a \times [CH_3COOH]$
        $[H^+] = \sqrt{1.75 \times 10^{-5} \times 0.100} = 1.32$
        $pH = -\log_{10}[H^+] = 2.88$

    iii $[H^+] = \dfrac{K_a \times [CH_3COOH]}{[CH_3COO^-]}$

        $= \dfrac{1.75 \times 10^{-5}\,mol\,dm^{-3} \times 0.100\,mol\,dm^{-3}}{0.125\,mol\,dm^{-3}} = 1.4 \times 10^{-5}\,mol\,dm^{-3}$

        $pH = 4.85$

6 b $[H_3O^+{}_{(aq)}] = 10^{-2.91}\,mol\,dm^{-3} = 1.23 \times 10^{-3}\,mol\,dm^{-3}$

  c $[acid] = \dfrac{4.62\,g}{168\,g\,mol^{-1} \times 0.250\,dm^3} = 0.110\,mol\,dm^{-3}$

    $K_a = \dfrac{[H_3O^+{}_{(aq)}]^2}{[acid]} = \dfrac{(1.23 \times 10^{-3}\,mol\,dm^{-3})^2}{0.110\,mol\,dm^{-3}} = 1.38 \times 10^{-5}\,mol\,dm^{-3}$

8 b i $K_w = [H^+{}_{(aq)}][OH^-{}_{(aq)}]$

    $\therefore [H^+{}_{(aq)}] = \dfrac{1.00 \times 10^{-14}\,mol^2\,dm^{-6}}{0.300\,mol\,dm^{-3}} = 3.33 \times 10^{-14}\,mol\,dm^{-3}$

    $pH = -\log[H^+{}_{(aq)}] = -\log(3.33 \times 10^{-14}) = 13.48$

    ii in $25\,cm^3$ there are $\dfrac{0.3 \times 25}{1000} = 7.5 \times 10^{-3}$

    $\therefore$ concentration of solution is $4 \times 7.5 \times 10^{-3} = 0.03\,mol\,dm^{-3}$

    $\therefore [H^+{}_{(aq)}] = \dfrac{1.00 \times 10^{-14}\,mol^2\,dm^{-6}}{0.03\,mol\,dm^{-3}} = 3.33 \times 10^{-13}\,mol\,dm^{-3}$

    $pH = -\log[H^+{}_{(aq)}]$
        $= -\log(3.33 \times 10^{-13}) = 12.48$

    iii in $25\,cm^3$ of $0.300\,mol\,dm^{-3}$ NaOH there are $\dfrac{0.300 \times 25}{1000} = 7.5 \times 10^{-3}$ mol of OH^-

    in $75\,cm^3$ of $0.200\,mol\,dm^{-3}$ HCl there are $\dfrac{0.200 \times 75}{1000} = 1.5 \times 10^{-2}$ mol of H^+

    since $H^+ + OH^- \rightarrow H_2O$ in neutralisation reaction, then
    $(1.5 \times 10^{-2}) - (7.5 \times 10^{-3}) \times 7.5 \times 10^{-3}$ mol of H^+ in excess
    this is present in $1000\,cm^3$ of solution
    $\therefore [H^+{}_{(aq)}] = 7.5 \times 10^{-2}$ mol dm^{-3}
    $\therefore pH = -\log(7.5 \times 10^{-2}) = 1.12$

9 a $pH = 2.69$ $\therefore [H^+] = 2.04 \times 10^{-3}\,mol\,dm^{-3}$

    $K_a = \dfrac{[H^+]^2}{[Hx]} = \dfrac{(2.04 \times 10^{-3}\,mol\,dm^{-3})^2}{0.15\,mol\,dm^{-3}}$

        $= \dfrac{4.16 \times 10^{-6}\,mol^2\,dm^{-6}}{0.15\,mol\,dm^{-3}} = 2.78 \times 10^{-5}\,mol\,dm^{-3}$

b i in 25 cm^3 of A there are $\dfrac{2.04 \times 10^{-3}}{40} = 5.1 \times 10^{-5}$ moles of H$^+$

$\therefore \dfrac{\text{volume of NaOH} \times 0.25}{1000} = 5.1 \times 10^{-5}$ moles

volume of NaOH $= \dfrac{5.1 \times 10^{-5} \times 1000}{0.25} = 0.20\,\text{cm}^{-3}$

pH of neutralised solution $= 7$

ii at this point $[\text{H}^+] = K_a$

$\qquad [\text{H}^+] = 2.78 \times 10^{-5}\,\text{mol dm}^{-3}$

$\therefore \text{pH} = -\log(2.78 \times 10^{-5}) = 4.56$

iii moles of NaOH in excess $= \dfrac{(25.0 - 0.2) \times 0.25}{1000} = 6.2 \times 10^{-3}$

(also moles of OH$^-$)

6.2×10^{-3} mol of OH$^-$ in 50.0 cm^3

hence $[\text{OH}^-] = 6.2 \times 10^{-3} \times 20 = 0.124\,\text{mol dm}^{-3}$

$K_w = [\text{H}^+][\text{OH}^-] = 1 \times 10^{-14}\,\text{mol}^2\,\text{dm}^{-6}$

$\therefore [\text{H}^+] = \dfrac{1 \times 10^{-14}\,\text{mol}^2\,\text{dm}^{-6}}{0.124\,\text{mol dm}^{-3}} = 8.06 \times 10^{-14}\,\text{mol dm}^{-3}$

pH $= -\log(8.06 \times 10^{-14}) \times 13.09$

10 b i $M_r\,\text{Na}_2\text{CO}_3 = 106$

$\therefore 13.4\,\text{g} = \dfrac{13.4}{106} = 0.126\,\text{mol}$

ii 0.126 mol of Na$_2$CO$_3$ contain 0.126 mol CO$_3^{2-}$

$\therefore$ mol of H$^+$ in 4 dm^3 $= 2 \times 0.126 = 0.252\,\text{mol}$

$[\text{H}^+] = \dfrac{0.252}{4} = 0.063\,\text{mol dm}^{-3}$

pH $= -\log[\text{H}^+]$
$= -\log(0.063) = 1.2$

iii 13.4 g of NaOH $= \dfrac{13.4}{40} = 0.335\,\text{mol}$

0.335 mol of NaOH contains 0.335 mol of OH$^-$

since H$^+$ + OH$^-$ → H$_2$O

then mol of OH$^-$ in excess in 4 dm^3 $= 0.335 - 0.252$

$\qquad\qquad\qquad\qquad\qquad\qquad\quad = 0.083\,\text{mol}$

$\therefore [\text{OH}^-] = \dfrac{0.083}{4} = 0.021\,\text{mol dm}^{-3}$

since $K_w = [\text{H}^+][\text{OH}^-] = 1 \times 10^{-14}\,\text{mol}^2\,\text{dm}^{-6}$

then $[\text{H}^+] = \dfrac{1 \times 10^{-14}\,\text{mol}^2\,\text{dm}^{-6}}{0.021\,\text{mol dm}^{-3}} = 4.76 \times 10^{-13}\,\text{mol dm}^{-3}$

$\therefore$ pH $= -\log[\text{H}^+]$
$= -\log(4.76 \times 10^{-13}) = 12.3$

● Chapter 6   Examination questions

5 a iv $2SO_2 + O_2 + 2H_2O \rightarrow 2H_2SO_4$
$2NaOH + H_2SO_4 \rightarrow Na_2SO_4 + 2H_2O$

number of mols NaOH $= \dfrac{20.8}{1000}$ dm$^3 \times 0.400$ mol dm$^{-3} = 8.32 \times 10^{-3}$ mol

number of mols $H_2SO_4 = \dfrac{8.32 \times 10^{-3}}{2}$ mol $= 4.16 \times 10^{-3}$ mol

4.16×10^{-3} mol in 25 cm^3, so $4.16 \times 10^{-3} \times \dfrac{1000}{25}$ mol in 1 dm^3

$= 0.166$ mol dm^{-3}
solubility of $SO_2 = 0.166$ mol dm^{-3}

● Chapter 7   Marginal questions

7 $CCl_4(l) + 2H_2O(l) \rightarrow CO_2(g) + 4HCl(aq)$
$\Delta H^\ominus = \Delta H_f^\ominus [CO_2(g)] + 4\Delta H_f^\ominus [HCl(aq)] - \Delta H_f^\ominus [CCl_4(l)] - 2\Delta H_f^\ominus [H_2O(l)]$
$= [-393.5 + (4 \times -165) - -129.6 - (2 \times -285.8)]$ kJ mol$^{-1} = -352.3$ kJ mol^{-1}

Examination questions

2 c ii moles of $PbO_2 = \dfrac{5.0\,g}{239.2\,g\,mol^{-1}} = 0.0209$ mol

the equation shows 4 times as much HCl is required, that is, 4×0.0209 mol $= 0.0836$ mol

∴ the volume of 12 mol dm^{-3} HCl $= \dfrac{0.0836\,mol}{12\,mol\,dm^{-3}} = 6.968 \times 10^{-3}$ dm^3 (or 7.0 cm^3)

volume of chlorine $= 0.0209$ mol $\times 24$ dm^3 mol$^{-1} = 0.50$ dm^3 (or 500 cm^3)

● Chapter 8   Marginal questions

6 Cu^{2+} Zn^{2+} Ag^+ Ag^+
7 a hexaamminecobalt(II)   **b** hexaaquacopper(II)
  c hydroxypentaaquairon(III)   **d** dichlorotetraaquachromium(III)
10 $E_{cell}^\ominus = +1.51\,V - (+0.77\,V) = +0.74\,V$
11 moles of MnO_4^- used $= 0.0100$ mol dm$^{-3} \times 0.02020$ dm$^3 = 2.02 \times 10^{-4}$ mol
  ∴ moles of Fe^{2+} in 100 cm$^3 = 5 \times 2.02 \times 10^{-4}$ mol $= 1.01 \times 10^{-3}$ mol
  ∴ moles of Fe^{2+} (and Fe) in 1 g powder also $= 1.01 \times 10^{-3}$ mol
  ∴ mass of Fe in 1 g powder $= 1.01 \times 10^{-3}$ mol $\times 55.8$ g mol$^{-1} = 0.0564$ g, that is 5.64%
12 $Cr(III) \rightarrow Cr(VI)$

Examination questions

3 d i Amount of MnO_4^- used in the titration $= 0.0200$ mol dm$^{-3} \times \dfrac{24.2}{1000}$ dm^3

$= 4.84 \times 10^{-4}$ mol

Mass of iron in steel sample $= 5 \times 4.84 \times 10^{-4}$ mol $\times 10 \times 56.0$ g mol$^{-1} = 1.36$ g

  ii Percentage by mass of iron in steel $= \dfrac{1.36\,g}{1.40\,g} \times 100 = 96.8\%$

5 d i 0.002 mol   **ii** 5   **iii** increases by 1   **iv** 2 (that is, $VOCl_2$, as this gives V as +4)

● Chapter 10   **Examination questions**

2 b i M_r of ester $= 136$
moles methyl benzoate $= 0.0198$

ii moles NaOH $= 4.0/40 \times 1000/50$
concentration $= 2.0 \text{ mol dm}^{-3}$

iii M_r of $C_6H_5COOH = 122$
maximum yield of $C_6H_5COOH = 0.0198 \text{ mol}$
maximum yield of $C_6H_5COOH = (0.0198 \times 1.22) = 2.42 \text{ g}$
% yield $= (1.50/2.42) \times 100 = 62.1\%$

iv hydrolysis not complete/C_6H_5COOH slightly soluble in water

5 f i **C** is

$$\begin{array}{cccc} & H & H & H \\ & | & | & | \\ H-&C-&C-&C-&C \end{array}$$

(skeletal: $H-\overset{\overset{\displaystyle H}{|}}{\underset{\underset{\displaystyle H}{|}}{C}}-\overset{\overset{\displaystyle H}{|}}{\underset{\underset{\displaystyle H}{|}}{C}}-\overset{\overset{\displaystyle H}{|}}{\underset{\underset{\displaystyle H}{|}}{C}}-\overset{\displaystyle O}{\underset{\displaystyle OH}{C}}$)

C is a weak acid, and so only partially dissociates
$$HA \rightleftharpoons H^+ + A^-$$
$$0.100 \qquad x \qquad x$$

$$K_a = \frac{[H^+][A^-]}{[HA]}$$

Substituting:

$$K_a = \frac{x^2}{[HA]}$$

$$x = \sqrt{1.52 \times 10^{-5} \text{ mol dm}^{-3} \times 0.100 \text{ mol dm}^{-3}} = 1.23 \times 10^{-3} \text{ mol dm}^{-3}$$
$$pH = -\log[H^+]$$
$$= -\log(1.23 \times 10^{-3} \text{ mol dm}^{-3}) = 2.91$$

ii buffer solution

iii molecular mass of sodium salt of **C** is 110 g mol^{-1}
5.5 g of the sodium salt of **C** is 0.05 mol
0.05 mol of the sodium salt of **C** in 500 cm^3 has a concentration of

$$0.05 \times \frac{1000}{500} = 0.100 \text{ mol dm}^{-3}$$

$$HA \rightleftharpoons H^+ + A^-$$
$$[HA] = 0.100 \text{ mol dm}^{-3}$$
$$[H^+] = y$$
$$[A^-] = 0.100 \text{ mol dm}^{-3} \text{ (from the sodium salt of } \mathbf{C})$$

$$K_a = \frac{[H^+][A^-]}{[HA]}$$
$$[H^+] = 1.51 \times 10^{-5} \text{ mol dm}^{-3}$$
$$pH = 4.82$$

7 a i From graph, end-point $= 15.5 \text{ cm}^3$ of NaOH

C	H	O	
40.0	6.70	53.3	% by mass
$\dfrac{40.0}{12}$	$\dfrac{6.70}{1}$	$\dfrac{53.3}{16}$	divide by atomic mass
$= 3.33$	$= 6.70$	$= 3.33$	
1 :	2 :	1	simplest ratio

empirical formula of **A** is CH_2O, empirical formula mass $= 30$

$$\frac{15.5}{1000} \text{ dm}^3 \times 0.050 \text{ mol dm}^{-3} \text{ NaOH reacted in titration} = 7.75 \times 10^{-4} \text{ mol}$$

7.75×10^{-4} mol of **A** in 10.0 cm³ of solution, so $[\mathbf{A}] = 7.75 \times 10^{-2}$ mol dm⁻³

7.75×10^{-2} mol of **A** is 7.20 g, so 1 mol of **A** is $\dfrac{7.20}{7.75 \times 10^{-2}} = 93$ g,

which is approx. $3\times$ empirical formula mass, so **A** is $C_3H_6O_3$

Chapter 11 Marginal questions

9 a $K_a = 10^{-pK_a} = 3.16 \times 10^{-10}$ mol dm⁻³ (units deduced from equation not calculation)

Chapter 14 Marginal questions

10 first process   60%
second process $90\% \times 78\% = 70\%$ this gives the best yield
third process $95\% \times 75\% \times 85\% = 61\%$

Examination questions

1 d mass of bromine = $(80 \times$ mass AgBr$)/180$
% by mass = (mass bromine $\times 100$)/mass bromoalkane

Chapter 15 Marginal questions

1 $A_r(\text{Ni}) = (58 \times 0.6776) + (60 \times 0.2616) + (61 \times 0.0125) + (62 \times 0.0366) + (64 \times 0.0116) = 58.77$

2 number of C atoms $= \dfrac{11}{1.1} = 10$

4 100g contains   40g C   6.7g H   53.3g O

$\dfrac{40\text{g}}{12\text{g mol}^{-1}} = 3.33$ mol C   6.7 mol H $\dfrac{53.3\text{g}}{16\text{g mol}^{-1}} = 3.33$ mol O

ratio   1   :   2   :   1
empirical formula is CH_2O for which $M_r = 30$
∴ molecular formula is $C_2H_4O_2$

13 100g contains   64.86g C   13.51g H   21.62g O

$\dfrac{64.86\text{g}}{12\text{g mol}^{-1}} = 5.405$ mol C   13.51 mol H $\dfrac{21.62\text{g}}{16\text{g mol}^{-1}} = 1.35$ mol O

ratio   4   :   10   :   1
empirical formula is $C_4H_{10}O$ for which $M_r = 74$
mass spectrum gives M peak of 74, ∴ molecular formula is $C_4H_{10}O$ (M + 1 peak of 4.40%
indicates 4C atoms)
IR spectrum indicates presence of an OH group, that is, C_4H_9OH (3300 cm⁻¹) as well as CH_3
and CH_2 groups (2800–3000 cm⁻¹)
NMR spectrum has 4 peaks, showing H atoms in 4 different environments:
$\delta = 0.9$ for R—CH_3 (integration 3)
$\delta = 1.4$ for R—CH_2—R (integration 4)
$\delta = 2.2$ for R—CH_2—O (integration 2)
$\delta = 3.6$ for R—OH (integration 1)
this confirms the structure as $CH_3CH_2CH_2CH_2OH$

14 100 g contains 45.86 g C   8.92 g H   45.22 g Cl

$$\frac{45.86\text{ g}}{12\text{ g mol}^{-1}} = 3.822\text{ mol C} \qquad 8.92\text{ mol} \qquad \frac{45.22\text{ g}}{35.5\text{ g mol}^{-1}} = 1.27\text{ mol Cl}$$

ratio   3   :   7   :   1

empirical formula is C_3H_7Cl for which $M_r = 78.5$

mass spectrum gives M peaks at 78 ($C_3H_7{}^{35}Cl$) and 80 ($C_3H_7{}^{37}Cl$), ratio of peaks is 3:1,

∴ molecular formula = C_3H_7Cl

IR spectrum indicates presence of C—Cl (610 cm^{-1}) as well as CH_3 and CH_2 groups
(2800−3000 cm^{-1})

NMR spectrum has 2 peaks, showing H atoms in 2 different environments:

$\delta = 1.2$ (integration of 6)

$\delta = 3.7$ (integration of 1)

hence indicates CH_3—$\overset{\displaystyle H}{\underset{\displaystyle |}{\overset{|}{C}}}$—$CH_3$ for H atoms, this confirms the structure as CH_3—$\overset{\displaystyle H}{\underset{\displaystyle Cl}{\overset{|}{\underset{|}{C}}}}$—$CH_3$

Examination questions

3 Linking together evidence from IR, NMR and mass spectrum suggests **G** is butanone,
$CH_3CH_2COCH_3$.

4 A 1715 cm^{-1} C=O group

   B 3350 cm^{-1} O—H group; alcohol

   A $CH_3CH_2COCH_2CH_3$
        ↑   ↑
        t   q

two environments or two kinds of proton

CH_3CH_2 adjacent or coupled

ratio 2:3 or 4:6

symmetric

$[CH_3CH_2COCH_2CH_3]^{+\bullet} \rightarrow CH_3CH_2CO^+ + CH_3CH_2{}^{\bullet}$

$m/z = 86$ $m/z = 57$

or M_r for **A**

$CH_3CH_2CHCH_2CH_3 \rightarrow CH_3CH{=}CHCH_2CH_3$
                |
B   OH   **C**

$\overset{+}{CH_3CH_2CHCH_2CH_3}$ and $\overset{+}{CH_3CHCH_2CH_2CH_3}$

both secondary

hydration gives **B** and $CH_3CHCH_2CH_2CH_3$
                              |
                    D   OH

about 50% of each

A → B reduction

B → C dehydration or elimination

C is an alkene; *cis/trans* isomers

D is a racemate or mixture of optical isomers

Periodic table

Index

Italic entries indicate references to figures and tables